PIERRE F. WALTER

THE SCIENCE
of Shamanism

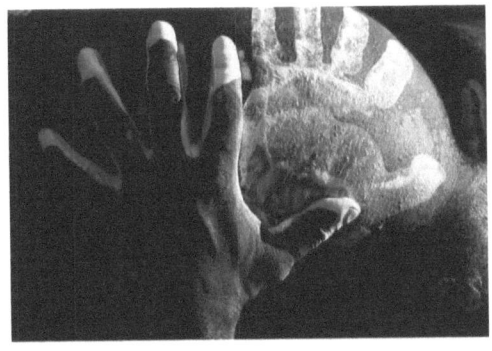

Millenary Model for an Integrative Worldview

Published by Sirius-C Media Galaxy LLC

http://sirius-c-publishing.com

http://siriuscmedia.com

http://ipublica.com

ISBN 978-1-45658-5853

Contact Information Pierre F. Walter

publisher@sirius-c-publishing.com

About Pierre F. Walter

http://drpfw.info

Quotation Suggestion

Pierre F. Walter, *The Science of Shamanism: Millenary Model for an Integrative Worldview*, Newark: Sirius-C Media Galaxy LLC, 2010

About the Author

Pierre F. Walter is an author, international lawyer, researcher, corporate trainer, and lecturer. After finalizing studies in German Law, International Law and *European integration* with diplomas obtained in 1981 through 1983, he graduated in December 1987 at the Law Faculty of the University of Geneva as *Docteur en Droit* in international law.

The doctorate was funded by scholarships from the *Swiss Institute of Comparative Law*, Lausanne, and from the *University of Geneva*, as well as a Fulbright Travel Grant for an assistantship with Professor Louis B. Sohn at *UGA Law School Department of International Law*, Athens, Georgia, USA, in 1985. Pierre F. Walter also served as a research assistant to *Freshfields, Bruckhaus, Deringer*, Cologne, Germany in 1983 and to *Lalive Lawyers*, Geneva, in 1987.

Pierre F. Walter writes and lectures in English, German and French languages; he has written *more than ten thousand pages* embracing all literary genres, including *novels, short stories, film scripts, essays, selfhelp books, monographs* and extended *book reviews*. Also a pianist and composer, he has realized 40 CDs with *jazz, newage* and *relaxation music*.

Pierre F. Walter's professional publications span the domains *International Law, Criminal Law, Holistic Science, Psychology, Education, Shamanism, Ecology, Spirituality, Quantum Physics, Systems Theory, Natural Healing, Peace Research, Personal Growth, Selfhelp* and *Consciousness Research*. 110 Book Reviews, thirty-eight audio books and more than hundred video lectures were realized in the years 2005-2010. Besides, Pierre F. Walter is author and editor of *Great Minds Series*, which features scientists, artists and authors of genius from Leonardo to Fritjof Capra.

Pierre F. Walter publishes via his Delaware firm *Sirius-C Media Galaxy LLC* and the imprints IPUBLICA and Sirius-C Media (SCM).

To all great shamans around the world

CONTENTS

Published by Sirius-C Media Galaxy LLC, 2010

Published by Sirius-C Media Galaxy LLC, 2010

CHAPTER FIVE 261
A Science of Pattern

Published by Sirius-C Media Galaxy LLC, 2010

Published by Sirius-C Media Galaxy LLC, 2010

Published by Sirius-C Media Galaxy LLC, 2010

Published by Sirius-C Media Galaxy LLC, 2010

If you study science long enough, and seriously enough, and dig deeply enough, if you don't come up feeling wacko about it, you haven't understood a thing!
– Fred Alan Wolf in 'What the Bleep Do We Know!?'

INTRODUCTION

What is Shamanism?

The Science of Shamanism

Why should we know anything about *shamanism*, the reader may ask, as it's after all an outlandish kind of behavior, not grown in our culture, and barely affiliated with our traditional scientific method? Another may argue it's something laid away for 'wild, primitive peoples', and still another may come up with: 'That may be normal and even necessary when you live in the forest, but not when you live in a civilized modern society.'

And yet, not only is the interest growing in our society for shamanism as a way to inner truth, but a growing number of people actually practice shamanic rituals without feeling offended by their admittedly otherworldly strangeness. Many may feel attracted to shamanism as a lifestyle that has something *poetic* and soulful about it, speaking in a *metaphorical language* thereby giving soul to even the most ordinary things and behaviors in everyday life. While shamanism has to be distinguished from *animism* and *paganism*, as we shall see further on, it is true for all natural religions that they affirm the ensouled nature of life, of trees, animals, lakes, planets – and humans. And further, they see the unity of life rather than, as our science for the last four hundred years, a universe cluttered with unrelated things or objects.

As we also shall see in this book is that what attracts many people to shamanism rather than quantum physics is that our science, while it admittedly has done a big step forward toward a science paradigm that acknowledges the existence of a unified field, or quantum vacuum, there is a still a strangeness about the vocabulary of quantum physics that is actually more estranging to the simple-minded than a natural shamanic science that talks about the 'spirits of nature'. And, let's not fool ourselves, not only Terence McKenna but a serious shamanism researcher such as Ralph Metzner, have recognized that when natives speak about spirits, they actually denote the unified field, the subtle energy that connects us all, and that pervades all, and which shamans know to use for their purposes.

I have myself done studies both on quantum physics and shamanism and therefore understand either point of view, or scientific point of departure. After several years of immersion in these matters, and having written several papers and books about both shamanism and quantum physics, I may say in public, without offending anybody, that I find shamanism *more coherent, more accessible, more comprehensive, and even more human* than quantum physics.

Before I continue to explain what I mean, let me clearly affirm here, right at the start, that I do believe that shamanism and quantum physics deal exactly with the same problems or scientific questions, and give exactly the same answers to them. The difference is the scientific methods that are used, and the different vocabulary to describe and evaluate the observations made.

This is not a final conclusion but rather a work hypothesis, and I put it up for mere rhetoric reasons. I believe the reader is entitled to judge if what I am writing here has some common sense to it, or is mere speculation. I do honestly not

Published by Sirius-C Media Galaxy LLC, 2010

like to indulge in speculation and if this book was only that, I wouldn't have written it. But for sure, I am not able to corroborate my theory here, to validate it, for this is much too large an assumption to be elaborated in a theoretical work. It needs thorough research and a certain pragmatism to either verify or falsify my hypothesis.

In the present book, I am not going to even attempt doing this, and I might even not be qualified for it. So please take this statement here as an *obiter dictum* and let me proceed with finding out why shamanism is of real interest today for many people in Western and international consumer culture.

I have actually talked to quite a few people about this subject and some were giving explanations along the lines of my theory. One person, let us call him Jim, said, for example:

Jim

I am looking since long for an *integrative worldview* after all those philosophies that have only dissected life and have thereby divided people. I have found this unifying worldview with shamanism.

I know that quantum physics comes essentially today to the same insights, but it formulates them in a bizarre manner, as if coded in a special language that is only accessible to insiders.

I think shamanism has kept a streak of simplicity that makes it actually more credible, while quantum physics with its overall lack of transparence seems to paint life more complex than it actually is. When the details are complex, that doesn't mean we need to lose sight of the whole.

I found this statement incredibly dense and accurate, and it curiously reflected my own inner chatter and intuition over many years.

Then I found Ralph Metzner's enlightening reader *Ayahuasca (1999)*, and was fascinated by the fact that Western shamans had received exactly the same insights and messages from the plant teachers, and had had the same kind of visions when taking the brew, as native shamans. This was actually a very important information in that it disproves the often-voiced assumption that only natives could experience trance states that open the doors of perception, thereby relativizing such experience as 'merely cultural'.

In truth, many Western people have in the meantime had visionary experiences of the same kind or a similar kind; the LSD research done by Albert Hofmann who discovered the consciousness-altering effects of LSD in 1938, and the even more extended research on LSD by the psychiatrist Stanislav Grof showed with clear evidence that the mythological content of the collective unconscious is universal, not person-specific, and not culture-bound.[1] And this, then, corroborates Carl Jung's earlier psychoanalytic research on the collective subconscious in an amazing way. Jung namely distinguished between culture-specific and universal content in our collective phantasmagory, but came to the ultimate conclusion that the cultural differences are minimal compared to the strength and boldness of the collective message in its universal, unifying character.

Now, here we speak about mythology, but I contend that *shamanism is more than mythology,* more than a poetic, soulful

Published by Sirius-C Media Galaxy LLC, 2010

approach to the facts of life, and the invisible world. It is that, too, certainly, but it is *more than that.* I contend that shamanism is a true science in that as with any other science, *observation* is its major focus and tool. Had I said 'speculation', then we would have to conclude that it's what we use to call 'philosophy', and had I said that it's magic, then we would have to conclude that it uses suggestions for healing.

Now, don't get me wrong. We all know that shamans also use suggestions for healing, and a whole pandorabox of actions to achieve their goals. But their approach is *not mythological,* and it's not merely art, not merely a technique. It is that also, but at the starting point, shamanism is a science because it observes nature, and this, in very meticulous detail.

Many people in our culture won't accept this *a priori* because they believe science is only valid as a truth-delivering endeavor when it's set within the framework of a technological culture, where all is mechanized, and where science thus uses machines for the observation of nature, or for processing that observation. The best example here is perhaps quantum physics where physicists use gigantic particle accelerators.

One of the next generation electron accelerators will be the 40 km long *International Linear Collider*, planned to be constructed between 2015 and 2020, supposed to start with a collision energy of 500 billion electron volts, and could later be upgraded to 1000 billion electron volts.

The New York Times, February 8, 2007, writes about it with the headline 'Price of Next Big Thing in Physics: $6.7 Billion'. Another just completed particle accelerator is the *Large Hadron Collider* at CERN, outside of Geneva. 'That ma-

chine', writes Dennis Overbye, will be the world's most powerful (...), eventually colliding beams of protons with 7 tril-
lion electron volts of energy apiece.'

This kind of research about the quantum field reminds
me strangely of *vivisectionist medicine* that was searching the
truth of life in the corpses of dead people. And I am wonder-
ing about the logic thinking abilities of scientists who are out
to find the laws of life in the realm of death, instead of looking at living organisms. We have that behind us, for the most
part, and systems research now fills this gap, but we still have
the *mechanistic approach* around when it goes to find the truth
about the subatomic world. Why do scientists have to demol-
ish matter to find the truth about antimatter?

I am not a physicist so I leave the question open, as I
would not be able to give a competent answer. But I do stay
with the question and find it valid. It may one day open a
door.

My intuition is that what in the subatomic world is called
the unified field, quantum field, or Planck scale, is what the
natives call the spirit world. The approach of the shaman to
this world is scientific, not mythological, as it's conceived of
as a reality, not a dream reality, as a real space, not a virtual
space. The shaman approaches spirits in a true quest of discovery; while this knowledge was traditionally an oral one,
transmitted only from master to disciple, it is scientific in the
sense that it is empirical and subject to revision. In addition,
shamanic experiences, when set and setting are as much as
possible identical, can be *duplicated and verified by researchers
around the world*. Hence, they are not subjective emanations of

Published by Sirius-C Media Galaxy LLC, 2010

the shaman's poetic soul reality, which would namely deprive them of the qualifier 'scientific'. Terence McKenna wrote in *The Archaic Revival (1992):*

Terence McKenna

The anthropological literature always presents shamans as embedded in a tradition, but once one gets to know them they are always very sophisticated about what they are doing. They are the true phenomenologists of this world; they know plant chemistry, yet they call these energy fields 'spirits'.[2]

Overview

I would like to give an overview now over the chapters of this book, shortly outlining what each chapter is about.

In **Chapter One**, entitled *What is not Shamanism?*, I show the divide between shamanism as a science, and *Animism, Paganism, Humanism, Parapsychology, Theosophy, Taoism* and *Zen Buddhism*. While the latter domains may have done cognitive contributions to our philosophical understanding of life and living, they *cannot semantically be described as sciences* in the strict sense. Why this is so, I explain it in detail in this chapter.

I also show the model function shamanism has gained in our days as a guidepost for the development of a truly holistic and integrative cognitive paradigm that actually expands science into any and every quest for knowledge that is rational, empirical, logical and driven by proper motives, and that tries to give answers to the question of humanity interacting with other realms of existence thereby gaining deeper insights as to the role and function of the human within the cosmos, and accordingly, the cosmic intention behind creation.

In **Chapter Two,** entitled *The Warrior-Scientist*, I inquire first into the roles of the shaman and here I introduce a notion that may sound queer in the ears of Western people; it is the notion of 'natural morality'. I namely argue that one of the roles of the shaman is to be a communicator between the realms of *true morality* and of *true humanity*. This brings me to look at a role of the shaman that so far was little if ever dis-

Published by Sirius-C Media Galaxy LLC, 2010

cussed in the literature, the role namely of an agent of order and right behavior. What has been traditionally emphasized over and over was that the shaman is a healer, a psychopomp and a communicator with the spirit world.

However, his or her role as an agent of social hygiene, if I may say so, was never really put in the foreground of the scientific discussion of shamanism. In the following, I discuss the enlightening research done by Dr. Alberto Villoldo on shamanism and its healing function, before I come to talk about my own, personal, shamanic quest and the development it took over the last twenty years.

In **Chapter Three**, entitled *The Shamanic Method*, I first question some common assumptions that our Western science tradition used to make and that show that our science is actually very little descriptive and very highly normative, as it projects those assumptions upon nature, instead of observing nature and deriving conclusions purely from that observation. It's as if by interpreting a poem, I project my own intellectual assumptions about it and about the poet, in my interpretation of the poem – which then becomes an *intropretation*.

I shortly retrace in this chapter how the initial *extremely prejudicial paradigm of ethnology* changed and became more able to actually see the uniqueness of the shamanic worldview and scientific paradigm without projecting cultural and intellectual assumptions upon it. I also present and discuss in this chapter the main detractors of shamanism, such as the philosophy of the 'Enlightenment', Cartesian science, the Reductionist science paradigm and, last not least, Catholicism.

In this chapter, I also discuss what I call the 'shamanic revival', which is a header under which I subsume the final recognition of shamanism as a valid epistemological, ontological and scientific tool for self-discovery, healing and renewal, but also a cognitive tool for learning the 'secrets of nature' from a scientific, not a mythological point of departure. This, then, leads me to inquire into what properly is to be called the 'shamanic method', and here I expand on what it really is that makes shamanism a science.

In **Chapter Four**, entitled *Shamanism and the Use of Entheogens*, I present and explore the world of hallucinogenic compounds and discuss their relationship to shamanism.

I show by perusing modern-day research on shamanism and the use of entheogens that Mircea Eliade's 'classical' opinion only decadent shamanic cultures used hallucinogens is not only outdated, but has little empirical value in view of the religious and deconditioning quest that goes along with the ingestion of these compounds. I then report in this chapter my own experience with Ayahuasca during my wisdom quest, back in 2004, with a Shuar shaman in Ecuador.

I use this experience for demonstrating, in a methodologically sound manner, that the theory that says the mind-opening voyage is a result of DMT contained in those compounds is mechanistic and untenable. I made several strong points in this report that demonstrate that what brings about the trance and the voyage is the consciousness of the shaman impacting on that of the client, via the DMT and other entheogenic compounds, as a *transmitter matrix*. This is to my knowledge a novel theory and has never been seriously ar-

Published by Sirius-C Media Galaxy LLC, 2010

gued in any publication on shamanism and the use of entheogens. The chapter is rounded up by a literature review.

In **Chapter Five**, entitled *A Science of Pattern*, I show that shamanic cultures have known since times immemorial what systems theory delivers us today to our desktop, the truth namely that all life is coded in energy patterns, and that all intelligence is patterned intelligence. Fritjof Capra has called it 'the systems view of life' and I demonstrate by my own research on the *Eight Dynamic Patterns of Living* that native cultures have set in place eight specific patterns of living that guarantee peaceful coexistence on all levels of the community. This is why I consider the eight patterns as a guide concept to be taken over and applied in social policy making by a future more wistful and more conscious society. These eight patterns are *Autonomy, Ecstasy, Energy, Language, Love, Pleasure, Self-Regulation* and *Touch*. With this research, I implicitly demonstrate that shamanic cultures were from the start scientific in their overall orientation toward life, and have therefore developed patterns of living that are in sound harmony with nature's own higher logic and holistic intelligence, while in dominator cultures these patterns have been more or less shunned, if they were recognized at all.

Now, for each of these patterns, I can quote a whole list of modern-day research that fully confirms it, and shows that we cannot do without it.

In **Chapter Six**, entitled *The Matriarchal Science*, I expand about the intrinsic quality of the science of shamanism and I come to qualify it as a 'matriarchal' science, explaining the term in detail. I show in this chapter the transition from matriarchy to patriarchy and what happened to the symbolism of *psychic representation* through this profound archetypical change. I also explain how we can call a science 'matriarchal' or 'patriarchal' without messing up the objectiveness of scientific observation with ideological considerations. I explain what quantum physics found out about *observer entanglement*, which in last resort means that man creates science, and not science creates man. Hence, science may well be objective, but the scientist will be to a lesser extent because he is human and more than just 'science'. It is through this natural bias, that we cannot deny because there is no such thing as a 'detached observer' that we can comprehend that science and scientist are entangled in mutually enriching relationship.

When we see this entanglement, we can see how science has changed over time, because scientists, the people who do science, have changed. And this insight in turn allows us to understand the intrinsic qualities of shamanic science, and its *imbeddedness in ecology and respect of nature*, while dominator science created disruption from nature and disrespect of nature. We can then eventually see why shamanism was never called a science, in the first place! It was a matter of ideological indoctrination for science to be defined over centuries only as the patriarchal vintage of it, while the rest of the equation was carefully wiped under the carpet.

In **Chapter Seven**, entitled *A Scientific-Shamanic Approach to Religion*, I develop my own completely intuitive, shamanic approach to the inner god, the inner creative realm. I contend there is *nothing comparable* to this hitherto unpublished

Published by Sirius-C Media Galaxy LLC, 2010

text that spontaneously came to mind, back in 1992, after a deep meditation, and in French language and that was originally entitled *Le jardin infâme* (The Infamous Garden). The present chapter is a recent translation I accomplished of this extraordinary production that I could never have realized by using my rational and intellectual mind. And yet, the procedure, the way to go, in this approach to religion, is scientific in exactly the sense that shamanism is scientific, methodologically sound, and to an utmost possible extent – objective.

It is namely through asking questions that this approach proceeds scientifically, not by giving answers. This is so because answers are closing doors, while questions open doors, and because answers are most of the time temporary – also *scientific* answers!

In **Chapter Eight**, entitled *The Integrative Function of Shamanism and Channeling*, I present intuitive insights on various subjects that I inquired into, in a trance-like condition, and which show how *integrative* the shamanic quest really is.

These spontaneous reflections that were written down in fast consciousness-style diction, which is a slightly more controlled form of *automatic writing*, reflect all basic areas of life from a strikingly different and partly novel perspective. One may consider them as cliff-notes, one may think of them as speculations, one may call them approximations to truth – these expressions are what they are, verbal distillations of something that cannot be properly put in words because the experience itself is beyond the level of verbal language and conceptual cognition.

CHAPTER ONE

What is Not Shamanism?

Shamanism and Animism

It is important to distinguish shamanism from *animism* and *paganism, parapsychology, humanism,* or *theosophy.*

The shamanic quest is different from these conceptions of life, or approaches to life, or scientific methods; it is important to realize this difference, and it is in my view because of the large grey area around these notions that shamanism is often confounded with speculation, superstition, primitivism or worse, an *entertainment* for Western tourists. It is none of that. It is not related to Zen either or to Taoism, it is not a philosophy. It is, to repeat it, a science.

To begin with, *animism* which is connoted to the old Latin word *anima*, the living spirit, is an intelligent religious attitude that sees all of nature as spirited, alive and intelligent, as well as interconnected by some kind of *overarching soul.* Some use the notion 'ensouled' for the way animists tend to look at the world and life as a whole. Animists, as in shamanism, believe that souls or spirits exist in animals, plants and other entities, in addition to humans. Animism may also attribute souls to natural phenomena, geographic features, and even manufactured objects.

However, the difference is that the animist is not a scientific thinker, while the shaman is. Typically, the animist thinker contemplates nature, while the shamans interrogates nature.

In other words, the animist sees nature as existential just like the shaman but stays with this *existential vision,* while the shaman goes beyond and enters into a direct dialogue with nature. When shamans say they have communicated with spirits, what they want to say is that they have been in touch with nature's intelligence, that they have been infused with natural energies, which in most cases are non-human forms of living that have their own vibrational code that comes over in their bioplasmatic irradiation or cell vibration.

The shaman, by letting his own cell vibration enter into resonance with the cell vibration of plants and animals, enters into a nonverbal and cross-species dialogue that can be highly revealing. When shamans do not get access to the energies they try to connect with, they search in a systematic manner for a way to achieve their goal.

The animist thinker does not typically become inventive to that point, and is generally not to that point pragmatically motivated. That is why animism while it's not a religion in the traditional sense, is often taken as a form of 'nature religion', because it does not generally inquire into the composite nature of creation and the revelatory means that are at the disposition of a shaman.

Published by Sirius-C Media Galaxy LLC, 2010

Shamanism and Paganism

While there are certainly parallels between shamanism and *paganism*, the latter is a term which, from a Western perspective, has connotations to s*piritualistic, animistic, and shamanic* practices or *folk beliefs.* Thus it is a very wide notion, or antithesis to clerical, dogmatic or fundamentalist Christianity.

As such there is nothing even remotely scientific about it. Typical for pagan beliefs is that they are *polytheistic* and worship the spirits of nature. Paganism is typically a ritual and often *nothing but a ritual.* Contrast with that the systematic and pragmatic approach of the shaman, who is a phenomenologist, not a believer in rituals.

In addition, it has to be seen that today neopaganism has become much of a belief system on its own, with its dogma, which is often but an anti-dogma, and there is very little scientific attitude within the neopagan communities as to delivering truth. There is much of worship in paganism, among which goddess worship is perhaps the most well-known. But worship certainly is not a scientific methodological approach to life. The shaman is not a worshipper; he dialogues with spirit entities on an equal level.

It has to be seen also that the term 'pagan' is a Christian adaptation of the *gentile* of Judaism, and as such has an inherent Abrahamic bias, and pejorative connotations among Western monotheists, comparable to *heathen, infidel, mushrik* and *kafir* in Islam. These negative connotations have precisely to do with the ritualistic and dogmatic stance in paganism, its almost ideological character. It has to be seen that this is a trait completely unknown to shamanism because who is active in shamanic communities is not the group, not the clan, not a herd of worshippers and grass hoppers, but the single shaman in his or her sacred solitude. Shamanism is a solitary experience.

I am particularly conscious of this because before I arrived in Ecuador for my voyage, I have been searching to participate in *Santo Daime* in Brazil which is a uniquely Brazilian spiritualist group ritual where Ayahuasca is ingested during dances, songs and incantations. But as I was not accepted in any one of these esoteric groups, I went, somewhat frustrated, directly to Ecuador, and recounted the experience to Esteban, the shaman, upon my arrival. He quietly and firmly replied to my little report:

– It is good that it did not work for you. I think you were guided. Ayahuasca is not to be ingested in a group setting, but only as a solitary journey. All Shuar shamans agree about this, and there are no exceptions to be made. The journey is a solitary one, the other rituals are mix-ups of various religious rituals, and as such they are *delusional*, not the way to gain real truth!

It is for this reason that ethnologists avoid the term paganism with its uncertain and varied meanings, when they refer to traditional or historic faiths, choosing more precise categories such as *polytheism, shamanism, pantheism* or *animism*. It also should be noted that the fact that neopaganism has its place today within popular culture and is currently gaining in popularity among the young, and with the fashion world, is

Published by Sirius-C Media Galaxy LLC, 2010

certainly not conducive to its developing one day a systematic and scientific method toward apprehending reality. This method, however, has been developed by shamanism, as I am going to show in the first chapter.

Shamanism, Parapsychology, Humanism, Theosophy

Further shamanism must not be confounded with *parapsychology, humanism* or *theosophy*. Apart from the fact that shamanism has in common with parapsychology and theosophy that the latter are equally scientific in their approach to reality, these sciences have their origin in the Western cultural world and tradition and cannot even remotely be conceived to operate within the shamanic set and setting. Let me explain this in more detail.

Shamanism and Humanism

To begin with, *humanism* is not a science, but an ethical quest, while this quest well affirms using science as a way to find truth.

Humanism is a broad category of ethical philosophies that affirm the dignity and worth of all people, based on the ability to determine right and wrong by appealing to universal human qualities, particularly rationality. It is a component of a variety of *more specific philosophical systems* and is incorporated into several religious schools of thought. Humanism can be considered as a process by which truth and morality is sought through human investigation. Focusing on the human capacity for self-determination, humanism rejects the validity of transcendental justifications, such as a dependence on belief without reason, the 'supernatural', or texts of allegedly divine origin.

Published by Sirius-C Media Galaxy LLC, 2010

Humanists endorse *universal morality* based on the commonality of the human condition, suggesting that solutions to human social and cultural problems cannot be parochial. Humanism rejects deference to supernatural beliefs in resolving human affairs but not necessarily the beliefs themselves.

According to humanistic agenda, it is up to humans to find truth, as opposed to seeking it through revelation, mysticism, tradition, or anything else that is incompatible with the application of logic to the observable evidence. In demanding that humans avoid blindly accepting unsupported beliefs, it supports *scientific skepticism* and the scientific method, rejecting authoritarianism and extreme skepticism, and rendering faith an unacceptable basis for action.

The difference between *science and humanism* is that humanism is not a science; it only points to science as a possible truth-giver, and says that religious dogma, by contrast, is an impossible truth-giver. As such, humanism is not a method by itself, but an appeal to ethical transparence and consistency. As a result, I conclude that science and shamanism can be compared because both use the scientific method, while they cannot compare with humanism as the latter is but an ethical quest or appeal to reason, not itself a scientific apparatus.

In addition it has to be seen that humanism is a concept that has grown on the soil of modern Western culture, and pretty much as an anti-reaction to Christian fundamentalism. As such it has no place in shamanic culture.

Shamanism and Parapsychology

Among the corner stones of my psychic research were the extensive studies on spiritism, in general, and ectoplasms, in particular, by Emmanuel Swedenborg, Charles Richet and Baron von Schrenck-Notzing, as well as the meticulous research done at Stanford University on the medium Uri Geller.

While ectoplasm research was never really taken serious before the team Richet-Notzing did meticulous and pedantic research, with all precautions taken against fraud, it is until today a topic that is controversial. The researchers came to the result that the phenomena exhibited are not to be explained with anything known from traditional science. Most of their research was done with the famous medium Eusapia Palladino.

However, the study of energy phenomena in general has given me a clearer picture of what material ectoplasms could be made of. It seems that this substance is a fluid emanation of the bioplasmatic energy that I call *e-force* and that has been called *ch'i, ki, prana, mana, wakonda, shakti, libido, animal magnetism, odic force, universion* or *orgone* over the ages, an energy-information-field that is at the basis of all life and that permeates all, penetrates all and is a major information transmitter in the universe.

What a group of open-minded researchers did for shamanism[3], one man did single-handedly for psychic research or parapsychology. This man is Dean Radin. Here is a quote

Published by Sirius-C Media Galaxy LLC, 2010

from his book *The Conscious Universe: The Scientific Truth of Psychic Phenomena (1997):*

Dean Radin

When modern science began about three hundred years ago, one of the consequences of separating mind and matter was that science slowly lost its mind. This split became painfully obvious about seventy-five years ago when psychotherapy began to intensely embrace the value of personal experience and behaviorism began to intensely deny the value of personal experience.

Parapsychology fits in this picture by straddling the edge separating the mind-oriented disciplines such as clinical and transpersonal psychology and the matter-oriented disciplines such as neuroscience and cognitive science. Parapsychology explicitly studies the interactions between consciousness and the physical world. It assumes that downward causation exists in some form, and it assumes that scientific methods can be used to study this middle realm in a rigorous way.

Thus, the persistent controversy over psi can be traced back to the founding assumptions of modern science. These assumptions have led many scientists to believe that the mind is a machine, and as far as we can tell, machines don't have psi. The problem is that most of the classical assumptions that originally spawned the idea of mind-as-a-machine have dissolved in the wake of new discoveries.[4]

In my research I soon found that ectoplasms were manifestations of dense bioenergy liberated by mediums in trance. This unique phenomenon, then, is at the root of an array of other research topics in parapsychology such as the materialization of ghosts and psychokinesis.

While psychic research is a very important and revelatory scientific approach to our higher human potential, it is entirely situated within the Western cultural setting and cannot be seriously compared with shamanism. This is so despite the fact that parapsychology also researches shamanism, but the opposite is not true. Shamanism doesn't need parapsychology for evidential backup of its scientific character. To see only that it is thousands of years older than the new science of *parapsychology* that actually only corroborates the perennial science tradition!

Shamanism and Theosophy

This argument is even more true with regard to Theosophy, which is but a *revival of millenary knowledge* that got lost in the last four hundred years of Cartesian reductionist madness. Theosophy certainly learnt from shamanism. Shamanism has nothing to learn from Theosophy. It's that simple!

Published by Sirius-C Media Galaxy LLC, 2010

Shamanism and Taoism

If there is one teaching that seriously be considered to have parallels with shamanism, it is Taoism, a philosophical school from ancient China. One of its foremost sources are the *Tao Te Ching*, by Lao-tzu.[5] Some of the foremost qualities that Taoism values are a non-biased mindset, acceptance of all-that-is, including the world, integration of all of our emotions, magnanimity, patience and tolerance toward the uneducated, brute and perverse majority of humans who are caught in innumerable projections due to their refusal to face what-is and their entanglement in possessions, status and time-bound concepts. Lao-tzu is considered, together with Chuang-Tzu, as the primary representative of Taoism.

Taoism was defeated by violent patriarchy in the same way as its Western homologue, the truly systemic *All-Flows* philosophy if Heraclites, was defeated by the aggressive, polemic and judgmental Platonic and Aristotelian thought that became the basis of the dogmatism of the Christian Church.

I came in touch with Taoism for the first time in 1991, when I was reading the *Tao Te Ching* and the illuminating poetics of Chuang-Tzu, through which I was transformed. The peace and joy that filled me when reading these works of wisdom are not to describe with words – it was sheer bliss, an experience, by the way, that I have never had with reading the bloody stories of the Bible or the bloodless social and sexual policies of the Koran that forever will leave me appalled and depleted of energy.

I will once have to write a larger essay about this transformative experience. So far, I have written an analysis of the *dynamic life patterns* contained in the I Ching that may serve as an entry point for the interested reader.[6]

There is one point that I would like to mention here, and where Taoism and Christianity really clash wide apart. It's the question of judgment, and of judgmentalism or *moralism*. One of the most salient characteristics of the three monolithic and theistic religions, Judaism, Islam and Christianity is that they are *judgmental*, and harshly divide life in black and white, good and bad, high and low, and so on and so forth.

Taoism really does the contrary. It finds unity and fosters growth as the major overarching pattern of living where the three major religions dissect, divide, split apart and fragment, and thus repress, regress and kill. It is not in any way judgmental; it simply is beyond any of these dualistic categories. It creates while traditional religions destroy, it is in favor of the young and free expression of emotions, where traditional religions *mutilate the young emotionally* and consider the free expression of emotions as a major sin. Thus, it is truly one with the Divine, while the major religions are truly one with the Devil.

But, to repeat it, Taoism is a philosophy or religion, not a science. Shamanism is.

Published by Sirius-C Media Galaxy LLC, 2010

Shamanism and Zen

Zen emphasizes *dharma practice* and experiential wisdom, particularly as realized in the form of meditation known as *zazen*, in the attainment of awakening. As such, it putatively de-emphasizes both theoretical knowledge and the study of religious texts in favor of direct, experiential realization. The establishment of Chan (Zen) is traditionally credited to the Indian prince turned monk Bodhidharma.

The emergence of Chan as a distinct school of Buddhism was first documented in China in the 7th century.

From China, Chan subsequently spread southwards to Vietnam and eastwards to Korea and Japan. In the late 19th and early 20th centuries, Zen also began to establish a notable presence in North America and Europe.

I began with Zen meditation in 1991, and have done it ever since. In the beginning it was rather painful because of my legs not having been used to bend, so it hurt quite a bit, and the blood circulation got messed up after about half an hour. But I realized that my body improved in many ways through the practice. To begin with, the legs got used to it, and I could sit around one hour, but more importantly, my blood circulation got a boost, and my spine problems ceased completely. I also felt that once the body was used to the posture itself, while it first was kind of disturbing, became a trigger of the relaxation response in the body. It was as if it was becoming a mold within which the body would feel totally at ease, and with every coming day of practice, even more so.

Zen is certainly important. It is a philosophy of life, a lifestyle even, a wisdom technique. *But it's not a science.*

We have here a striking similarity with animism, as I have already pointed it out. Both animism and Buddhism are contemplative in nature, that is, their regard upon nature is descriptive, and besides, meditative, in the sense that nature serves the Buddhist as a metaphor for life. While the Buddhist view upon nature is metaphorical, the shaman's view is realistic; it is not contemplative, but inquisitive. This is why it is at its very root scientific, while the Buddhist view is at its very root religious, but of course not in the sense of dogma or of indoctrination but in the sense that contemplating the beauty and perfection of nature helps us to connect with our inner beauty and our inner perfection.

This is what I mean when I say Buddhists see nature as a metaphor. Generally speaking, the spiritual view of life takes nature, and the paradoxical and unfathomable in nature, as a *metaphor for the divine,* for the unspeakable, for the unknown. This is also the Taoist view of nature that is deeply spiritual in the sense that the deepest ground of verity in humans is derived, metaphorically from the complexity that is inherent in nature, and in living systems.

Published by Sirius-C Media Galaxy LLC, 2010

Shamanism's Model Function

What I would term the 'model function' of shamanism is perhaps the most important element in the slow but definite migration of knowledge that roughly since about a century is taking place; it will hopefully result in modern science integrating the millenary wisdom about life that it has formerly ruthlessly discarded out, first through the ecclesiastical, and then through the Cartesian blindfolding of scientific reality.

We always need models, a mold where we can go back to when we lose track, a map that shows the territory, while it's essential to not take the map for the territory. That we have Western shamans now, men like Michael Harner or Alberto Villoldo, is certainly a good start, but it would probably be hubristic at that point to talk about an *integration of shamanic knowledge* into our own scientific thinking and doing.

We are not yet there. We are at the pioneer level, at best, in the starting holes of a new adventure that may lead us very far, and at the same time far back into *our own perennial science traditions* that were, all of them, holistic, before they were dissected and ostracized, declared heresy, and, later on discarded out from science as 'vitalistic nonsense' or 'superstitious beliefs'.

Not long ago, as Mircea Eliade, Jeremy Narby and other authors on ethnology and shamanism remind us, shamanic science was still taken for a pathology of the senses, or 'psychotic delusions', or a 'schizophrenic perception' of reality.[7] Similarly clear-cut judgments were applied to another human sensory ability that exists within our own culture, but was always considered an *aberration of the senses*. I am speaking of *synesthetic perception*, described by Merleau-Ponty in *Phenomenology of Perception (1945/1995)*. Alberto Villoldo considers synesthetic perception as the rule, and was confirmed in his intuition by what he saw and experienced over more than twenty years with the Laika and other native peoples he encountered. He writes in *Shaman, Healer, Sage (2000)*, p. 116:

Alberto Villoldo

As the philosopher Maurice Merleau-Ponty wrote in Phenomenology of Perception, 'Synesthetic perception is the rule, and we are unaware of it only because scientific knowledge shifts the center of gravity of experience, so that we have unlearned how to see, hear, and generally speaking, feel, in order to deduce, from our bodily organization, the world as the physicist conceives it, what we are to see, hear and feel.' Synesthesia grows as we bring awareness to touch, taste, sensation, and sound. One of my favorite synesthesia exercises involves 'tasting' your emotions. Become aware of the taste in your mouth. Is it sweet? Sour? Woody? Metallic? Now recall an incident that made you feel sad. Notice if the taste in your mouth changes. Recall a pleasurable situation, and notice again how the taste in your mouth changes.

While synesthetic perception is known to be a characteristic many genially gifted people possess, to name here only the famous composers *Alexander Scriabin* and *Olivier Messiaen*, where the feat is well-documented, it has never received sufficient attention in our mainstream culture. I know however through personal encounters with a high-class Bulgarian family that the psychiatrist Georgi Lozanov, famed for invent-

Published by Sirius-C Media Galaxy LLC, 2010

ing *Superlearning*[8], was widely experimenting with synesthetically gifted children. I had the opportunity to meet such a child, who, a seven-year old girl, was able to recognize colors with her hand palms, while her eyes were covered with a scarf. Her mother was telling me after the demonstration that the activity had been taken up by the girl's teacher who, after reading Lozanov's books, had begun to introduce the activity in the art class. To everybody's surprise almost all the children, after being introduced to the idea, were found to be synesthetically gifted.

By the time I am writing this book, in 2010, there is a real fashion to be made out especially in the United States that shows that almost all children are synesthetic in some or the other way, a discovery that has the dimension of a landslide! I consider this fact as a particularly striking example for the extent shamanic culture, which always recognized synesthetic perception and that encourages children to develop synesthetic abilities, has become a cultural model for us.

CHAPTER TWO

The Warrior-Scientist

The Shaman's Roles

The shaman is a catalyzer of the numinous experience, not a magician. The phenomena he[9] invokes are not his own, and he has but limited control over them.

Eliade sees the shaman as a manipulator of the sacred, almost as a technocrat of ecstasy, as the prime religious experience. But he's not the originator of that experience, he's not even the trigger of it. In my view, the shaman is but a communicator between the realms of true morality and true humanity.

When I say 'true morality' I do not mean *moralism* in the sense that moralism stands for enforced and compulsive morality. I am speaking of *natural or cosmic morality*, which rules human conduct that deliberately avoids to attract a negative karmic response.

Here, the shaman differs from the Priest, the Imam, and the Buddhist monk in that he doesn't *moralize*. He doesn't tell people that what they do ought to be 'right' or 'wrong', from a fixated morality scale. What he does is asking the spirits who then will render their verdict, through his mouth. The verdict may be surprising, as it may not comply at all with human moralistic rules and laws.

A good shaman, for this purpose, needs to be immoral, or rather *amoral*, for if he complies with a fixated concept of morality, he won't be able to *objectively communicate* the verdict of the spirits. And the morality of spirits is certainly not the one human patriarchy has brought about, with all its sense-less compulsion, and all the violence that flows out from it.

So, the *first* role of the shaman is clearly to be a *communicator*, an intermediary between the eternal world of the spirits, gods and demons, and the perhaps rather *temporary* world of human beings. The *second* role, for the purpose also of the first, is to be a *traveler*, which includes to be a *time traveler.*

The shaman travels for three distinct purposes. He travels for receiving information from certain spirits he knows might have the answer he is looking for. He also travels to certain sacred places or energy vortexes for meeting spirits there, or for receiving information through telepathic channels; he also travels for collecting medical remedies for healing a particular disease of a client.

In addition to these feats, powerful shamans can time-travel as well, which means they can travel way back into the past or forward into the future to test a certain outcome, or receive a karmic feedback in advance of its actual realization. To give an example, if a man intends to marry a certain woman, and asks a shaman for guidance if it will be a good union, the shaman can travel both in the past, to connect with that woman's energy vortex, inquiring about her karma, and in the future, to see how the marriage is going to turn out.

This brings me to the shaman's third and major role, the role of the *healer*. What does that imply? It does *not* imply, first of all, to administrate medication from a known reservoir of remedies as the mainstream doctor does, or even the modern homeopathic healer. It means to administrate no medication

Published by Sirius-C Media Galaxy LLC, 2010

at all. *The shaman doesn't heal the client at all.* What he does is to take over the sickness of the client in his own organism, and then heals himself, thereby healing both his own cloned sickness and its original version in the client.

This is obviously a very sophisticated approach to healing, and not to compare with the mechanistic 'medical' healing approach practiced by major civilizations. In fact, it's not a healing approach at all, because the very concept of illness is different in shamanic cultures and in non-shamanic cultures, such as ours.

In shamanic cultures, all illness is attributed to a violation of an eternal *morality code* imposed by the spirit world, not as a purely organismic deregulation. This morality code varies a little from one shamanic culture to another, but when you look closely at it, the differences are minor. It's more or less *always the same kind of interference* in healthy evolution and constructive human relations, through either the fact of committing incest, or murder, or both, or violating secondary rules, such as procreating a child with the wrong woman, or procreating it through rape, or by hunting the wrong animal, or the right animal at the wrong time, thereby risking to attract misfortunes to the clan by the animal spirit world, or else to violate the quietude of ancestors through lacking out on reverence for the ancestor cult rituals that are ongoing ceremonies and that require a considerable investment of time and effort in every single shamanic culture.

While some of these rules of conduct have been taken as the result of 'magical thinking', it may well be that those who put up such a judgment are simply lacking out on the knowledge that would justify each and every of those rules in their cultural context. It may not surprise to see such judgments coming from a culture that denies the existence of the spirit world because it denied, until very recently, the existence of the human energy field and cosmic energy fields that interact with it, or are superimposed upon it.

Holger Kalweit, a German ethnologist who is specialized on shamanic healing, describes the shaman's role in rational terms, explaining each and every of his tasks as *functional* and *comprehensive* within the particular set and setting of a tribal culture. In his book *Shamans, Healers and Medicine Men (2000)*, Kalweit writes:

Holger Kalweit

Shamanic therapy means the healing of an entire life rather than just healing failing functions and disruptive pains. For shamans, healing involves philosophy, a view of life. In this regard, giving careful account of ways of healing is only one aspect of our task.

The other is to depict the shamanic manipulation of energies of a higher dimension that for us are invisible and even unthinkable.[10]

Energy healing or *vibrational healing* as a sophisticated technique of manipulating cosmic energies, and the human energy field, is something that only very recently was being recognized within the Western scientific context. While it has roots that reach back to Antiquity in Europe, Persia, Egypt, India, China and Japan, to name only these traditions, and can be considered as an integral part of perennial science, it was first forbidden as *heresy* by the Christian Church and then

Published by Sirius-C Media Galaxy LLC, 2010

was discarded out as a 'vitalistic theory' by Cartesian science, from about the 17ᵗʰ century. It was only from about the second half of the 20ᵗʰ century that this age-old effective healing technique was being introduced into modern Western science, within the framework of so-called 'holistic science'.

At the same time, systems research corroborated it because all living systems were found to be information fields, where energy, consciousness and information are blended into functional units that can be understood not by focusing upon its parts, but upon the *relationships* between those parts, which are *energy patterns*.

However, while cutting-edge science now can effectively embrace and explain *vibrational healing*, our culture is far from accepting such a reality; this knowledge, then, is reserved to a tiny minority of systems theorists and otherwise, to energy healers, vibrotherapists, Reiki practitioners, phytotherapists, homeopathic healers and those who practice *orgonomy*, the energy science that Dr. Wilhelm Reich created and that is founded on the same principles.

Kalweit considers our culture and tradition lacking in understanding of the principles that shamanic healing, and the functional use of higher energies are based upon. It has to be seen also that the notion of *initiation*, which is fundamental to the path of the healer in all shamanic cultures, has never been considered in our culture without a negative connotation, approaching it to realms of danger and abysmal sorcery. Our culture's approach, in this respect, is such that it *suffocates natural spirituality*, while it is strangely open toward the artificial, merely ornamental and dogmatic organization of the religious experience. Kalweit writes:

Holger Kalweit

The path of initiation is branded in the West as degenerate; by contrast, in tribal society the initiation of the shaman is accepted, even encouraged and supported by everyone; and the teacher helps the student to decipher his experiences by means of cultural symbols. But in our culture the symbols of transformation are negative: they include hospitalization, schizophrenia, brain-wave tests, stupefying psychotropic drugs, and ostracism from society. How many unrecognized shamans, mediums, and saints fill the madhouses of rationalism? How many powers have been mangled and cut off during the long history of psychiatry? How many people has psychology reduced to mindless robots through its abasement of the psyche? The spiritual climate in our society shuts down shamanic experience in its incipient stages, distorts it and desacralizes it as neurosis and psychotic deception. But psychic transformation cannot be extirpated by societal taboos. Spiritual experience is a transhistorical, transcultural phenomenon and can break through in individuals at any time.[11]

Accordingly, Kalweit sees the role of the shaman primarily as a *functional, effective, pragmatic manipulator of higher energy fields*, with the intention to bring about a transformation of consciousness, and thereby, a form of total healing. Healing, in this connotation, must embrace the entire human, as it is a *transformative process*, not just the curing of some or the other symptoms or ailments. The research Kalweit appears to have done on this complex question is admirable, and has delivered detailed knowledge of how the various tribal cultures around the world understand and describe the cosmic energy

Published by Sirius-C Media Galaxy LLC, 2010

fields their shamans have witnessed. I must admit that despite my own twenty years of research on the *energy nature* of human emotions and sexuality, I came across these specific notions of the *cosmic energy matrix* only through Kalweit's book. He writes:

Holger Kalweit

All tribal cultures live in unity with nature, in unity with the universal laws. Their sense of the world is a symbiosis of God, world, and ego. Thus, they have a worldview for which our 'high culture' is not yet ready. If the symbiotic, synchronistic, synergistic is the acme of the natural-mystical worldview of shamanism, another high point in the natural philosophy of natural societies is their belief that a power permeates all being. Each tribe has its own concept of this universal energy. Pacific peoples call it *mana*; the Crow speak of *maxpé*, the Dakota of *wakan*, the Hidatsa of *xupa*, the Algonquian of *manitou*, he Hurons of *oki*, the Tierra del Fuegans of *waiyuwen*, the African Sotho of *moya*, the Masai of *ngai*, the Bantu of *nzmbi*, the Pygmies of *megbe*, the Australians of *joja*, the Dajak of Indonesia of *petara*, the Batak of Sumatra of *tondi*, the Malagasy of *hasina*. The same is meant by the ch'i or the ch'i gong of the Chinese, the ki of the Japanese, the Hebrew ruach, and the *prana* or *akasha* of the Indians - *akasha*, the matrix of the universe, the *mysterium magnum*, the primordial ocean.[12]

As I already mentioned, the astonishing difference between the way shamans cure and Western doctors cure is that the shaman takes the medicine, while in Western medicine it's the patient. The shaman, through the trance, *enters the vibrational field of the patient*, and can thus detect the real problem of their illness, by screening their luminous body. This is all the secret, or the most part of it. No medicine is needed when you can alter vibrations within the aura, an insight that today has been made useful for medicine, and that is at the basis of what we call *vibrational medicine*.[13] Mircea Eliade observes in his classic, *Shamanism (1964)*:

Mircea Eliade

The morphology of shamanic cure is the same almost throughout South America. It includes fumigations with tobacco, songs, massage of the affected area of the body, identification of the cause of the illness by the aid of the helping spirits (at this point comes the shaman's trance, during which the audience sometimes ask him questions not directly connected with the illness), and, finally, extraction of the pathogenic object by suction.[14]

A particularity almost unknown in Western culture, that however can be found in many tribal nations is that illness is often attributed to the interference of the spirit world. While Western consciousness evolved from a merely palliative and mechanistic medical paradigm to one where the patient is seriously asked how he or she may have contributed to bring about their disease, the spiritogenic etiology, method used by shamanic cultures, would by most doctors probably be qualified as schizophrenic. Not so in tribal cultures. Rule and exception can be seen as reversed in the sense that in most native cultures, illness is primarily seen as a form of *superimposition* of malignant spirit power, and only in second instance as a possible result of an individual's condition, weakness, or fragility, or corruption to have let it happen. Eliade describes a healing ritual:

Published by Sirius-C Media Galaxy LLC, 2010

Mircea Eliade

Throughout Melanesia treatment of a disorder begins with sacrifices and prayers addressed to the dead person responsible, so that he will remove the sickness.

But if this approach, which is made my members of the family, fails, a mane kisu, 'doctor', is summoned.

By magical means the latter discovers the particular dead man responsible for the sickness and begs him to remove the cause of the trouble.[15]

Shaman, Healer, Sage

Dr. Alberto Villoldo, a highly acclaimed modern-day expert on shamanism, teaches in his workshops techniques for *self-healing and awareness-building* that he himself needed almost twenty years to acquire during an apprenticeship with Laika shamans high up in the Andes. Villoldo has written captivating and bestselling books on the matter that I have reviewed, considering them as prime literature for any intellectual who wants to know why our Western culture failed to bring about integrated humans, and how we can heal our collective inner split and become whole again, and *holy*.

While another who went that far would perhaps stayed with those who loved him enough to share their most ancient secrets about life, healing and peaceful living, Villoldo came back with the intention to share what he had learnt, and organized a series of ongoing workshops that are held at various destinations all over the United States. In studying and reviewing four of Villoldo's major books and his DVD *Healing the Luminous Body (2004)*, I was intrigued enough to include his insights in my own ongoing research on energy science and vibrational healing. To take up the thread of thought from the last sub-chapter, I would like to quote Villoldo on the main cultural difference between the shamanic and the modern-day perspective because Villoldo expresses it particularly well. He writes:

Published by Sirius-C Media Galaxy LLC, 2010

Alberto Villoldo

In the West we identify with the side of matter, which is by nature finite. The shaman identifies with the side of energy, which is by nature infinite.[16]

One of the strikingly novel insights that Villoldo gained on psychosomatic medicine was that the age-old dichotomy between body and mind eventually turned out to be a gigantic misunderstanding of the functional unit that the human body represents.

Here, modern science eventually meets the oldest of traditions that always affirmed there is no such split, except a culture brings it about through putting the mind 'over the body', thereby separating the mind from the body.

It was through the new science of psychoneuroimmunology that Western scientists could eventually grasp the notion of an *'intelligent energy field'* that is upheld through *flow*, and the enormous liquidity of the human body, which consists of 98% of water. Villoldo writes:

Alberto Villoldo

In the last few decades the field of psychoneuroimmunology (PNI), which studies how our moods, thoughts, and emotions influence our health, has matured. PNI investigators discovered that the mind is not localized in the brain but rather is generalized throughout the body. Dr. Candace Pert found that neuropeptides, which are molecules that continually wash through our bloodstream, flooding the spaces in between each cell, respond almost instantaneously to / every feeling and mood, effectively turning the entire body into vibrant, pulsing 'mind'. Our body as a whole experiences every emotion we have.[17]

Villoldo makes a complex field of research easier to understand by associating the realm of spirits and the vibrational field that is at the origin of life, as *infinity*. He explains to the reader that infinity in this sense doesn't mean eternity as eternity implies the notion of time, while infinity is outside of time, a realm of existence that knows no beginning and no end. In fact, our *higher self* belongs to this infinite realm of pure beingness. Villoldo explains that the powers the shaman receives during his psychedelic travels are coming from the fact that he encounters infinity.

For all of us, Villoldo explains, meeting with infinity is a transformative and rejuvenating experience because we realize that we are not defined by our history, by our past, by our accumulated karmic experience:

Alberto Villoldo

We realize that we are not our stories. And the experience of infinity shatters the illusion of death, disease, and old age. This is a not a psychological or spiritual process only; every cell in our body is informed and renewed by it. Our immune system is unbridled, and / physical and emotional healing happen at an accelerated rate. Miracles become ordinary, and spontaneous remissions, those mysterious and baffling cures that confound medicine, become commonplace.

And a spiritual liberation or illumination takes place. In the presence of infinity we are able to experience what we were before we were born and who we will be after we die.[18]

I mentioned already several times that the main reason I wrote this book was my old idea that, contrary to what traditional anthropology asserted, the shaman is essential a scien-

Published by Sirius-C Media Galaxy LLC, 2010

tist, a discoverer, and only randomly a magician or, as it was believed, or a trickster.

It is not my own idea but was found out already by the late Terence McKenna that when shamans speak about 'spirits', the convey *energy fields* or energy patterns that in one way or the other become *superimposed* upon our own human energy field, thereby triggering various kinds of responses.

These responses range from simple psychedelic visions, over various pathological states to highly dangerous paranoid delusions that can lead to sudden and dramatic suicide, as affirmed by Michael Harner in his revealing book *Ways of the Shaman (1980/1982)*.

Sorcery, the malevolent malignant kind of work done by shamans can be understood, under this scientific perspective, as an application of energy principles for deliberately triggering pathological conditions in the body of the person that the sorcerer intends to harm. The methodology of both doing harm and good is the same, a deliberate logical approach to manipulate energy fields in the *luminous body* of the person that the shaman is focused upon.[19] This has thus very little to do with tricking somebody out, with creating illusions or with doing 'magic' in an attempt to manipulate reality without knowing what the factors are that are really in play for doing so. In this sense, therefore, the shaman is *not a magician* for he knows what he is doing and why he is doing it; he knows what those factors are that are at the basis of reality creation.

He also knows that 'reality' is *not a fixated universal condition* but a state of mind that is individually upheld through *consciousness*, and influenced by energy fields. Logically then, it is by manipulating these energy fields in the luminous body of the client that the shaman changes the reality of that person, for *eternity*, through having that person consciously experience *infinity*.

Now, the knowledge about the human energy field is not a dubious vague matter, but a *body of detailed knowledge* that was not understood by Western official science for the simple reason that it was secret knowledge, not revealed to the world until first the Toltec[20] and then the Laika broke the long silence. Dr. Villoldo explains:

Alberto Villoldo

The Luminous Energy Field is shaped like a doughnut (known in geometry as a torus) with a narrow axis or tunnel, less than a molecule thick, in the center. In the Inka language it is known as the *popo*, or luminous bubble. Persons who have had near-death experiences report traveling through this tunnel in their return voyage to the light. The human energy field is a mirror of the Earth's magnetic field, which streams out of the North Pole and circumnavigates the planet to reenter again through the South Pole.

Similarly, the flux lines or *cekes* of the Luminous Energy Field travel out the top of the head and / stream around the luminous body, forming a great oval the width of our outstretched arms. Our energy fields penetrate the Earth about twelves inches, then reenter the body through the feet.[21]

It is important to realize that wherever in the universe, there is consciousness, there is energy, and where there is energy, there is consciousness. But also, where there are consciousness and energy, there is *memory*. Memory, thus, is a function of energy fields or energy patterns, and not just of

Published by Sirius-C Media Galaxy LLC, 2010

'chemical structures in the brain'. The chemical structures that were detected by Western neuroscience are real, but they are created by *energy streamings in the luminous energy field*, which are the true beholders of the memory interface. Villoldo explains:

Alberto Villoldo

The Luminous Energy Field contains an archive of all of our personal and ancestral memories, of all early-life trauma, and even of painful wounds from former lifetimes. These records or imprints are stored in full color and intensity of emotions.

Imprints are like dormant computer programs that when activated compel us toward behaviors, relationships, accidents, and illnesses that parody the initial wounding.[22]

Studying the once secret energy science of native cultures, such as the Laika, but also the Kahunas in Hawaii, first discovered for the West by Erika Nau[23] and Max Long[24], we realize that all true knowledge about life and death, and about health and sickness, begins with the conscious perception and manipulation of energy fields. And interestingly so, this science, just as Wilhelm Reich's orgonomy[25] and the old Chinese science of *Feng Shui*[26], make a clear distinction between the human energy field and other, planetary and cosmic energy fields.[27]

If native energy science was, as still today many mechanistic scientists assume, but a 'vitalistic' belief system combined with delusions produced by 'a psychotic delirium' that is believed to be a 'suggestive healing' ceremony, and effective only because of *belief*, it couldn't be learnt by a Western doctor and applied with clients who *do not harbor the cultural beliefs* that it was originally based upon. Dr. Villoldo writes:

Alberto Villoldo

Over the years I developed the ability to perceive the streams of light that flow through the luminous body, and read the imprints of health and disease. I believe that this is an innate ability that we all possess but either do not develop or lose after the age of seven or eight because we are taught to believe that the material world is the only 'real' world. Shamans throughout the Americas rely on their ability to perceive the energetic realm.[28]

I personally consider a witness report originating from a rather ignorant outsider *higher* than such report written by a member of that wistful culture itself. This is so because *we are all biased,* if we want it or not, and, while this sounds queer, the danger to distort reality is *higher* when we judge *our own cultural reality* than when we try to learn and comprehend a *different* cultural reality.

It is exactly the strangeness and queerness of the experience that straightens and reinforces our awareness level; when we deal with everyday life, we tend to be shallow and unaware. When we enter novelty, we are fully in the present!

This is one of several reasons why Castaneda's and Villoldo's accounts of shamanic reality are primed sources of knowledge because the cross-cultural researcher intends to *strip research of its cultural bias and of its innumerable projections.*

When we penetrate into shamanic energy science, we gain detailed knowledge about the human body's energy sys-

Published by Sirius-C Media Galaxy LLC, 2010

tem and its rotating chakra wheels that are primordial energy vortexes, managing each a certain *clearly definable frequency range* of the total energy rainbow that is the base layer of all life in the cosmos – for *life is light*. This knowledge is not reserved to tribal cultures only, but has a long tradition in our own culture through *clairvoyance*.[29] Alberto Villoldo explains the chakras as follows:

Alberto Villoldo

The chakras are the organs of the Luminous Energy Field. They are swirling disks with wide mouths that spin a few inches outside the body; through which they drink in the radiant fuel stores in the luminous body to nurture us spiritually, emotionally, and creatively. The narrow, funnel-shaped tip hooks directly into the spine.

The chakras transmit information of past trauma and pain, contained in imprints in the Luminous Energy Field, into the nervous system. The chakras inform our neurophysiology, affecting our moods and influencing our emotional and physical well-being.

The chakras also connect to endocrine glands that regulate all of human behavior. [30]

Now, this is the entry point for understanding how shamanic healing works, how the shaman really effects the total healing of the organism, through a *deliberate interference with the luminous energy field* of the client. Alberto Villoldo explains:

Alberto Villoldo

The blueprint that shaped and molded us since we were inside our mother's womb contains the memories of all of our former lifetimes – the way we suffered, the way we loved, how we were ill, and the way we died. In the East these imprints are known as karma,

> forces that sweep through our life like a giant tide that
> we cannot swim free of. These imprints contain in-
> structions that predispose us to *repeating certain events from
> the past.* We want to learn where these energy imprints
> are located in the Luminous Energy Field and how to
> erase them so that the body, mind, and spirit can re-
> turn to health.[31]

The science of *vibrational healing* is complex, too complex for quacks and charlatans who are, according to common Western media culture, the only participants in the ecstasy of the participatory healing experience that is defined by the fact that the healer is but a catalyzer of cosmic energies that are *already present* in the client's human energy field – but that are blocked by the particular disease that is at stake.

This being said, I insist that energy healing is not, as commonly believed, a transmission of cosmic energy from the healer to the patient, but *activation of the patient's own energy resources* that are stalled or in a state of stagnation that is the result of neurosis – which in turn may be the consequence of abuse suffered in early childhood. Villoldo explains compre-hensively how distortions in the luminous energy field impact negatively upon our emotions:

Alberto Villoldo

Imprints etched into the emotional-mental layer of the Luminous Energy Field predispose us to live in par-ticular ways and to become attracted to certain people and relationships. These imprints dictate the course of our emotional lives. It is very difficult to change our lifestyle without clearing the imprints in this layer.

Imprints stored in the etheric or soul layer inform and organize our physical reality. Imprints in the causal or spiritual layer choreograph our journey through life,

Published by Sirius-C Media Galaxy LLC, 2010

including the kind of spiritual peace and fulfillment that we will attain.[32]

In my own long-term research on human emotions that goes back to the 1970s, I came to the same result as Villoldo, namely that *emotional memory* is not stored in the brain but in the aura or luminous energy field.

Alberto Villoldo

The language of computers consists of magnetically charged zeros and ones. / The Luminous Energy Field is similarly coded. Childhood abuse is not recorded as an image of a child being battered. Likewise, a cancer does not appear like a blob in the energy template. They both appear like pools of dark, stagnant energy to those who can see.[33]

Imprints are formed when the negative emotions that accompany trauma are not healed. (...) Crises or emotional stress would trigger the script contained within the imprint, which would begin to play itself out again.[34]

Psychologists believe that the subconscious motifs and behaviors we inherit from our parents might be encoded into the circuitry of our brain, and that the only way we can reprogram these circuits is through psychotherapy. I'm convinced that these energy patterns and habits are encoded in the Luminous Energy Field as well and that the Illumination Process can accomplish in one session what can often take years to heal through psychotherapy.[35]

Our medical and mental health industries probably have an interest in psychotherapy remaining long and costly, while there is a clear trend to make for a shift toward shorter and lesser costly therapies, such as *hypnotherapy* and other, recently developed therapies.

Our mainstream scientific and medical authorities are *still far from recognizing a therapy approach that directly works with repatterning the luminous energy field,* thereby tackling the primary memory surface of the human mindbody unit.

Another aspect to healing in general is to see it as a learning experience. Our traditional medical science has this 'bullet approach' to it, that lets us see disease as an enemy that has to be attacked and killed, while in native cultures there is a lesser judgmental tone regarding the fact that humans turn ill and well again. This has been seen already back in the 1970s by the Simontons as a major handicap in our understanding of health, and as a factor to keep people from healing, because they themselves, and the whole medical establishment behind them, is completely focused on disease.[36]

It's a fact of life, and a feature of human consciousness that we reinforce what we direct our attention upon; so if it's *not health but disease* which is the main focal point of our attention, we *reinforce disease,* and when we do that on a group level, then we as a society will have a hard stand on 'fighting disease'. In fact, then we as a group or nation will be preoccupied with nothing but disease and as a result, our health statistics will worsen, and not improve, with every year. And this is exactly what's going on and what the Simontons have already found thirty years ago.

While for shamans this is *the only valid approach to healing,* this kind of direct experience of the energy nature of life is considered by most Westerners as esoteric, and outlandish. It's almost as trying to explain to a blind man how a multicolored bird's feather looks like. Villoldo goes as far as assert-

Published by Sirius-C Media Galaxy LLC, 2010

ing that 'Westerners have not developed the neural pathways to sense energy'.[37]

Another factor may be our primarily intellectual understanding of life and living, in our Western scientific tradition, while the approach of most native cultures is holistic, intuitive and kinesthetic.

This in turn has of course an impact upon our readiness for change, our else *our rigidity to avoid change.* Villoldo writes:

Alberto Villoldo

Change happens first at a core energetic level, and then the intellect gets it. In contrast, in the West we insist that understanding must precede healing. We first rehash and rationalize how our mother or father was not emotionally available for us before we embark on change. In luminous healing, the mind can have its insight after the energy field and body change, but true transformation *can never be preceded by the intellect.*[38]

It may well be that in so far phylogenetic factors come into play, especially when we consider the recent cutting-edge research on *morphogenetic fields,* by Dr. Rupert Sheldrake and others.[39] If a mouse in New York City learns to get out of a maze because a mouse in Los Angeles was able to do so, while the two mice are *connected with each other through nothing but a general 'morphic' resonance pattern,* then humans certainly, and even more likely than mice learn their basic abilities not only through individual development, but also through the *memory surface of the entire species* or large groups within the species, such as tribes, races, nations. And here we have the phenomenon that the Inka have the ability to see the vital energy streamings, as we know it from cobra snakes and other animals, but most humans on this planet, including the author of this book, have not developed this ability. We know it well from singular exemplars within our own culture that often are taken for charlatans, who well possess this ability. I am speaking of mediums, paranormally gifted individuals, and clairvoyants.

Also, what might stand against developing these abilities is our left-brainism, our group fantasy of 'total reason' which disconnects us from our bodies, our hypertrophy of *deductive logic* to the detriment of *associative logic*, and our almost political focus on 'yang' values in education and the general social code. It is our *moralism* that stands in the way, our judgmental attitudes, our constant obsession with 'good' and 'bad' judgments that pervade our media world, our religions, and our talk shows. It seems that in tribal cultures, this is very different, if not completely absent. Villoldo writes:

Alberto Villoldo

In the Inka shamanic traditions there are no 'bad' energies. There are only energies that are 'light', and so support life, and energies that are 'heavy', which cannot be digested.[40]

This is what Wilhelm Reich termed the 'functional regard' upon nature, well aware that at that time, more than half a century ago, there were very few scientists ready to see that this holistic view was going to be the leading paradigm of future science, and not what we know today was 'mechanistic science'.

Published by Sirius-C Media Galaxy LLC, 2010

We can only hope that the energy view of life will soon be part of our own science, and will after a while even become the mainstream part of it. There are many road signs *to be made out already now* that show that this is indeed what's most likely going to happen, in a few decades from now.

More detailed information can be obtained through my monograph *Energy Science and Vibrational Healing (2010)*.

My Shamanic Quest

My mother and my grandmother have been psychic. None of the males in the family, apparently, none of my uncles, nor my father and his family had this ability. But my mother and even more so, her older sister, were psychic from early childhood. My grandmother often told the story how, during World War II, she got in touch with all her four children, who were far from home, and with whom from a certain moment, all contacts had broken off. My mother, after she had finalized her art studies in Leipzig, moved to Berlin for working as a news reporter for Berlin Radio. My mother's sister had been in Munich, and my two uncles were at the front in France.

My grandmother said that in a dream she got the information that my mother was in Berlin, working at Berlin Radio, Masurenallee. My grandmother knew a fire police man who knew my mother well, as he had courted her for some time. Now she called him, asking him to drive to Berlin to get my mother out of the bombings that were particularly heavy at that time. While it sounds like a miracle, the fire police man found my mother, unconscious, as she had almost suffocated in the bunker under the radio station that was bombed down to ashes, and while all other people had left the bunker already. He was able to save her life and to drive her back to her home town, where my grandmother received her in great relief and with a shock as she did not recognize her at all.

Published by Sirius-C Media Galaxy LLC, 2010

Through other circumstances she got in touch with my aunt in Munich, and received a clear dream that her two sons were okay, and would return home soon. It was actually shortly before the end of the war. From that moment, she said, her fears had vanished off and she was confident that her four children would be saved. And so it was.

My grandmother used the pendulum for getting answers to any questions when an important decision was to be taken. And she was always fortunate, a rich, blessed and most respected woman all her life, while she comes from a modest bakers family. Both my mother and my grandmother knew instantly who was calling when the phone rang. It was like a family game later in which I participated and saw I could with the same ease predict who was the caller. It was always correct. I still vividly remember the evenings when I was around twelve, thirteen, when we saw Uri Geller on German television. The interest my grandmother and my mother had in paranormal matters was amazing. We would discuss about it until late at night and both my grandmother and my mother would tell me many anecdotes from their lives that involved paranormal experiences. But somehow, through school and my law career, my psychic abilities had fallen into oblivion. In addition, I had developed a revolt against religion as I had been cruelly mistreated and even tortured in a Catholic home early in my life, in the name of Jesus. So I became an atheist.

It was not before I was in my thirties that I experienced anything that one would qualify as 'unusual', 'paranormal', 'psychic' or 'mysterious'. After finalizing my doctorate in in-

ternational law, deeply unsatisfied with the law career, I re-
treated from professional life for an extended sabbatical. My
purpose was to find out about my true mission. I also was
searching for my true religion for in the meantime I became
painfully aware of my inner vacuum. So in January 1989 I
engaged a psychotherapy, and started a parallel auto-therapy
for healing my inner child. In addition, I practiced spontaneous art and Zen meditation for almost five years, and became
a complete vegetarian. It was at that time that my interest in
Oriental religions arose, especially Zen, but also Taoism.

I began studying, and divining with, the I Ching. Several
friends of mine, coincidentally so, also turned toward a spiritual life at the time, but they focused on a more or less fundamentalist Christian dogma; and while I generally respected
their involvement, I could not sympathize with their ideas.

Prophetic Dreams and Spirit Visions

The *first* unusual experience was a dream I had in February 1989, and that shocked me. It was a dream with many
details and that was about a huge world war, a real Arma-
geddon that was somehow related to the Bible's Apocalypse.
The dream was so complex and emotionally so heavy that I
woke up in a real trance and had to recover from the impression for several hours. Unfortunately, the lethargy that had a
grip on me held me back from writing the dream down at
once, and I had forgotten many details when I eventually
wrote it down in the late afternoon. I retained that one of the
strongest symbols used in the dream was the number 666,

Published by Sirius-C Media Galaxy LLC, 2010

and that the 'Russians' had been winning the battle, and that the Americans had lost it. It was only through a visit in the university library and checking out several books about the Apocalypse that I got to know the meaning of the number 666, the number that stands for the Antichrist.

The *second* unusual experience was a prophetic dream I had in March 1989. In this very clear dream I saw the Berlin wall falling and masses of people applauding, full of joy and relief. Six months later, the Berlin wall really fell, to my surprise.

The *third* unusual experience occurred during a spontaneous art session, on a sunny day in summer 1990, in the late afternoon. I was completing the third drawing when suddenly, there was a shadow in the room. I felt a freeze, and a strange silence that at once filled the space. There was a presence! I did not move. From the angle of my left eye, I saw the silhouette of a little girl hiding under the table. I was deadly shocked and afraid, and uttered:

– Please leave … I cannot bear the experience, I am not ready for it. But please later, tell me by your thoughts why you came here.

The presence was gone at once, and I felt relieved and sat down on a chair. I prayed for the spirit of that little girl, and asked again aloud what her mission was to see me. A little later that day, a clear message suddenly formed in my mind and I wrote it down:

– You were not aware that I was in one of your recent drawings. I had tried to get in touch with you to tell you my story, and the story of my death. For I was murdered by a

man who had raped me, and it took me very long to understand where I was once I went to the other realm. But my
grief is so strong for I had loved that man, and that is why I
cannot go. I felt strong love for you when I felt your presence.
I felt safe around you. I was not aware that you could be so
afraid to meet me.

I became very pensive and prayed for her spirit all over
the next weeks, getting dozens of books from the library
about the meaning of death, the actual death experience
people go through, as far as it was known, the *Tibetan Book of
the Dead*, and Raymond Moody's famous study on our-of-
body and near-death experiences.

The *fourth* unusual experience was another prophetic
dream in February 1991, equally very clear and real. In this
dream I saw Mikhail Gorbachev holding a speech in which
he resigned his office. A voice in the background murmured
'The Fall of Gorbachev'. Again, about six months later, the
event actually occurred, much to my surprise.

The *fifth* unusual experience was in 1994. I had a very
clear and detailed dream that was obviously a guiding
dream. This dream that I wrote down in my dream journal
under the header 'Train to Shanghai', was as follows:

Train to Shanghai

I was standing at a platform, waiting for a train. Initially I had with me five suitcases but a voice had told
me to leave them all behind and take only one plastic
bag with me. So I had with me only a white plastic
bag. The train arrived and I saw a board on the loco-
motive showing the name 'Shanghai' as a destination.
I stepped in the train and was appalled by the smell of

Published by Sirius-C Media Galaxy LLC, 2010

the young German people standing there. Not thinking long, I opened the door to a compartment to the right, and was surprised when I entered it, for it was very luxury, the walls being covered with Mahogany wood, the floors with Persian carpets, and in the middle of the first class compartment were standing two man next to a table. It was obvious that it was a high-class Chinese man and his assistant. He was wearing a beige-colored elegant suit and had shining black hair that was pomaded and brushed backward, while his assistant was wearing a black suit and had a moustache. Both men smiled at me with much courtesy and invited me to enter the compartment and close the door behind me. I did so and only then noticed a huge golden clock standing in the middle of the table. Our conversation immediately was about this clock that I could not enough study as it contained many wheels that I have never seen with a clock. There were obviously wheels indicating cosmic time zones, not just time zones regarding our planet, and this was what the man more or less told me. I felt much love irradiating from this man toward me, and was full of admiration for him, for it was obvious to me that he was a being of high nobility and spiritual development. He then handed me over a gift, a precious stone contained in a little box, and I was speechless about so much hospitality and friendship.

The dream had been so vivid that I woke up full of energy, and immediately went to tell it to my mother and her friend. I was asking them if they thought the dream meant I should really move to Shanghai? They were skeptical, as always, and even negative about my 'crazy ideas'. I then consulted the I Ching which was not every positive about the idea either. Exactly one week later, I had a dream that I instantly felt was a follow-up dream, and that dream left me depressed. In that dream I was running toward a quay where I wanted to catch a ferry boat. But I arrived too late, in the moment namely when the boat was taking off. I could have jumped still on the platform, but a guard on the ferry waved his arm with a gesture to stop my attempt.

After that dream I was very angry with myself, thinking I had refused to follow a very clear call. Shortly thereafter I moved to Holland for writing a book, but then got in business, but that business was not successful; I lost more than one hundred thousand dollars. But nonetheless, there was a positive side about it. I had seen the end of that international discount system approaching, through those hunches, while nobody believed me.

The Turning Point

Six months before that international discount network broke apart, I namely had a strong intuition that something was going wrong within the management team of 'Premier Club International'. As a result, I began to give the subscription fee back to the strongest members in my down line.

Some of them thought I had gone mad as all was looking so brilliantly, among them two Indonesians, a wealthy business lady and a man who later revealed to be one of the eight brothers of a leading Chinese business family in Indonesia. As my intuition turned out to be true and after the downfall of the company, because of gratitude, this man acquainted me with his family in Jakarta where, six months later, I gave my first business presentation for my newly setup *Autohuna Institute* in Jakarta.

Published by Sirius-C Media Galaxy LLC, 2010

The coming four years in Java, Lombok and Bali were providing me with the rich texture of a *truly shamanic culture* that opened my eyes of wisdom even further.

The new challenging goal and the experience to work in a new culture triggered hard work. I began to formulate a training concept and worked out training schedules, work sheets, games, creative sketches, produced brochures and promotional CDs, constantly on tournée between Jakarta, Yogyakarta, Surabaya and Denpasar, Bali. I saw that all my past life experiences had been extremely valuable for this new orientation. Living in hotels was from then on a feature of my daily life and I tried to cope best with it – choosing good hotels and primarily those that supported me in return, and getting the best corporate rates possible, such as first of all the Radisson hotel chain as well as Accor hotels. My training methods and seminars were taking off in Indonesia and in 1998 I gave a train-the-trainers program to an elite corps of the Indonesian government.

I later conducted a successful mid-management and top-management seminars for several five star resort hotel outlets belonging to international hotel chains.

As for a long-term visa, I needed a so-called 'sponsor' or business partner, I was approaching a man I saw each time when I had to do my international calls. He was running that Telecom agency and twelve others in Jakarta. We really became friends, and there was really strong sympathy on both sides. He had two very gifted children, a boy who was a black belt in Karate and a girl who was awarded as the best Bali dancer in Jakarta. We got very close to each other, he invited me to his home often times. He never asked anything in return to his services for me, so I just spoilt his children with gifts, as he never wanted to accept anything. He was a small man from Sumatra and one day I learnt he was one of the local kings of that area and had a 35 acre property, and a real palace.

Psychopomping Baginda

One day I had an alarming dream about him, a dark dream, and I woke up apprehensive something could happen to him. I called him after breakfast, telling him about the dream, and he invited me to his office for later that day.

When we were sitting in his very simple private office, shielded from the noisy hall of the agency where there were always many foreigners for doing their calls and sending their faxes, he looked at me with his usual generous smile and smoked a cigarette:

– Thank you, he began, that you care so much about me. This is very unusual for me. I have my customers, I have my family. My children are busy, they have their lives now, I have my life, as you see, and hardly a minute for myself. And the situation with my wife is not easy because my former wife messes everything up. She doesn't want to let me go and spoils my new relationship. What can I do? I feel I should die, then the problem is solved...

Shocked, I said his thinking was self-defeating and destructive and it was not his fault that his former wife had such a demanding and possessive attitude and tried to bog down

Published by Sirius-C Media Galaxy LLC, 2010

his new marriage with all her female powers. I gave him a positively affirmative prayer and explained him how to repeat that affirmation over and over again, on a daily basis, so as to give freedom to his former wife and let her go.

– Do you mean, he asked, that she is so entangled with me because I am still entangled with her?

– Exactly, I replied, otherwise it couldn't happen, you see?

He looked a moment up to the ceiling, in a pensive mood and took a long draw from his cigarette.

– That's a smart idea, he said, softly, and in fact, it's true. The problem is that half of my heart still hangs with her ... and yes, by the way, as I am talking about my heart, my doctor says I should be careful with smoking. Actually I get those edemas on my arms ...

He showed me his right arm where I saw a swollen area that was blueish is color. And suddenly I had a hunch what my dream had been about. I left him very apprehensive, after giving him advice to seek out a heart specialist.

Two months later, my mother called and asked me to come back to Germany for some time to help her with our family business, so I went to see him, and had some gifts for his children. But to my surprise, he was not in the office and his staff said he had been hospitalized for his dropsy. I left the agency that day in a very apprehensive mood.

I was at home perhaps a week, when a fax arrived from Jakarta. It was from his family who informed me he had suddenly passed over and that the whole family was mourning, as he had left behind five children, three of his former marriage and two of his new marriage. He had never told me that he had had three children with his first wife, and even a baby, so he actually had ongoing relations with both his former and his new wife. While, I guessed, he had to keep the old relation entirely secret in front of his new wife and perhaps also in front of his two bigger children. One may imagine the psychic stress this man went through for years of his life!

I was to a point knocked out of my way and thrown in a depression that I wondered why? After all, he was but a business relation, I reasoned. But in that moment, I felt that that was just a way of talking. He had been a real friend to me, all of his family. I sent a long fax back to his wife that was never replied to. And as I just had finished reading the *Tibetan Book of the Dead* for the second time, I took the sudden decision to psychopomp him. Following the instructions of the book, I scheduled regular sessions for every day, over a month, during which I was reading the special passages of the book that were destined for accompanying the soul on its way to the bardo, and again out of the bardo, to its final destination in the afterworld.

I did this work with a strong sense of devotion, and did not for a minute worry about the fact that he had been a believing, however not practicing, Muslim, and that that book comes from the Tibetan Buddhist culture. I understood that the truth contained in the book is universal. I reasoned that my work may be important for him as I knew that his whole family was what in Indonesia is called 'liberal Muslims' which means they do not practice, and what is worse, they do

Published by Sirius-C Media Galaxy LLC, 2010

not practice any rituals to accompany the dead on their last voyage, after they have passed the tunnel and arrive in the astral dimension, the so-called 'Bardo'.

Each time when I did this work, I was thinking of him with love and devotion, and after the month was over, I had a feeling of relief, a sudden joy, and all grief was gone. I knew in that moment that the work was coming to its end and that it had been successful. I could have rejoiced was there not a new depression approaching. While my mother was in her seventies, and while she had been operated at the heart already several times, I had no idea that that year, 1997, I was seeing her for the last time. Yet several dreams left no doubt that this was going to be the case and that she was going to die, and not just that, but that she was going to die in a rather unpleasant manner.

The dreams I received were shocking, most of them containing the motif of suicide. In one rather absurd dream she had driven her car on top of the garage and then stepped out of the car without knowing she was stepping in the air...

Dreams Regarding Mother's Death

I had in total around five dreams where she was suffering a rather violent death, and as I felt so oppressed about the dreams I was going upstairs one morning after breakfast and told them. My mother did not say a word, but her friend viciously attacked me and went as far as suggesting that I wished my mother to die. From that moment, I knew that this man was my enemy, while already before, the tensions between him and me were not minor. The next day, my mother came to see me in the evening, when her friend was drunk and watched television. She said she wanted to have a serious talk with me, if I had the time? I said I had of course always the time, as I had come back from Asia for being at her disposition when she needed me.

She sat down in an armchair with a gesture of exhaustion, and said nothing for a long time. I did not push her to talk. I just waited, and brewed some tea. Finally, when I was back in my armchair and the tea on the table, she said:

– Your dreams did not surprise me. I know you are psychic. We all are, I mean you, me, and my mother was, and my sister. But *he* is not, as your father was not, nor anybody of his family, nor were my brothers. That may be the reason why he hates you, and I come to believe now he was not the right partner for me. Unfortunately, I found out too late. I actually would like to leave him and live alone. Better alone than with the wrong partner. But I think I do not have the force for doing such big a step now. I feel so weak after the last operation, when they put me the pacemaker. I am going to die, and I just want to make sure you come back here when I call you.

I felt very sad, could almost not get a word out, and my tears came up. I assured her I would come whenever she was going to call me. And I asked her why she possibly thought there had been that suicide motif in most of my dreams about her? She was not surprised at all, and replied:

– Oh, that's an old matter. I wanted to suicide myself since the time when all was going south with your father,

Published by Sirius-C Media Galaxy LLC, 2010

when he drank more and more, cheated me more and more, and was more and more violent. You were about one year old when he finally left and we divorced, but the aftermath of that divorce was terrible as I had to pay all the debts he had made for our shop, and poor as I was at that time, I had to pay all that money back to the creditors for the next seven years. And even later in my life, the idea of suicide was recurring, it was like a recurring intention during all of my life.

She apologized that she had never told me about it and what was even more important, she apologized for all the mistakes she had committed in her educating me within a real co-dependent relationship that denied me any autonomy, including her totally intransigeant attitude regarding my professional choice and the constant ultimatums she had faced me with in my younger years.

That evening I went to bed with an immense feeling of relief and gratitude as I needed no more information to pardon and forgive her all the hurt I had been suffering during my childhood and youth.

It was all buried from that historical moment, and I felt my mother was a great person, to eventually, and before she died, being able to leap forward in her self-perception to such an extent! I felt whole, then, and integrated, and in peace with her and myself.

The Sabdono Connection

About a week later I flew back to Jakarta and that coming year was going to open me many more doors of perception. Destiny somehow had its hand in the box, as my second sponsor, a very gentle elder, turned out to be the president of the *Parapsychological Society of Indonesia* in Jakarta, Mr. Sabdono, PhD. The only request he had in exchange for the service he provided me regarding my long-term visa was asking me to write a letter to Professor Bender in Freiburg, Germany, the international capacity on parapsychology, to get him interested in the psychic research Dr. Sabdono had conducted over more than twenty years. I did this service for him with much pleasure and devotion and my respect for him was unlimited. He was a royal person, connected with only the finest and most reputed intellectuals in the country, and as he felt I was in search of some spiritual guidance, he recommended me three honest long-term friends of his, reputed paranormals and clairvoyants who had celebrities as their clients. To my great astonishment, and presumably because I was recommended by Dr. Sabdono, their consulting fees were very moderate.

The events that I am reporting from now will show that somehow the energies around me became more and more focused, more and more condensed, for there was from that moment a whole avalanche of more or less unexplainable events happening in my closer surroundings for the next three years, until the death of my mother.

It started with a Chinese paranormal coming to visit me in my new flat, on the 30th floor of *Batavia Residence* in Jakarta, an apartment that was almost entirely covered with mirrors; the bedroom really was one mirror cabinet. The small quiet man sat down and looked around, saying:

Published by Sirius-C Media Galaxy LLC, 2010

– Do you know that it's not very lucky, according to Chinese Feng Shui, to live in an a place with so many mirrors?

I said I did not know but admitted that I was sleeping not very well since I had moved to this flat, while before, when I was having a flat on the 23rd floor, which was much less equipped with mirrors, I had slept much better and felt more peaceful. Once of a sudden, when we were having a cup of Chinese tea together, his eyes fell upon a doll I had, an *Arlecchino* doll I had bought in Holland, during my stay there. I saw his face becoming pale, and after a pause, he said, softly:

– Sorry, I must tell you something about this doll. You should get rid of it as soon as possible …

I wondered why, and asked him to explain. He said:

– It's a long story. I can see that the doll is inhabited by three spirits who are probably following up on you. There is a blond woman from Holland. The doll was in her possession first, before you bought it. She has committed suicide. The other two are men from Brazil.

I was speechless. He invited me to sit down on the floor with him for a 'scan'. He was taking a seat close to my right, and put his left hand on my navel, my hara point, while closing his eyes. I felt there was suddenly a deep silence in the room. He did not speak for about five minutes, then he said:

– I must warn you. Your hara point is wide open. This is probably a scar you carry from a childhood trauma. When the hara is wide open like that, all kinds of spirits can enter who may haunt you, obsess you or may suck your vital energy.

I was telling him that I was suffering sometimes from an unexplainable fatigue, that was at times so strong that I had to lie down in bed at any hour of the day when it happened. I also told him that I had frequent diarrheas that sometimes lasted for several months. He replied all this were just symptoms, but that the root cause was a lack of vital energy that was related to my open hara point. He advised me to stop eating meat and drinking wine, and to do every morning an hour meditation, preferably between 4 am and 5 am. In addition, he recommended me to get rid of the doll as soon as possible.

Then, I asked him about my former lives and he replied:

– You were a sage at the time, totally dedicated to a spiritual mission. But one day, you met a woman and that was the end of your spiritual career. For you got so dependent on her that you abandoned your mission.

The week was not yet over and I had another surprising encounter regarding the doll. I had become acquainted with a marketing expert, a woman who was running an advertising firm, and I had wanted to try some of her services for promoting my seminars. So one of the next days, she sent her secretary to take some notes of my needs, and when that young woman entered the flat, she saw only one thing: my doll. She was shrieking, while walking backwards to the door, and when I saw where she was looking, I was trying to calm her down, saying:

– Oh yes, please, don't get in such a frenzy, I know the story of the doll already. A paranormal told me.

Published by Sirius-C Media Galaxy LLC, 2010

She calmed down somewhat, and sat down at the table where I had a cup of hot tea for her. She thanked and asked me why, if I knew the story, I had not got rid of the doll? I replied I had no idea what to do, that if I gave the doll to some children for play, they might be affected, and if I burnt the doll I might harm the spirits, so I was really in a dilemma. But curious, I was asking her then, what she saw in the doll, intrigued if what she saw was coinciding with the story the Chinese had told me. (Silently I doubted it).

She said, without thinking long, there were three spirits in the doll, a blond woman and two men, one of whom had a moustache. She said the woman was from a Western country and had committed suicide. She also said the woman had fallen in love with me and did not want me to find a girlfriend or partner. That she did all to prevent it. I was dumbfounded. Then she said, the men were from a country in Latin America. I was asking her since when she had clairvoyant abilities and she replied that already as a little girl she had seen spirits and that sometimes people asked her what to do when they have to take an important decision. But she said, she didn't want to do any professional consulting using her talent, and preferred to make her money in a 'simple business job', which got us to talk about my promotional needs.

Renata

The second clairvoyant who came to see me was equally a secretary. She was actually the secretary of the owner of the residence and five other residences in Jakarta, a well-known multimillionaire. Her name was Renata. In about half an hour she gave me a presentation about one of my former lives which I video-taped. Here is the entire text:

Renata

Good Day,
I am Renata, I am one of the clairvoyants from Jakarta, Indonesia, and I would like to present my message, of course through my spiritual guide.

Good Day Doctor,
You are now in the era of a new world, and have a mission to assist people in lost regions. Your past was way back in the 13th century. You are a full descendant of the Ronenberg family. You were named Kul Manerheim von Ronenberg, and you were a talented musician and served as a full member of the Royal court, of the Steinem-Strauss clan. She is [a queen] who was also the reigning Queen who gave birth to the German kings. In the past, you were very famous and summoned to the court. You will in the future be the head, a chairman, of the *Inspirational Classical Arts*. There is going to be a gentleman from Indonesia who will contact you through a colleague who will be interested in your business. You will combine business and arts, and travel, full force into many regions, in Indonesia, and there is going to be a stage concert in your honor, together with be a surprising charity event in this near coming future. Please continue through meditation your journey in this very interesting world. There will be a chariot to bring you to a spiritual world; you will feel fulfilled and rewarded, and surely have to come with your relatives and friends. A man who is trying to be in contact to promote your business will all along take you to competent people and perhaps you will sign a contract. In your midlife, you will gradually setup your house, your office. It's a tower and you will have a full collection of music, arts and system where there will be extra-inspirational power to help the people through your system. I must congratu-

Published by Sirius-C Media Galaxy LLC, 2010

late you then. May this world come to peace, no war, and music will bring mankind together to peacedom.

I few days later, I met Pak Sujanto, equally a long-time friend of Dr. Sabdono, and we recorded four video sessions in which we dialogued. While I was not really trying to research the veracity of Renata's clairvoyance, I found the dialogues with Sujanto to the point, helpful and interesting. As to the writing of this text, I have done research on Renata's oracle but could not confirm any of the details. I have not found a Steinem-Strauss clan to be existent, but well two powerful Jewish families, the Strauss family which well goes back to the 13th century, and the Steinem family, both families being very famous in the USA today. However, I could not find a 'Ronenberg' family nor an individual with the name *Kul Manerheim von Ronenberg* in all of French or German history.

For the rest, her 'predictions' of the near future were completely off-track. What happened over the next weeks were violent street riots and attacks against the Chinese minority, during which many Chinese women and children had been brutally gang-raped and burnt to death by hordes of barely clothed men who later turned out to have been hired by the military for the pogrom against the Chinese minority.

A few months later the Soeharto regime was definitely facing its end, with all the violence that broke out in the aftermath. It was only at that time that I got informations about the genocide that was committed by the Indonesian military in East-Timor and that was funded by the United States government.

I had to virtually flee to Yogyakarta because it was no more safe to stay in Jakarta, and my staff arrived one week later with my luggage, in a totally confused and anguished condition. The man had to keep the bed for one week, to that point he had been traumatized by the street attacks and the constant assaults and threats to be shot. He had been a real hero to defend my luggage at the danger of his life, paying all the money he had to the street militia. And of course, I took good care of him and payed him the money back. I never again went to Jakarta, but settled later on in Lombok and in Bali, until my mother died in 2000. It was a time of turmoil not only for me but for the whole of Lombok island.

A tragic incident was the trigger of major violence. On a nearby island, fanatic Christians had killed a Muslim family that was praying in a Mosque. As an act of revenge, gangs of Muslim fundamentalists swamped Lombok for weeks, destroying all the churches and killing many Christian families. The family next to my villa who had two small children, was burnt down during a tempest and monsoon-like rain, using a whole tank of gasoline, and my staff told me to have found dismembered bodies at the beach early the next morning. The country was in a horrible condition and the airport was closed. I could not fly back to assist my mother in any way. I could not even fly back at the time of the funeral, a Hindu funeral that she had expressly ordained in her testament.

Published by Sirius-C Media Galaxy LLC, 2010

Only in the first week of March the airport opened and I could fly back to Germany. After I sold my villa project in Bali by 2002, I never went back to Indonesia again.

Black Magic on Lombok Island

However, the two years I spent on Lombok island were *highly initiatory* from a spiritual point of view. I went through many paranormal experiences, some of which were painful. At two occasions, the strange tale of the 'gang of three' that persecuted me was confirmed in a way that dumbfounded me. It began with the two little daughters of my staff who one night wanted to sleep at the entry door because they claimed the house was haunted, and wanted to entrap the 'ghosts' with a mirror. These girls were very courageous after all, to be so unafraid about matters they knew much more about than I. The next morning I saw them at breakfast, quite mute and pale. Upon my question how their night had been, the older girl said:

– It was quite frightening but I did not waver. They were three, two men and one woman. The woman was blond, the men had black hair. They were very fierce-looking and were asking for you. They asked me where you are, where you sleep? They said they wanted to penetrate in your bedroom and I held them the mirror which shocked them and threw them back.

I was so astonished that I couldn't say a word because a few days earlier I had hired a gardener who turned out to be the son of a quite famous natural healer. He had said already the first day there was something strange in the house and that I had to be careful. Asking him, what he meant, he said if I'd allow him to, he would sleep the night at the entry door to see what was going to happen. I thought why not, and let him pass the night in the big hall from where a small stairs leads to the frontyard. The next morning, I woke up quite early, and went to see him. I found him in an acute state of excitation, almost out of his mind. He stuttered. He said three spirits had penetrated from the frontyard, one blond woman and two men who had rather black hair. One of them had looked like someone from South America, with his strong moustache. He had tried to stop them, he said, as they announced they wished to penetrate into my bedroom, but they had viciously attacked him and threw him on the floor, he said. Upon which they had quickly moved through the air to my bedroom where they had opened my belly and had eaten my intestines.

A few weeks later, after horrible dreams that I experienced over this period of time, I had to dismiss one of my staff as she had so much trouble with her husband, that one day he had come with a knife in his hand, to kill her. I had defended him to do so and the next day had dismissed her with the argument that I wouldn't tolerate violence in my house and that if she couldn't handle her husband, I would prefer to look for somebody else. A few days after this discussion I got a very bad diarrhea that did not stop, and I became meager and meager, and suffered terribly from fever and an ongoing diarrhea that nothing seemed to stop. Eventually, in my despair, I went to a Hindu priest at a sacred

Published by Sirius-C Media Galaxy LLC, 2010

source, quite far from my house, and among the eight priests in this place, he said he was the only one who understood how to counter Muslim magic. He said I had to sleep a night in a small temple that was built for the honor of a particular Goddess, and so we did. In the evening, after a long prayer in that temple, he was going into a meditative state and said after a long silence:

– Now I can tell you what is the cause of your illness. It was revealed to me.

I said, I was curious to know it and he continued:

– She is a quite ugly woman, right? She has very irregular teeth, especially her front teeth. She has one son and one daughter and her marriage is very bad now. Her husband beats her often. She doesn't see a way out. This woman had a grievance with you and went to a black magician. This sorcerer did some sort of black magic against you; it is specific Muslim magic and that one is really dangerous. If you had not come here to see me, you would die in a few months, for sure.

I was speechless. All he had said was true, her features, her ugly teeth, her family details. I felt humbled and we went to sleep very soon and the next morning, in the cool early morning, we woke up and had some tea together before I went home. He had walked down the hill with me, and drew me holy water, after having bought five plastic canisters that he filled with this famous water from a sacred mountain river. Then he gave me instructions how to take my shower in the morning with that water. I had to wash myself first and then poor that water over me, but not dry me with a towel. I had to let the water dry up by itself. I did exactly what he had said and already two days later I felt significantly better. The fewer was gone, my head was for one time no more burning hot but felt normal, the headache was gone, but the diarrhea persisted. And it persisted for four more years, it was healed eventually, after many visits to natural healers in various countries, by a Chinese doctor in Phnom Penh, Cambodia, using traditional herbal tea. I had ingested that tea every day, according to precise instructions, for three months and the diarrhea stopped and never recurred.

Now, I will precisely recall the dialogues with Pak Sujanto that were fortunately video-taped, for my memory would not suffice to report those details here. We met, to repeat it, in 1998, in my Jakarta flat, shortly after the Chinese paranormal and Renata had seen me. He was coming for four afternoons, and each session had something like a header or a theme.

Sujanto's Psychic Readings

Session One

Sujanto

What is your question?

Pierre

What is my mission?

Published by Sirius-C Media Galaxy LLC, 2010

Sujanto

First, it is to correct your previous lives' karma? That is the first.

Pierre

What kind of karma was that?

Sujanto

You have abandoned your mission at the time, and now you must continue!

Pierre

At the time, what mission was it?

Sujanto

Spiritual, as a guru.

Pierre

For children, or also for adults, in the general sense?

Sujanto

Adults. At the time, you stopped because you were attracted by a woman. So now you have the chance to renew your mission – to go further.

Session Two

Sujanto

So concentrate on teaching, concentrate on assisting people in their stress. Then, in this process, you will find power – without asking, without willing …

Pierre

Yeah. I recently became to organize a social project with some people, business people. We wanted to do it for children, orphans, and maybe also old people, poor people. And I see, it's coming very quickly, whereas when I was looking for more business, it was *not* moving. So I see the energy is more on that side.

Sujanto

That is parallel with your spiritual path, because … business is okay as long as you do not attach to the result … it's okay, you may study it, cooperate with local people, setup a factory and so

Published by Sirius-C Media Galaxy LLC, 2010

on … it's okay, but don't attach to the result, and don't use … bad tactics to make the project a success.

Session Three

Sujanto

Once you have been … not in this world but in that world, you were an angel. It is crazy. For the first time I see that a human being was an angel. How can? It is for the first time. I must laugh because you were an angel, in that world. Also to serve people.

Pierre

It is very strange because I have this attachment to angels. When I was living in Bali, I had a new project to make a film about the angels. And I am looking for people to help me making this film.

Sujanto

This does not astonish me because you have had this experience, to be an angel.

Pierre

It's incredible …

Sujanto

It was for the first time that I see a human being that had an angel's life in that world ... So I don't know why you came back here, to pay your karma or by your own choice, or because God or the spiritual unity was fighting you to go back to this world ...

Pierre

Actually, throughout my whole childhood I thought I do not belong to the world here. I didn't want to go out until I was 33, when I committed suicide, and it did not succeed. And then, my whole life changed. And I had then the intuitions coming for the mission. And then I went to astrologers and I studied, and I saw, yes, this and that, and I began to develop my art. Because I was never supported, my whole childhood was traumatic. I never had a backup with my family, it was just absolutely awful. Nobody was appreciating what I was doing ...

Sujanto

Hopefully, now, you don't feel that way, because they have their own mission, they have their own karma. They still don't support you, but it's okay, because you have a strong will and your spiritual knowledge will become better and better. So, without your family's support, you will be supported by other humans.

Published by Sirius-C Media Galaxy LLC, 2010

Session Four

Pierre

You have to detach yourself from the appearance; well, they are jealous, and so on. This is only the outside level. Inside, it is all this indoctrination, through the religion and all that, which brings people to that point … It is an ideology. Religion, if you take the word, it comes from Latin 'relinquere', which means 'link back' – to the source. So you don't need any organization to do that. You have to do it yourself …

Sujanto

That's right. We are of the opinion that there is no need for an institution …

Pierre

Because all institutions seek power …

Sujanto

Yeah …

Pierre

So the power actually gets in the foreground, power becomes the motivating thing, and they forget about the mission they have. The Christian Church is the same ...

Sujanto

Religions defy the people. They don't see the other truth ...

Pierre

Yes, and that separates human beings ...

Sujanto

Yeah ... and yet we know that we are *one*, humankind is one, so there is no border, spiritually we are the same.

Psychopomping Mother

After the return, my life changed, and while I inherited a fortune, everything and everybody seemed to be against me. I was facing a whole lot of problems, and missed my friends in Indonesia and the wonderful climate there, the magic, the silence, the respectful way people behave – except when they are mad with their leaders and pass through a psychotic episode.

Published by Sirius-C Media Galaxy LLC, 2010

Because of that exposure to a basically shamanic culture, facing Germans once again was even worse as all through my younger years. It was an awful experience in every respect, and in the worst moments, when I was cheated, betrayed and attacked in the most brutal ways, I was taking a vow to leave Germany forever, and never return – much as Arthur Rubinstein did. Today, a decade later, I have not changed my mind, and still live in Asia.

Yet I had to stand my man for those two years, from the moment I inherited the family business, until the moment I could eventually sell it, for a price that was neither lucky nor unlucky. I managed it as good as I could. But my preoccupation was lesser of a business nature, it was rather a religious endeavor; it was about psychopomping my mother who had died in January 2000, and came to visit me regularly in my dreams. These visits were not 'dreams' in the ordinary sense, and to explain this, I have to begin three years before.

In 1997, when I was living in my mother's house for a few month for preparing and finetuning my new career as a sales trainer and coach in Asia through writing brochures and conceptual introductions for my training business, I had some very deep and unique conversations with my mother. To repeat it, I also had premonitory dreams about her death, a fact that upset her friend and partner to a point he got angry at me. He, a typical German man from a low class family that my mother regretted over and over to have lived with for the last twenty years of her life, just suspected me to wanting my heritage, interpreting those dreams as 'wishful thinking'.

But they were of a totally different kind. In fact, these dreams clearly indicated that my mother had been mal-treated by her doctors with her recurring heart problem and that she was so unhappy in her love relation that she really wanted to pass away. It happened two or three times during these three months that she went downstairs to my apart-ment, with her suitcases packed in her hands, saying:

– I cannot live with him anymore, please let us rent a house together, and give up in Asia. I must leave him, he is stupid and insensitive, and I should never have met him.

I agreed each time, but the next morning she would come again and say:

– Please forget all I said last night. I cannot restart my life, it's too late. I should have listened to you when you said he's not the right partner for me. He's really not, even your father, while he was a drinker, had more brain than this man. But I feel I cannot make it to a new turn in my life, I am too old. I do not have the force anymore …

I accepted her decision without questioning it. Only she could know what was best for her. Her partner was a man with whom I didn't get along at all. He was still married with his wife and had five children from her, but lived with my mother saying he 'loved her'. He was from a poor family, a small government employee who had some savings, but it was not to overlook that he had chosen 'a well-situated' mis-tress like my mother to get some extra bonus from life. After my mother's death, I was able to see that my negative feelings about him had not been *negative enough!* He had given me a credit of fifty thousand euros for starting my business in Asia,

Published by Sirius-C Media Galaxy LLC, 2010

which I paid him back right after I inherited the family fortune, but then he came with the idea to take about twenty thousand euros interests while we had agreed upon the default 6% p.a. interest prescribed by the German Civil Code, which I had paid him. It was only after I threatened him with a lawsuit that he gave up his ludicrous plan.

We had long talks during those months, while my mother never stayed long with me. She was terribly afraid of his, and anger against me for any slight occasion, for example when he scheduled to cut the grass and I was not immediately assisting him. So she was always restless and squirmy when she was with me, but in one of those talks we touched the most important subject. She had been in touch with her mother for several years after she died, back in 1973, when I was just about to leave the boarding. She had told me often about those 'encounters'; she never spoke about 'dreams' with her mother and I was not too sure what she meant to say. When asking her, she would say something like:

– It was not a dream. She was really there. And she said always something for introducing the communication. She would speak about that 'other world' and that she just wanted to tell me that she is always near to me …

Well, in the meantime I had done my research on parapsychology and I knew what she had been talking about. So I offered her a contract. I said:

– Whoever of the two of us dies first, promises the other to connect back from the 'other world' to tell important things or whatever, but to keep that communication alive for a certain while, okay?

And she immediately agreed; that was the point of departure. I am sure she never told a word about our agreement to her friend. This is how the story started that I was psychopomping my mother for about six months after her passing over, and it was the most important time in our relationship, for various reasons that I will try to elucidate as I am going through the material. Destiny made it that I could not assist her when she approached the point of her passing out of her body. I was living in Lombok at that time, and it was a time of revolt and chaos, after the fall of the Soeharto regime, with murders and robberies committed virtually every day. To make it full, a gang in Maluku, an island with mostly Christian population, murdered a Muslim family in the Mosque while they were praying. It was, as one can imagine, one of the most horribles crimes that a Muslim or any other person can ever think of. And while Muslims are a tiny minority in Maluku, so are Christians in Lombok. And the revenge was terrible. For more than a month it was like hell had been, all Churches were blown up by grenades, and next to my house, a Christian family was burnt down including their children. The owner of my house, a merchant from Qatar, feared very much for my safety and wrote on the wall of the property in big black letters 'MUSLIM' so as to signal to everybody that the inhabitant of the villa was a Muslim. For one week I could not leave the house. And in that very week, my mother's friend called and said a critical point was reached, my mother was hospitalized and 'this time it looks she will not leave the hospital alive'. And I was stuck there in Lombok. It was in January 2000. And the airport remained

Published by Sirius-C Media Galaxy LLC, 2010

closed until March, when I finally could leave the island and return to Germany.

The first thing I inquired about was if her testamentary instructions had been followed, and he confirmed it to me, handing me over a document that attested that her body was burnt and her ashes thrown in all four winds on the North Sea. There was an official stamp, seal and signature of the captain of the ship. And I was relieved, as the notary first had refused to put that clause in her testament. Why my mother wanted to be buried in the 'Hindu' way, I have no idea. Since her childhood, she told me she had loved India and among the books she has written and that unfortunately she lost through a bomb attack on our family property, was a novel that played in India. My mother was all her life fascinated by India but never went there, and the last who could have understood her wish was her friend, the prototypical German 'Biedermann' who was interested in nothing but watching TV, drinking beer, talking his 'Rheinland' dialect and tell his little life story every night after about the fifth and definitely after the tenth bottle of beer.

And not even in our conversations, when she began to visit me regularly in dream visions, while this definitely were not dreams, she lifted the veil over this issue. I suppose she had a strong previous life karma in connection with India. In her childhood she had wanted to learn classical dance but her father replied:

– Either you become a *Prima Ballerina*, or you don't even start. I do not want my daughter to be 'ein Balletthäschen' (a ballet rabbit).

Hence, my mother died after a basically missed life. She had spent her last years in a *totally unproductive* way, reading boulevard papers every day. All the books about human potential and self-transformation I had given to her remained untouched in her bookshelf. With choosing a partner about 20 years before she died, who was a total misfit, both intellectually and from an aesthetical point of view, she had buried the last rest from a once creative and positive life track; she had not only buried her dreams, she also buried the last years of her life in an utterly boring and nonsensical daily routine of just 'passing her time'.[41] And instead of staying in her region, the *Saarland*, with all its French influence and lifestyle, she had followed her partner, moving to the *Rheinland*, worse, to the *Westerwald*, a typical German region where she felt like an alien. All the food she once liked, like French-style snails, seafood, fish, her friend abhorred. He could not even smell a normal French vinaigrette, saying the mere *smell of vinegar* made him 'vomit'. So she adapted to her partner's residual lifestyle, renouncing her sensual appetites, thereby dying inside of her with every day to come.

He recounted endless times how brutal his father was. He must have been a totally insensitive nerd and 'house tyrant' who used to throw his large fist on the head of any of the children who was not eating properly or didn't eat a certain kind of food, so that the face of the poor child landed in the often broken plate. One may easily explain why this man was appalled of such a variety of good and healthy food, virtually traumatized from childhood against culinary pleasures. And the story goes on and on and the detail of it are so

Published by Sirius-C Media Galaxy LLC, 2010

appalling that I spare then out here for they do not really add to the essence of what this story is about. (They rather fit in my memoirs). This chapter is more precisely about how my dear mother experienced her death and the about six months she passed in the astral dimension, the dimension that Michael Newton calls 'life between lives' and that the Tibetans call the 'Bardo'. I already mentioned that besides me, my mother had nobody to ever talk about her premonitions and her knowledge about the after-death experience. First of all, she was not afraid of death. She knew about the afterlife, and she knew it from *first-hand knowledge* gathered from her own mother. She also knew it intuitively. I remember that as early as about six, I was asking mother questions about religion.

She used to answer that what they tell me in school is ridiculous and that her religion teacher had thrown her out of the class because of her 'queer questions'. With my father it was about the same. But while my father became an atheist, my mother kept her own genuine spirituality.

She used to say that the Hindu teaching about spiritual matters was basically true and could be relied upon, by and large, that what we call 'god' was but spirit, an all-pervasive energy that guides us toward right behavior, and that karma is but the effect of ignorant behavior, without any judgment connected to it, simply cause-and-effect. She said she regretted that the religious teachings in schools had not changed since her own childhood and that teachers told children 'the same nonsense' today as before, but that I should not believe it but make up my own mind about it all. Which I did. My mother's view made sense to me. She also basically believed in reincarnation but added always:

– But I know that I myself will not reincarnate. I know this intuitively while I know it's not the rule. But I feel I will not come back to this dimension but to another place.

However, when I asked her why she was so utterly negative about her life and others, while she had so much of spiritual knowledge, she used to say:

– It all went wrong in a way and I think I cannot change it or it is too late to change it. When I lost all my writings in the war, I knew I lost the most precious while I have saved all the valuable items of my parents. I never thought of myself and my own art, my own writings, as something of value, and that was a big mistake. I was raised to be 'unselfish', but in truth, it was a self-defeating education. I never learnt as a child to believe in myself, my brothers ranged first, and my older sister, while I was their maid, their servant, that is how I was raised. And while my father was a loving soul, in this respect he was an absolute patriarch! I was his cutie, as long as I had no thoughts and longings for myself, as long as I was nice and obedient …

The first session was about four months after her death, one month after my return, when I was living in the apartment she had built in our family property, on the first floor, actually for herself, as a kind of 'safety shelter' to come home to in case she really wanted 'leaving him forever'. It was conveniently located in the center of our little town, with shopping malls all around and no car needed to go anywhere. It was a place where she could have had contacts every day, not

Published by Sirius-C Media Galaxy LLC, 2010

a golden cage as that huge villa she had bought with him in that little town in the Westerwald. She could have visited our clothes stores, our jewelry, or have an icecream in our Italian gelateria. Everywhere she could have spoken to the people, or have a glass of champaign, as my grandmother used to have it. Or she could have spent an afternoon in Café Löwe, the owner of which was a student of my grandmother and where she would always have got her free coffee and cake, as I used to get it later on. But not only did she fear people, her attitude was so negative that she was not received in the same manner as my grandmother was, or as I was later on. People instinctively reacted with a silent fear when they saw her, in awe she could again fault-find them, as she used to, or even threaten them with a court action for anything that didn't go the way she wanted to. She hardly had a relaxed chat with anybody, always 'in a hurry' (for reasons that were irrational and neurotic), while my grandmother would take all her time to really talk constructively with her tenants and all the business people around. And it was worse when, as usual, she came to visit them with him, who was playing the 'big mafia boss' in a property where he had absolutely nothing to say. It was embarrassing to witness it, it looked all so 'cheap' how they were coming over to people, so 'important' but utterly 'amateurish', making up some kind of 'professional' attitude, while my grandmother had been one of the most respected business women in town without needing that kind of show.

In that first vision, she came to apologize. Her face was swollen and red, and she said she was suffering from great heat. It was like burning all around her, she said, and I instinctively thought she was in what we call 'hell'. I asked her how she felt about it? She said she felt terrible and got some kind of announcement that she would have to stay quite long in this transitory world as she had done so much bad to me and inflicted so much suffering also upon other people. She said she had been told to get in touch with me and that if I'd pray for her she might be able to leave that state of suffering earlier on. I then promised her to pray for her and she was showing gratitude and a feeling a love irradiated from her to me. So I prayed for her, fervently, every day, and cried often. My weight was at that time just 48 kg, I was just a bundle of flesh and bones.

Some time later, she came again, this time I did not see her but she was showing me scenes from her childhood, as if watching a film, and commented about it. First she thanked me to pray for her and that she felt greatly relieved but that she had to tell me some 'important things' about her childhood. In this quite shocking vision, I saw her naked, as an about eleven year old beautiful girl, with her blond curly hair, in a room with old furniture. And she said she had been constantly sexually abused by her older brother but that she told nobody about it, but had to get it off her soul once forever. I felt pain and compassion for her, and suddenly understood all her terrible hatred for men, all her harshness and her bitterness. The information was not a surprise for me as I since long had suspected and intuited something of that kind, knowing how screwed-up my uncle was, and given that her relationships with men all bore that typical trait of 'abused women' I had observed with many 'incest survivors'.

Published by Sirius-C Media Galaxy LLC, 2010

This also explained to me why she was to that point 'sexually obsessed' in the terms of my father, who once told me she had slept with all his colleagues at university including the professor and just could not 'get enough of it'. In fact, my mother was very sexual all her life through, which for me was just normal, but I felt there was more to it without knowing the truth. Not even my father or any of her friends, she said in this vision, had known about it, as she had never revealed it to anybody because of the shame she felt all about it. In my incest research I actually found that with siblings incest, the shame may be greater for the girl as with father-daughter incest, as the girl may secretly consent to have her first sexual experiences with her brother. It also may not be rape but a consensual play. But psychologically, in the patriarchal setting where incest is simply the 'greatest sin', the shame is inevitable and typically comes up later on, once the incestuous relationship is entirely cognized, and the shame, then is *also on the boy's side*, but tends to be rationalized away with 'macho' arguments and behavior. In fact, my uncle later in life opened a bar where, as it was the rumor in our little town, he let his wife and daughter, my beautiful cousin, prostitute themselves for money. This in fact was the major reason why my grandmother disinherited him and gave the fortune to my mother, as by law, it was him who was deemed to receive it, as the first-born son.

In the next vision, my mother appeared in a different mood, and seemed relieved and happy. She said she felt relieved of a great burden that she was able to have told me 'her secret'. Her face ywas no more swollen and she said she had good news. It had been announced to her that within a month or so she could leave the astral realm and move on to a higher realm from where she could not come back to visit me. She also said that her mother was on that plane already, and that she was looking forward to meeting her there. She said that she would return for one time before departing.

I felt very happy for her, glad also that I was able to shorten her time in the Bardo through my constant prayers. The last vision was the most beautiful one. She was not alone that time but came together with my father. Both looked very young, radiant and beautiful.

They said they were *together again* and I replied this was the greatest gift they could ever make me. (In fact, I had broken with my father years before that time and I had never any information about his death). They handed me over a framed photo of a very beautiful little girl and my mother said:

– I understand your love now, sorry that I never did during my lifetime because of ignorance. Also your father embraces your feelings for little girls. This is our gift for you, and it should help you embracing your love and living it constructively.

When I asked who this little girl represented, I got the following answer:

– She was once a little prostitute, under the reign of Napoleon III in France. You may have some karmic connection with her. We only know that she dearly loves you and will never forget you, that is why we give you her photo. Bear it in your heart and always think lovingly of her and us!

Published by Sirius-C Media Galaxy LLC, 2010

It was the greatest gift I ever got in my life and I thanked dearly for it! I was completely reconciled with both my parents, and for weeks, was in a state of beatitude, of gratitude and of inner peace. I never ever in my life had expected such a blessing could happen to me.

This experience was the greatest trigger and motivator in my life to act, think and pray always in alignment with universal love, while I sometimes departed from that way, and committed sins.

My parents never came back in visions to admonish me, but higher spirits did from about that time, to regularly hold me a mirror so that I could see in which ways I acted toward others in ways that hurt them. And this is an ongoing experience. I believe it was unconditional love for my parents (that admittedly I developed rather late in life) together with the fact I really completely pardoned them the suffering I went through as a child and youngster that had opened this guidance for me for which I am infinitely grateful.

In fact, I do not believe in Nirvana, nor in any state or condition where there is no sin, where a human always acts in total alignment with the cosmic purpose. I believe that we all have a major weakness that from time to time lets us fall back in the 'old pattern' and sin again in the sense of doing harm to another.

I am no exception here but these experiences have held me open for receiving guidance in any case where some of my antisocial desires for a moment get the overhand. Having faith doesn't mean to be flawless, it means to holding one's heart open for guidance, for advice, and then adjusting one's behavior accordingly.

Published by Sirius-C Media Galaxy LLC, 2010

CHAPTER THREE

The Shamanic Method

Common Assumptions

The first real science humanity has developed, it has developed not under the pulpit of scientists, but of native shamans. This is my proposition.

Of course, when I assume that shamanism is a science, I must be able to show that it uses a method, a scientific methodology, that is, a set of tools that serve to look at nature in a truthful and possibly objective manner. Is that the case with shamanism?

Let me look at the literature first. Most authors' assessment of shamanism coincides with saying that it is a way of apprehending reality, a set of insightful techniques, rituals and patterns, as well as a natural and organic lifestyle centered not at dominating nature or cosmos, but at participating in and understanding nature and the cosmos.

Stanley Krippner and Alberto Villoldo, in their study *Healing States (1984)*, define shamanism as 'an attitude, a discipline, and a state of mind that emphasizes the loving care and concern of oneself, one's family, one's community, and one's environment.'[42]

Most authors agree with this view, in that the shaman is having a regulatory function within tribal society, for bringing inner peace and healing to the clan or the whole of the tribe. But this is not his only function. This quote also does not give flesh to my theory that shamanism is essentially a science, even though this science is used for doing good to people, for healing or for divining future events. What I am

Published by Sirius-C Media Galaxy LLC, 2010

saying is that the shamanic method or technique is not just a fancy ritual, not just an experience of ecstasy, but essentially, a *scientific investigation* into the nature of things, the nature of the cosmos, and the role that the human being plays within this cosmos.

Most authors also agree that the most important to find out about shamanism is its *use of entheogens*. These are plants that contain psychoactive compounds, such as DMT, which, when taken at appropriate doses, produce a consciousness-altering effect upon our psyche and perception. There are various names for such plants, and the name that is given often reflects the state of mind of the researcher. Mircea Eliade states in his book *Shamanism: Ancient Techniques of Ecstasy (1972)* that any given shamanic culture was at its decline or caught in decadence when their shamans began to use psychedelics for the shamanic voyage. However, it has to be noted that today this opinion is clearly contradicted by the large majority of researchers, such as for example Metzner, Harner, Gottlieb, Schultes, Hofmann, Rätsch or McKenna who agree in considering Eliade's bias here as a myopic view and a basic misconception about shamanism. For example, contrasting with this view, Terence McKenna writes in *The Archaic Revival (1992):*

Terence McKenna

While Eliade asserts that the use of narcotic substances as an aid to ecstasy invariably indicates a decadence or vulgarisation of the shamanic tradition, there is reason to doubt this.[43]

The shaman typically is the one who stands out because of his unique capability to explore, and travel into different realities and levels of consciousness. The second point where modern researchers largely coincide in their opinions is that shamanism cannot be defined under the exclusion of entheogens.

While there are methods to alter consciousness without plants, using esoteric breathing techniques, body postures or ecstatic dance, drumming, prayer, fasting and other techniques, researchers agree that from a point of view of effectiveness there is a large gap between those latter techniques, and the use of entheogenic compounds. Entheogens are *several hundreds of percent more effective* than non-plant based methods.

Several researchers have seriously tackled the question why this is so, and one of the most persisting on this specific point was Terence McKenna. In *Archaic Revival (1992)*, he affirms that entheogenic plants contain the very essential genetic code, the basic information about the evolution of life on earth, and that for this reason the ingestion of the psychoactive compounds they contain leads to an immediate *opening of consciousness,* which was something much broader and much more intelligent to experience than mere colorful visions.

I believe that McKenna's visionary and illuminating books would never have had such a powerful impact on the consciousness change of Western society if they only talked about some nice hallucinogenic visions. Anthropologists or generally researchers who try to understand the unique phe-

Published by Sirius-C Media Galaxy LLC, 2010

nomenon of shamanism and reduce the entheogenic experience to a mere social game, a distraction or a search for some kind of artificial nirvana are deeply misled. But for sure, when I say shamanism is a true science, actually the first science humanity used for reality assessment, I go even beyond the visionary message of McKenna, and I certainly go beyond Eliade's assumption that shamanism was merely a set of techniques for achieving ecstasy.

It is therefore not surprising that most anthropologists, and especially those of them who really do not understand shamanic culture, tend to employ expressions such as hallucinogens, narcotic drugs, narcotics or psychedelics when talking about entheogens.

Apart the fact that these plants are not narcotics, because a narcotic drug, such as for example opium, renders somnolent but does not alter consciousness, the important thing to know is that entheogens are not understood, in shamanistic cultures, as leisure drugs, but really are considered as assets of the religious and numinous experience. That is why the only expression that comes close to the shamanistic mindset is the term *entheogens*, facilitators for getting in touch with the inner god.

It has been equally affirmed that entheogens, apart from their helping us to reach the inner mind, also dissolve habits such as alcoholism, and generally help in a process of social deconditioning. In clear text, entheogens help us lift the veil of the normative behavior code in any given society as they show us options of *different behavior*. They actually show us the *immanent potentiality* in all of nature's setup, and especially in how nature has setup the human being, that is a basically free creature, who is not a priori bounded by a preset program. What we can thus learn from taking these plants as a sort of 'social medicine' is to recognize the patterns of 'normative behavior' we are caught in and that obstruct our creativity and self-realization.

People who are socially oppressed, racial, ethnic, religious or sexual minorities, may want to inquire into the possible dissolution of rigid behavioral rules and oppressive normative standards in society. They may thus look for the ultimately most intelligent catalyzer that exists to see all the options reality offers and, as a result, might want to engage in a consciousness-opening voyage.

Another important observation regards mental health. It has often been wrongly stated that indigenous shamanic populations were psychotic or at least pre-psychotic and it comes to mind that usually pedophiles, in our society, once convicted and subjected to 'psychiatric expertise', are labeled in exactly the same way. When we remember the times of communism in Russia or read books by Aleksandr Solzhenitsyn and others, we learn that under that totalitarian regime the same murderous psychiatry with exactly the same vocabulary had been used to eliminate intellectuals who were treated as system enemies because they defended human rights and democratic values. All this may not surprise any informed individual. The effective mechanisms to defend a given societal 'standard behavior' code are all founded not upon natural pleasure-seeking behavior but upon adaptive perversity. Hence, the necessity to look beyond the fence of

Published by Sirius-C Media Galaxy LLC, 2010

behavior patterns and inquire into realms that seem apart from it but aren't.

The human soul expresses its originality always in paradoxes and sometimes in extreme behavior and the very attempt to classify human behavior into rigid 'standards for all' is in itself an ideology, or political program. The more a given society puts up general standards, the more it is alienated from life and its creative roots and the more it is subject to decay and alienation from nature.

This being said, there is common agreement among researchers that shamanism is an effective guidepost for revisiting the realm of nature's wisdom and true connectedness with all-that-is. As far as I can see, people caught in minority groupings and the social fight involved with minority lobbying hardly ever come up with beyond-the-fence solutions such as studying shamanism and experiencing entheogens, which makes the extreme poverty of many of those movements, not to say their ultimate system-obedient stupidity. And this unconscious system-obedience can be seen in many a limitation that social activists impose upon themselves and that are, ultimately, still system-prone. As Krishnamurti said, repeating an old wisdom: the revolt is still within the same frame of mind as the society it revolts against.

The entheogenic quest is therefore an *inner quest*, not necessarily something like a defeatist approach on a social level, but certainly an important add-on to any social activism for any possible social or humanitarian cause.

The Detractors of Shamanism

The detractors of shamanism were the Enlightenment, Cartesianism, Reductionism and Catholicism. Let me point this out in more detail. It is perhaps the most putatively known fact that Catholic missionaries had a particular grip on native peoples, and their shamanic rituals, as the Church considered such practices as 'devil's domain'. It is for this reason not surprising that colonialism together with missionarism was doing great harm to shamanism, in many parts of the world. This was direct physical harm, that often resulted in malady or death of the concerned native populations.

That is today standard school knowledge. It is however much less known how Cartesian reductionist science, which was a fruit of the rationalistic thoughts of the Enlightenment, was largely prohibiting knowledge about shamanism to percolate into Western society and culture. Here the effect and the harm done was not direct and physical, but indirect. It was something like intellectual or scientific racism. Ethnology, psychiatry and even certain branches of psychoanalysis were initially treating native shamanic wisdom as 'primitive' or 'barbarous' practices. Shamanism was not considered to effect valid healing of disease, but fake healing, or 'magic'; and even less was it considered to be a science.

The Age of Enlightenment

The spook of *rationalism* began in the second half of the 17th century, which is not surprisingly also the time when two

Published by Sirius-C Media Galaxy LLC, 2010

other large movements started out, industrialization and child protection.[44] This rationalist streak in human philosophy advocated so-called 'Reason' as a means to establishing an authoritative system of aesthetics, ethics, government, and logic that would allow human beings to obtain pretendedly 'objective truth' about the universe. I simply call it the *Age of Darkness* because it is now firmly established by both quantum physics and systems theory that the values of the Age of Enlightenment were bringing us widespread intellectual and emotional narrow-mindedness, rampant functional disease, spiritual confusion, fragmentation, racism and worldwide ecological destruction. The typical concern for the enlightenment was mechanics; so was its understanding of the world, that is, as a gigantic clockwork.

Most of the intellectual avantgarde today agrees with this critical view, as for example Fritjof Capra, one of the greatest exponents of today's intellectual elite; many further references and other authors are to be found in Capra's books.[45]

All that didn't fit in the mindset of those total rationalists was *ruthlessly discarded out* and labeled as 'mysticism', 'paranoid delusions', 'freakish daydreaming' or 'charlatanism'.

The sciences that were particularly hit by this myopic paradigm were parapsychology, shamanism and astrology. It is interesting to see that today adherents of this outdated and judgmental paradigm are to be found in the rings of mechanistic science and skepticism. An example is the fight of Randall James Hamilton Zwinge alias *James Randi* against so-called pseudoscience, and his personal fight against Uri Geller, a medium who was tested by Stanford University and found to be not a fraud. For James Randi, Goethe's 'school wisdom' theorem literally applies that says that what mustn't be, cannot be.

Cartesian Science

The Cartesian or Newtonian worldview is a life and science philosophy marked by left-brainism, a hypertrophy of deductive and logical thinking to the detriment of the qualities of the right brain such as *associative and imaginative thinking*, and generally fantasy.

It's also a worldview that generally tends to disregard or deny dreams and dreaming, extrasensorial perception and ESP faculties as well as genuine spirituality. The term *Cartesian* has been coined to mark a similarity in reasoning of Cartesian-minded people with the reductionist philosophical theories of the French philosopher René Descartes (1596-1650). Historically, and philosophically, it was not Descartes who came up first in world history with this schizoid worldview, but the so-called *Eleatic School*, a philosophical movement in ancient Greece that opposed the holistic and organic worldview represented by Heraclites; but it was through the Cartesian affirmation and pseudo-scientific corroboration of the ancient Eleatic dualism that in the history of Western science, the left-brained reductionist approach to reality, which is actually a fallacy of perception, became the dominant science paradigm. Fritjof Capra, in his bestselling book *The Tao of Physics (1975/2000)*, observes:

Published by Sirius-C Media Galaxy LLC, 2010

Fritjof Capra

The birth of modern science was preceded and accompanied by a development of philosophical thought which led to an extreme formulation of the spirit/matter dualism. This formulation appeared in the seventeenth century in the philosophy of René Descartes who based his view of nature on a fundamental division into two separate and independent realms: that of mind (res cogitans), and that of matter (res extensa). The 'Cartesian' division allowed scientists to treat matter as dead and completely separate from themselves, and to see the material world as a multitude of different objects assembled into a huge machine.[46]

Presently, even mainstream science gurus declare Cartesianism to be overruled by the new physics and the emerging *holistic sciences* that are presently breaking through as a preparation for a completely new worldview in the West, while in Eastern culture this organic, holistic worldview was always the prevailing one.

Quantum physics has demolished the classical Newtonian worldview with its strict determinism. As Fritjof Capra concludes, a careful observation of subatomic particles shows that these particles give meaning only when seen not as isolated entities, but when understood as interconnections between the preparation of an experiment and the subsequent measurement.

Quantum physics reveals a basic oneness of the universe at least at a subatomic level of observation, which is exactly what perennial science and mystical traditions of the East and West always have assumed as the main characteristic of reality.

In his second bestselling book, *The Turning Point (1987)*, Fritjof Capra then concludes this insight and extrapolates it beyond the realm of physics:

Fritjof Capra

In contrast to the mechanistic Cartesian view of the world, the world view emerging from modern physics can be characterized by words like organic, holistic, and ecological, It might also be called a systems view, in the sense of general systems theory. The universe is no longer seen as a machine, made up of a multitude of objects, but has to be pictured as one indivisible dynamic whole whose parts are essentially interrelated and can be understood only as patterns of a cosmic process.[47]

Thus we can conclude that Cartesianism, which is actually rooted in ancient Greece, as it was the dominator science paradigm for about four hundred years in Europe, the United States and other Western nations, was deeply hostile to shamanism, declaring shamans to be either psychotic and delusional, or to be charlatans. It has to be seen that the power of the Church in suppressing and rooting out shamanic cultures all over the world was backed up by science, by natural science, by psychiatry and by 'colonial' ethnology. To see the holocaust committed against native peoples only as religious fanaticism overlooks the much more important fact that this fanaticism was largely backed by Cartesian murder science.

Published by Sirius-C Media Galaxy LLC, 2010

Reductionism

Reductionism is a typical modern-day phenomenon. It is something like a thinking habit that results from a hypertrophy of the left brain. Historically it has taken root with the French philosophers René Descartes (1596-1650) and La Mettrie (1709-1751) who were considering humans as machines and nature as a complex yet mechanical machinery.

Thus, the nature of complex things is reduced to the nature of sums of simpler or more fundamental things. This can be said of objects, phenomena, explanations, theories, and meanings. More and more, with a holistic view of the universe as it is emerging from about the 1980s, the mechanical reductionism of Darwinian evolutionary psychology is overcome and science presently changes many of its fundamental paradigms because of this shift in understanding nature, human nature and the cosmos at large.

Let me give a few typical examples for reductionism in scientific texts and popular imaging. For example, it is written by Rupert Sheldrake in his book *A New Science of Life (1995)* that the old idea of a cosmic life energy, life force or vital energy was but a 'vitalistic theory'. What Sheldrake means is that there is no such cosmic life energy, and he thus was reducing the whole idea of a cosmic energy to the term 'vitalism'.

It has to be seen that often in science and also in political scripts and writings, reductionism is used for belittling, or outright downplaying important concepts and phenomena of life, thereby manipulating public opinion. A reductionist argument against shamanism would be the affirmation that shamanism 'is but a set of wild rituals that put primitive peoples in a state of trance, in which they do all kinds of things they wouldn't do when they are sober'. Typically, reductionists would deny shamans to be real healers and to have a scientific approach to knowledge gathering. They would downplay shamanism as a 'barbarous ritual' that 'may appeal to primitives but is to be rejected by civilized society'.

Catholicism

While Catholicism has ravaged shamanic cultures, especially in South America, and as a result of the *Conquista*, it could have had a better understanding of the shamanic quest because, after all, the esoteric Christian tradition is highly 'shamanic' in the sense that it values the inner experience over the outer ritual.[48] However, just as with Buddhism, this inner quest for enlightenment is not really scientific in nature, but contemplative. There are many phenomena that saints produce, for the most part involuntarily, such as stigmata, that cannot be rationally or scientifically explained.[49] It is for this reason in my view lesser a problem of an inner contradiction between the Christian dogma and shamanism, but a general problem of *power*, and power politics, with all organized forms of religion.

The same inner congruence but split on the outside level is to be seen in Islam, between the official dogma and the Sufi tradition. It is important to see that in most shamanic cultures, such as Siberia, or South America, there are relig-

Published by Sirius-C Media Galaxy LLC, 2010

ions in place that are neither in contradiction to shamanism, nor are they in any way in alignment with it.

Interestingly, shamans, when questioned if shamanism was a 'religion', tend to answer that it had nothing to do with organized religion but that the inner quest, the quest for real knowledge was a form of true *religio*, in the sense that it brings us closer to our inner god, and thereby, in a condition of cosmic alignment. It is in this sense also that shamans use *entheogens*, which is why these plants or compounds have been called that way – that is, inner god plants, as we are going to see further down in this study.[50]

Again, as with the other detractors of shamanism I discussed here, we see that shamanism stands out not because of an inherent conflict between shamanism and religion, but because the shamanic quest is scientific, religious and *teleological* at the same time. The latter element is important, for it makes exactly the divide with the purely contemplative or existential quest that is at the basis of esoteric religious traditions. If I was to compare shamanism with any other ancient science tradition, I can only think of *alchemy*.

The particularly destructive thrust that Catholicism is to be reproached regarding shamanic cultures, and shamanism in general, is that the Church preached that *non-believers had no soul*, that they were soulless and accordingly, were lacking the essential quality of being human. That was of course a hubristic view that is today largely contradicted even within Church circles. In fact, both the esoteric Christian tradition and the shamanic tradition deal with what they call 'loss of soul', which has a totally different meaning than the polemics of the Christian Church.[51]

Published by Sirius-C Media Galaxy LLC, 2010

The Shamanic Revival

The shamanic revival may coincide with the publishing of Terence McKenna's book *The Archaic Revival (1992)*, but it actually began in the 1970s. To understand the turning of the tide, let us consider the first honest information sources about shamanism that penetrated in the West, and that date back to the 1920s. It was the writings of the American-Polish anthropologist Bronislaw Malinowski (1884-1942), the American ethnologist Margaret Mead (1901-1978), and the writings of Sigmund Freud (1856-1939) and Carl-Gustav Jung (1875-1961).

Before I come to summarize these early glimpses into shamanic culture and lifestyle, let me give an example for the level of semantic confusion that reigned until very recently in this field of research, since exactly that time that I would qualify as the 'early colonial adaptation of shamanism to the reductionist mindset of Western researchers'.

The example I have chosen is taken from the article *Lévi-Strauss on Shamanism*, by Jerome Neu, published in Andrei A. Znamenski's reader *Shamanism: Critical Concepts in Sociology (2004)*. The author observes:

Jerome Neu

Lévi-Strauss actually speaks of the sorcerer abreacting for the silent patient (p. 183), which is without sense in psychoanalytic terms. And again, he does not explain why symbolic thoughts should provide a lever for producing physiological changes, except that the thoughts run 'parallel' to the physiology. But do they? If they did, would that *explain* anything?[52]

This is a striking example of a scientific misunderstanding of gigantic dimensions. First of all, the author here criticizes Lévi-Strauss for observations he has made, and thus as a mere messenger of a phenomenon neither Strauss nor the author seemed to understand. Second, it is true that the shaman effects changes in the physiology of the patient by performing acts on himself, or within his own spiritual oversoul, instead of the patient doing anything about them, or taking any remedy against them. This is about the most important fundamental difference between shamanic healing and the healing concepts in larger dominator civilizations. It has been pointed out clearly and with much detail in a study by Sabine Hargous, *Les appeleurs d'âmes (1985)*. In addition, a quote from Terence McKenna's book *Food of the Gods (1993)* leaves no doubt that shamanic healing uses what we today know as the quantum field or quantum interconnectedness for effecting healing:

Terence McKenna

Usually, if drugs are used, the shaman, not the patient, will take the drug. The motivation is also entirely different. The plants used by the shaman are not intended to stimulate the immune system or the body's other natural defenses against disease. Rather, the shamanic plants allow the healer to journey into an invisible realm in which the causality of the ordinary world is replaced with the rationale of natural magic. In this realm, language, ideas, and meaning have greater power than cause and effect. Sympathies, resonances, intentions, and personal will are linguistically magnified through poetic rhetoric. The imagination is invoked and sometimes its forms are beheld visibly. Within the magical mind-set of the shaman, the ordi-

Published by Sirius-C Media Galaxy LLC, 2010

> nary connections of the world and what we call natu-
> ral laws are deemphasized or ignored. /6

It doesn't actually surprise me that the psychoanalytic framework is used to deny shamanism its intrinsic scientific novelty, when compared to pre-quantum physics Western scientific methodology. I have shown in my book *Normative Psychoanalysis (2010)* that psychoanalysis at no point in time was a science, but represents a collection of myths that ultimately were forged and are upheld for providing a pseudo-scientific and ideological roof structure for Western consumer reality.

Terence McKenna has anticipated this scientific novelty while he explains the fact of the co-emergence of healing both in the healer's and the patient's organisms in poetic rather than scientific terms. But his language is accurate in that the quantum field is indeed 'an invisible realm in which the causality of the ordinary world is replaced with the rationale of natural magic'. The same is true for the last sentence of this quote, where he says that within the magical mindset of the shaman, 'the ordinary connections of the world and what we call natural laws are deemphasized or ignored'. The truth is that indeed on the subatomic or quantum level of reality, these natural laws that we know from Newtonian physics are invalid. Regarding the vocabulary McKenna uses, I can only refer to the old truth that humans consider as 'magic' all they don't really (yet) understand. Had McKenna anticipated cutting-edge research on the quantum field, he would have used a scientific instead of a poetic vocabulary to express this truth.

Sigmund Freud

I am at pains to qualify Freud's opinions and speculations about tribal cultures in any even remotely positive way. What Freud writes in *Totem and Tabu (1913)* about the 'primitive' Australian aborigines will not lead the interested reader to a comprehensive grasp of shamanism. In keeping with Freud's personal style, not to say his personal obsession, he uses examples mostly from the Australian Aborigines, gathered and discussed by anthropologist James George Frazer.

In his first essay, entitled 'The Horror of Incest', Freud points out, with some surprise, that although the Aborigines do not seem to have any sexual restrictions, they exhibit an elaborate social organization whose sole purpose is to prevent incestuous sexual relations.

In the second essay, 'Taboo and Emotional Ambivalence', Freud considers the relationship between taboos and totemism, using his concepts of 'projection' and 'ambivalence' developed in his work with neurotic patients, concluding, somewhat precipitously, that 'primitive peoples' feel ambivalent about most people in their lives, but will not admit it to themselves.

In the third essay, 'Animism, Magic and the Omnipotence of Thought', Freud draws another parallel between primitives and, this time, *early libidinal development*. He asserts there is a belief in magic and sorcery that derives from an

Published by Sirius-C Media Galaxy LLC, 2010

overvaluation of psychical acts whereby the structural conditions of mind are transposed onto the world: this overvaluation, he sees in both primitive men and neurotics, concluding that the animistic mode of thinking is governed by an 'omnipotence of thoughts', a projection of inner mental life onto the external world. This imaginary construction of reality is also discernible in *obsessive thinking, delusional disorders and phobias.* Freud comments that the omnipotence of thoughts has been retained in the magical realm of art.

In the final essay, 'The Return of Totemism in Childhood', Freud argues that combining one of Charles Darwin's more speculative theories about the arrangements of early human societies, Freud located the beginnings of the Oedipus complex at the origins of human society, and postulated that all religion was in effect an extended and collective form of guilt and ambivalence to cope with the killing of the father figure (which he saw as the true original sin).

It is almost incredible how today anybody can invoke Freud as a cultural or psychiatric innovator, as what he was standing for is simply nihilism and a total ignorance in front of the primacy of spirit over matter that aboriginals do know about since millennia. That neurotics in their confusion sense something true is not surprising, but obviously Freud was not up to match their level of evolution. He was on a level below 'primitives' and 'neurotics', ignoring about everything about spiritual laws, stating literal nonsense in most of his books and theories, which, to make it worse, were adopted as eternal truth by subsequent generations of psychoanalysts and even lay people. More is not needed actually to understand to what point our culture is 'intuitively ignorant' to a point it begins to border ridicule; for otherwise, none of Freud's arrogant assumptions that have no backup even in common sense would never have been adopted as bearing any ontological truth value.

But still, there was an opening effect somehow on Western society when an authority such as Freud took note, even in a distorted manner, of aborigine culture, religion, lifestyle or sexual customs, for this attention, and I almost would add, *nothing but that attention*, despite his display of superiority and 'colonial' arrogance, was enough to validate these tribal peoples as not only somehow 'interesting and original', but simply human, plain and full of rich and structured emotional intelligence.

Bronislaw Malinowski and Margaret Mead

As early as in 1929, Malinowski published his report on the sexual life of the Trobriands in which he draws the reader's attention particularly to the *sexual life of children and adolescents.*[53] Malinowski observed, not without surprise, high sexual permissiveness toward children's free sexual play.

More generally, he noted the total absence of a morality that condemns sexuality in children. Instead, he observed, children engage in free sexual play from early age.[54]

Initiatory rites, Malinowski found, were absent with the Trobriands since children were initiated from about three years onwards, generally by older children, in all forms of

Published by Sirius-C Media Galaxy LLC, 2010

sexual play. This play is completely nonviolent and includes, with the older children, coitus.

The most interesting finding for Malinowski was that in this culture violence was as good as non-existing and that there were equally as good as no sexual dysfunctions. Trobriands were found to be almost ideal marriage partners and divorce is a rare exception. Violent crimes are non-existent and incest strongly tabooed and inhibited by social norms.

Other researchers found similar phenomena with the Muria tribe in South India where children stay until their maturity in so-called ghotuls where they live their sexuality freely and in utter promiscuity. Older children initiate younger ones progressively into sexual play.[55]

These researchers found that after a phase of total promiscuity, the children, from the age of sexual maturity, form strong bonds and partnerships that are based not on sexual attraction, but on love. They further found that these first steady relationships formed the basis for later marriages that, regularly, last life-long.[56]

This field research conducted by Malinowski and Margaret Mead[57], while it is certainly of high importance for sexology, cognitive psychology and research on emotions, has not given any information about shamanism.[58] This is simply so and my remark is in no way a value judgment. Their research was not intended to provide information on shamanic culture and lifestyle. But …, historically speaking, it is quite uncanny to see that before researchers came up with looking at native peoples' spiritual life, they were looking at their sexual life. I guess it says more about the typical obsession of the observers than the subjects observed! And the parallel with Freud's early regard upon aboriginal culture is obvious and not coincidental.

What does this mean? It means that what quantum physics says is really true, and also on a practical level: the observer is always entangled with the object of observation. When researchers focus on sexuality and physical reality, they will see emerging sexual properties, when they focus on spirituality and metaphysical reality, the will see emerging spiritual or religious properties. Both Malinowski and Freud were focused on the former, Jung was focused on the latter. All of them were one-sided in a way, considering a partial spectrum of life, not life in its holistic total quality, as shamanism does.

Carl-Gustav Jung

Carl Jung was not a representative of psychoanalysis because he was *different*, so different that I say he was a shaman himself! And I am not the only one who says this. C. Michael Smith writes in *Jung and Shamanism in Dialogue (2007)* that Jung, during the time of the break with Freud, his mentor and 'father figure', became seriously depressed and close to psychosis, which led to as it were his shamanic initiation. The author writes:

C. Michael Smith

Shamanically speaking, soul loss has traditionally been associated not only with a loss of will, such as we find in depression, or with a loss of vital powers, such as we find in pathological dissociation, but also with a loss of connection to community, to the social sphere. In soul loss, one may be so lost in the 'realm of imagination',

Published by Sirius-C Media Galaxy LLC, 2010

in altered states of consciousness, that there is little relatedness to the outer world. In this respect, Jung must have felt very 'lost' indeed.

He heard voices, had visions, dreamt of rivers of blood, talked with spirits as he walked in his garden. He was so absorbed in the altered states of consciousness associated with psychotic, mystic, and shamanic realms that he had to remind himself that he was a doctor, a really existing person with a family, patients, and responsibilities.

During the shamanic initiatory crisis, the initiant has available ritual elders, master shamans to whom he or she can go to guide / and safely contain the transformative process. The wounded healer learns to heal himself or herself partly through the encounter with the spirits, and partly under the necessary structuration and guidance of the ritual elders. Jung had no ritual elder, no analyst to help him sort through and understand the emerging material, and no professional therapeutic containing vessel was available.[59]

From a scientific point of view, it seems daring and shaky to call Jung 'a shaman' even though he might have experienced phenomena that are usually reserved for initiants of shamanism and psychic research.

The difference between original shamanic initiation and Jung's psychic experiences is that Jung lacked a genuine *intention* to become a shaman; his experiences were involuntary for the most part, a result of his extreme psychic tension during these times of trial. He was closer to a psychotic who kind of manages to make senses of his delusions and psychic extravagances.

The difference to shamanism is that the shaman has mastered this initial phase that however, he entered with a firm intention to become a shaman, to enter the shamanic tradition, usually after having received a guiding dream in his adolescent years.

But despite this precaution, I can conclude that Jung's struggle with shamanism was certainly an honest opening to the influence of shamanism upon the Western psychological and psychiatric tradition. It is well documented that Jung left his initiation, though an involuntary one, unharmed and become a major spokesman for nonordinary states of consciousness and shamanic reality in the Western mythopoetic and psychiatric traditions. C. Michael Smith observes:

C. Michael Smith

Jung had found the key to his own healing, to his own psychological theory, and to others in his tribe of western society. The Self is the goal of his personal quest and simultaneously the goal of his mature psychological theory. In this insight we have Jung the shaman becoming healed, and returning with the boon of his tribe: the individuation process is a path towards self-realization. The understanding of the Self and its realization as the goal of the life process, the individuation process, became the program and mission for the second half of his life. From this point on, Jung had a clear sense of his mission (purpose), and a valuable psychology to offer the modern western world.[60]

Michael Smith asks 'Was Jung a Shaman?' He first says he was probably not a shaman in the classical sense, as suggested by Eliade, Harner, and others, but that he was certainly a wounded healer. Then he goes on reasoning:

C. Michael Smith

The wounded healer is a fundamental aspect of the shaman. It is through the tended wound that the shaman is able to see, to empathize, and heal. Jung pos-

Published by Sirius-C Media Galaxy LLC, 2010

sessed empathic abilities in a high degree, and his abilities increased immeasurably after the resolution of his midlife crisis. Like the traditional shaman, Jung was a loner, an individual who preferred solitude and absorption in the non-ordinary or imaginal dimensions of *what he later came to call the collective unconscious.*[61]

The Grand Opening

The grand opening, as it were, for shamanism in Western society occurred not before the 1960s. It can quite accurately be seen coincident with the publication of Mircea Eliade's book *Shamanism* in 1964, followed by the books of Michael Harner, Richard Schultes, Ralph Metzner, and Adam Gottlieb.

Newer research eventually recognized that shamanism is the oldest healing practice, a sort of religious medicine that originated more than twenty thousand years ago in the Paleolithic hunting cultures of Siberia and Central Asia. Mircea Eliade observes that the word shaman is derived from the Siberian Tungus word *saman*, which is defined as a technique of ecstasy. The shaman is considered a great master of trance and ecstasy. He is the dominating figure in certain indigenous populations.

Most early cultures' healing practices stem from a shamanic tradition. For instance, when visiting the sick, Egyptian magicians often brought a papyrus roll filled with incantations and amulets in order to drive out demons. It is further recognized today that the shaman is regularly the religious leader or priest of the tribe. He is believed to have magical powers that can heal the sick. The shaman is called upon to mediate between the people of the community and the spirit world to cure disease, exorcize evil spirits, and to promote success in hunting and food production and to keep the tribal community in balance. Traditional shamanic rituals include singing, dancing, chanting, drumming, storytelling, and healing.

The shaman also is a psychologist, or psychoanalyst in tribal society, a sort of specialist in human souls. He is able to see them and know their form and destiny. The shaman controls the spirits. Rather than being possessed by them, he communicates with the dead, demons, and nature spirits. The shaman's work is based on the belief that the soul can forsake the body even while a person is alive and can stray into other cosmic realms where it falls prey to demons and sorcerers. The shaman diagnoses the problem, then goes in search of the wandering soul and makes it return to the body.

Shamanism is still practiced all over the world, although each culture's shamanic tradition has evolved in different ways. Native American medicine men perform soul flights and vision quests to heal. North American Inuit shamans undertake undersea spirit journeys to ensure a plentiful supply of game. Tibetan shamans use a drum to help them in spirit flight and soul retrieval. Central and South American shamans often use hallucinogenic plants to invoke their shamanic journeys. Australian aborigine shamans believe that crystals can be inserted into the body for power.

Despite the variety of these practices, I stress in this book that the shaman has a methodology, a more or less precise, and 'computable' set of techniques that are orderly, logical and sound, and that are based upon a *scientific* outlook upon

Published by Sirius-C Media Galaxy LLC, 2010

reality. This fact is hardly ever mentioned in the admittedly large literature on the subject today.

The Shamanic Method

The shaman is not a theorist, but a scientist, not a theologian, but a pragmatist. He is practical, a solution-finder and his first rule is effectiveness. He is something like a highly effective Zen manager in his universe of natural laws, and he is a communicator. He communicates with the spirit world, the world of the ancestors and the world of the animal and plant spirits.

Now, how did I come to speak of shamans using scientific method to explore the spirit world and learn healing in the trance state? I was studying how shamans learn their 'techniques of ecstasy' and found there is very little ecstatic about this learning process, but that it actually is rather strict, methodologically sound, logical, and empirical.

Shamans observe the living, nature, the human organism, they observe every little details, and this usually starts when they turn into adolescence. They are rather introvert and often bold and persistent people, men or women, who have in common that they do not accept the common folk wisdom, nor the common lies and superstitions but set out to inquire by themselves. They spend years in solitude, occasionally meeting their tutor, and often have no families in their years of learning. That means that most shamans live rather ascetic lives, that are turned toward their science, toward their discoveries, just like any great Western scientist in his or her younger years of scientific achievement, lab work, experimental studies, and publishing of papers.

Published by Sirius-C Media Galaxy LLC, 2010

A shaman receives his basic education from the entheo-genic plant teachers, and only at a minor degree from an-other, elder, shaman tutor. Shamans around the world, asked why they knew this and that secret about healing, about cer-tain *hidden connections*, as Fritjof Capra calls holistic knowledge or about specific illnesses, answer they knew it directly from the plant spirits. They tend to affirm that they themselves knew very little and that they just humbly asked the plant spirits every time they could not solve a problem or not find a remedy for a certain illness. And the effectiveness of a sha-man, then, is exactly to maximize the response ability he has for all possible problems is is asked to solve (sickness, counter-magic, right timing for harvest or even political questions regarding tribe relations) by maximizing his unique commu-nication with the invisible world.

By the same token, shamans around the world, when asked about *reality* tend to affirm that our visible reality, the one most city dwellers think was the only one, is a very minor and rather insignificant form of reality and that the real real-ity is the hidden one, the one that is unveiled during the en-theogenic visionary experience.

If we refuse this bias of a *more-or-less* in terms of reality assessment, we can still enrich our mindset with the option that there might be parallel realities and that all realities, visible or invisible, or visible only through facilitating com-pounds or other consciousness-altering devices, are equally valid and equally important. Such an opening of science to-ward parallel universes and acknowledging the option of a multitude of possible realities that are not conflicting each other but may or not be intersecting would be a great advance and evolution of Western science.

I am serious when I allege that an experienced shaman is able, through scientific method of exploration into the unknown, to use the laws of the subatomic world, thereby directly connecting with the quantum field for time travel, for rendering himself temporarily invisible, or for traveling with lightning speed to remote places, clothed only in his auric body, while leaving the physical body behind on a bed, while being in trance.

Western society's notion of reality and that of most native populations clash worlds apart. Let me quote from Michael Harner's *The Way of the Shaman (1990)*, which represents a mark stone in shamanism research.

Michael Harner

Shamanism represents a great mental and emotional adventure that implies both the patient and the healer. Through his voyage and his heroic efforts, the shaman helps his patients to transcend their normal, ordinary, definition of reality as well as their self-definition as being sick.

This is obviously a psychological explanation of the healing experience. I can say that a shaman helps his patient to reframe his illness, just as a psychotherapist does; but I can also say that the shaman doesn't really impact upon the perception of the client, but directly upon his quantum field, using the zero-point field as the connecting agent, and scientific means to regulate the client's energy body.

Published by Sirius-C Media Galaxy LLC, 2010

That is a completely different explanation, because it is *epistemologically different!* When I say the shaman is a healer, that is one thing. When I say the shaman is a healer using scientific method for healing, that is another thing.

In the second case, there is namely no more difference between a shaman and a Western physician as both use scientific method for healing. While the scientific tools they use are obviously very different, they both proceed empirically, using observation and verification/falsification of a 'medical theory' for achieving their healing goals.

Now, obviously, my view here is going way beyond the accepted paraphernalia of shamanism research; it is today no more doubted that the shaman impacts upon the perception of the client. I can even say, it's popular knowledge after the movie *What the Bleep Do We Know!?* When the shaman in the movie touched Amanda's third eye or frontal lobe, this was clearly meant as a 'reality changing perception opener'.

But it was hardly meant to denote the shaman as a doctor who proceeds scientifically to heal Amanda's emotional stuckness, sexual neurosis and hysteria that was in part the result of her traumatic marriage experience.

Interestingly, this same statement could be made about hypnotherapy, especially the method applied by Milton H. Erickson (1901-1980) by replacing 'healer' and 'shaman' by hypnotherapist.

And in fact, Erickson certainly has learned many of his secrets by studying shamanic theory and practice. In a way, we can say that our modern psychotherapists are something like Western shamans. They are in fact borderline figures in a worldview that is almost hermetically closed toward recognizing the universal existence of soul values. Or, to remind the saying of Carl Jung, psychotherapy begins with the study of our dreams (individual unconscious) and of our myths and cultural sagas (collective unconscious).

A native would qualify somebody with a narcissism problem as a person who has lost a part or the whole of his soul. A psychotic patient who, in his delirium, says that he's Jesus Christ would be qualified by a native shaman as somebody whose soul is occupied by a spirit who, for whatever reason, speaks through him.

While both worldviews are quite opposite, the fundamental principles of healing, in shamanism, on one hand, and in Western psychoanalysis, on the other, are not very different. Where the split opens much farther is where we talk about Western medicine, as it is applied still by a majority of physicians. The sometimes sharp opposition between physicians and psychoanalysts has its deeper reasons here! The fundamental incompatibility, today, of shamanism and Western society is that the latter lacks almost totally out on acknowledging and integrating the *ecstasy pattern*, which I have identified as one of the *Eight Dynamic Patterns of Living*.[62] Terence McKenna was asked by Jay Levin to define shamanism.

Terence McKenna

Shamanism is use of the archaic techniques of ecstasy that were developed independent of any religious philosophy — the empirically validated, experientially operable techniques that produce ecstasy. Ecstasy is the contemplation of wholeness. That's why when you experience ecstasy – when you contemplate wholeness – you come down remade in terms of the political and social arena because you have seen the larger picture.[63]

But what is ecstasy, then? Terence McKenna explains:

Terence McKenna

Ecstatic is a word unnecessary to define except operationally: an ecstatic experience is one that one wishes to have over and over again.[64]

Terence McKenna was probably right in not complicating something that is basically so easy but that most of us have unlearnt it through our educational conditioning, while we originally, as small children, possessed the gift to connect with all-that-is. I have called it in one of my writings the capacity 'to feel good without reason'.

Most authors in the literature on shamanism emphasize the shaman's role as a manipulator of consciousness, an expert in states of ecstasy, a visionary, a healer, a communicator with the spirit world, and a psychopomp. But I have found only two authors who, like me, are convinced that all this, while it is much already, is not the essential, and that there is more. These authors are Stanley Krippner and Alberto Villoldo. They relate in *Healing States (1987)*, a book they authored together:

Alberto Villoldo and Stanley Krippner

In other words, shamans represent the world's oldest profession. Their roles probably varied from one society to another, but it is likely that they served a number of functions: artist, healer, magician, priest, psychotherapist, seer, storyteller. In so doing, they assisted the evolution of human consciousness. (…) Shamans were also the world's first scientists. Their discoveries of medicinal and sacred plants were made through observation and trial-and-error, both honored scientific procedures. (…) They provided humankind with the first tangible clues that there was order in the universe, because these observations could be replicated hundreds of times and still yield uniform results. If the shamans did not produce reliable data, their role was endangered and their days of honor were numbered. Thus, humanity owes a massive debt to shamans for their pioneering work in the accumulation of knowledge and the development of human capabilities. /161-162

In his later books, Alberto Villoldo stays true to his position and provides many more details for us to see that shamans work in a *methodologically correct* and conscious manner for gathering the data they need for their work. The specific knowledge they gather for healing has primarily to do with the states and the condition of what Villoldo calls the 'Luminous Energy Field', and which connotes the human energy field that pulsates both within the protoplasm and the surrounding aura.

In truth, Villoldo's books are an invaluable and important source for this knowledge to expand to a greater extent within Western culture. As I have shown in my *Idiot Guide to Emotions (2010)* and my audio book *The Science of Emonics (2010)*, this knowledge was once hermetic in our ancient traditions, and as long as it was hermetic, that is accessible only to a *chosen few* of sages and natural healers, things were okay. From the moment however that first alchemists and later natural scientists tried to vulgarize this knowledge, problems

Published by Sirius-C Media Galaxy LLC, 2010

arose. It was not a minor confrontation, as a number of scientists lost their lives, Giordano Bruno perhaps the most prominent on the list. This knowledge taboo persisted even in the 20th century, and we need only to think of how Wilhelm Reich ended his days to be reminded of how fierce, blind and brutal the opposition was.

The Western tradition that was for more than a millennium under the knowledge denial of the Church, was not ready to absorb this knowledge. Interestingly enough, things got even worse with the Cartesian science tradition, that is roughly during the last four hundred years, while this science was declaring itself separate from religion and agnostic. This science paradigm completely rejected the *concept of the ether*, the human energy field and the very existence of a creator energy that is the origin of all life.

Now, things look quite different. I have shown in my book *Do You Love Einstein (2010)* that it was first of all the emergence of quantum physics, and here especially its two proven base assumptions, *uncertainty* and *nonlocality*, that made for a wide opening where formerly there were only barriers and fears. Actually, the impact of quantum physics upon the Western scientific paradigm was so dramatic that scientists felt the ground had moved away from under their feet. But it was exactly this dramatic shift in consciousness, and the insight that the observer is inevitably entangled with the object of observation that made for novelty and new ways of scientific thinking. Within the last two decades, then, and given the unwavering and almost tumultuous progress of quantum physics beyond the borders of what Krishnamurti called 'the known', there was something like a grand opening not only in science but also in society, in the whole of our Western consumer culture, for this millenary knowledge to be eventually accepted and integrated.

What actually changed in this tremendous paradigm shift was not so much science itself, but the way we are perceiving reality; as a result of this shift in perception, our scientific processes and methodology changed accordingly. It is important to see this difference, for it is the very way of thinking of native peoples that *our perception conditions reality* and all we do and achieve within this reality. Once I look at the world in a different way, I will do a different kind of science. Once I see that my old science was destroying the planet, I will be able, from my new holistic perspective, to conceive and design a new science that is sustainable, and that respects and integrates spiritual values.

It is on this fertile ground that Villoldo's books could take root in our culture while just some decades before the knowledge they bring would perhaps have been violently rejected, with the result that the books would simply not have been accepted for publishing. This is how it was, and I know this from first-hand information, as around the same time, I tried to publish some of my own books and was constantly, and persistently, rejected. Now, things look quite different; there has not been a time in our culture that was more vibrantly interested and motivated to learn the most possible amount of knowledge that was formerly rejected and banned, as part of the forbidden tree of knowledge. A first breakthrough was happening with the books of Carlos Castaneda in the 70s

Published by Sirius-C Media Galaxy LLC, 2010

and 80s, then, at around the same time, with the books of Terence McKenna.

But it has to be seen that at that time, the idea of a shaman being a scientist was still so daring and outlandish that even a popular author such as Terence McKenna carefully guards against possible attacks when he states in the *Archaic Revival (1992)* that he was 'not a scientist, but an explorer'.

To reduce shamanism to being a catalyzing method for bringing about ecstasy, while this is certainly true, is not giving justice to the science of shamanism. It would be as if saying that medical science is to bring about painless operations, and generally, was a methodology not for healing people, but for effective painkilling. (That our Western medicine actually used to have this reductionist approach to healing, I won't discuss here as it's off-topic, and also because things are quite dramatically changing now).

While I admit that it was daring to state shamanism is a science, just two decades ago, now such a view cannot be dismissed as groundless or speculative. Why? Because there is evidence, a lot of evidence to corroborate my theory. Villoldo's books are a *good starting point,* and they contain much of the knowledge I have found in other recent publications about shamanism, but explain matters in more detail, and in a way that is *comprehensive* not for a few select field researchers only, but for many people.

To begin with, the shamanic medical system and methodology differs in several ways from the Western medical science model. The first fundamental difference, I have mentioned already earlier on; it is the fact that the shaman impacts first of all *upon his own neuronal and bioenergetic network* for effecting changes upon the neuronal and bioenergetic network of his client. This means the shaman uses *quantum interconnectedness* for bringing about any beneficial changes in the complete organism of the client.

The second important difference is that the shaman doesn't directly impact upon the physical body; he impacts upon the subtle, auric body, the luminous energy field; this is systemically sound as any changes effected in the quantum field automatically will trigger changes in the physical body as well. This is actually very smart as a healing methodology because the spirit body creates and maintains the denser physical body; as a result, any changes in the physical body are always preceded by changes in the subtle energy body.

Now, the interesting thing is that shamanic medical science in this point fully resonates with the insights and practices of intuitive and clairvoyant healers within our own culture. I have found important references in the writings of Paracelsus, Franz Anton Mesmer, Carl Reichenbach, Wilhelm Reich, Charles W. Leadbeater, Shafica Karagulla and Dora van Gelder Kunz, to name only these, that actually show in minute detail that shamanic healing wisdom exactly coincides with their Western esoteric correlates. It is important to note that both the shamanic and the traditional healing systems of large civilizations are very old, much older namely than our modern medical science. Another parallel, I have found in the oldest healing tradition of India, Ayurveda, and both Chinese and Tibetan medicine. These traditional healing practices equally know about the luminous

Published by Sirius-C Media Galaxy LLC, 2010

energy field and effect healing primarily by manipulating anomalies in the field.

Science and Ecstasy

Now, how can the experience of ecstasy go together with a scientific attitude? On first sight, the two experiences or attitudes toward life seem to be contradictory. But our Western scientist also returns home from the lab and has dinner, and perhaps watches television and goes to a party to have some fun, and we would not for that reason deny to him that his basic approach to life is *scientific*. And the shaman amuses himself with ecstatic dances, right? He is stamping and hopping like a fool, to show that he can relax and let go, right?

No. This comparison is an entire non-sense. The shaman uses ecstasy for sharpening his perception, as a scientific tool, so to speak, for accessing perception channels hidden to us in the normal waking state. So science and ecstasy are by no means contradictory experiences, attitudes or methods. They are complimentary! They go hand in hand when it goes to register a most-possible slice of the cake of life, and of experience, means, first of all, of nonsensory or extrasensory experience.

Ecstasy is one of the *Eight Dynamic Patterns of Living*, one of the basic ingredients of life with all native populations around the world. Contrary to common belief, ecstasy is not to be confounded with overindulgence or debauchery, or any form of 'bloody ritual' you may think of spontaneously when encountering the notion for the first time. To repeat what McKenna once said, ecstasy is the 'contemplation of wholeness', thus in other words it's an individual and a collective

Published by Sirius-C Media Galaxy LLC, 2010

form of meditation. When I contemplate wholeness, I am in a state of bliss, or blissful awareness. *Now I can't see why this heightened state of awareness should clash with scientific observation?* I assume the two ways of approaching the world are valid, and are non-conflicting, even complimentary. They sustain each other.

Let me give some example from our Western world. When Albert Einstein was working on relativity theory, he had a dream vision in which he saw the whole theory written down in precise mathematical equations, accomplished once forever. And he would accomplish it later in exactly the way he had seen it in that vision. Another well-known example is the discovery of the benzol ring. After benzene had been the subject of many studies, inter alia by Michael Faraday and Linus Pauling, it was still unknown which molecular structure the molecule had while the empirical formula for benzene was long known.

In 1865, the German chemist Friedrich August Kekulé published a paper suggesting that the structure contained a six-membered ring of carbon atoms with alternating single and double bonds. Only in 1890, during a honorific event organized to his honor by the German Chemical Society, Kekulé spoke of the creation of the theory. He said that he had discovered the ring shape of the benzene molecule after having a reverie or day-dream of a snake seizing its own tail (this is a common symbol in many ancient cultures known as the Ouroboros). This vision, he said, came to him after years of studying the nature of carbon-carbon bonds.

These inner visions that are reported by scientists and inventors all over the world, are moments of contemplation where the mind meets itself, and contemplates the wholeness, the hologram of a specific creation. Typically, those visions are preceding the accomplishment of the theory or invention, and thus they can be said to be *premonitory*.

There are many other examples that show that ecstasy is not bounded within tribal cultures but a phenomenon of the human mind in its greatest dimension that goes beyond space and time, and is connected with the creator force. Seen from this perspective, it was never questioned that science and ecstasy could not go together, or that a scientist might be non-scientific when he has such kind of visions. Experience shows that the contrary is true, it is the greatest, most reputed and most honored scientists who have these holistic experiences, not the mediocre vintage of the Cartesian, blindfolded and doctrinaire scientist who takes as truth only what he has 'under his two eyes'.

If this is true for our highly intellectual culture, then I think it's so much the more true for shamanic culture. Compared to our cultural setup, within shamanic cultures, values like intuition, nonsensory experience and nonordinary states of consciousness encounter less of stigma and are more easily accepted. In shamanic culture, one would not even think that ecstatic experience could probably interfere with logical, scientific scrutiny. This concern is typical for our Cartesian science tradition.

Published by Sirius-C Media Galaxy LLC, 2010

Let me explain this with two examples, the relationship between science and divination and why thinking in terms of a Gestalt is scientific thinking.

Science and Divination

When I talk about *divination*, I include all possible devices, methods and traditions that are used for getting a glimpse of truth for decision-making, or potential outcome of specific events around a chosen developmental theme. Thus, divination can mean *astrology*, it can mean *Tarot* and it can mean *geomancy*, and it definitely also can mean using the *I Ching*.

Cartesian science never cared about explaining divination and why it works, while archetypal and transformational psychology, especially the Jungian branch of it, has been a pioneering and thought-provoking pathway for opening the depth of the psyche and its divinatory potential to the modern researcher or psychologist. One of the leading publications in this context is Sallie Nichols' *Jung and Tarot*.[65]

Here, shamanic traditions surely have been showing the way, when our own science paradigm was still stuck in denial and prejudice about the alleged 'psychotic character' of native shamans, and the practice of divination as 'a god-forbidden devil's play'.

It is important in this context to realize that divination is not deterministic in the sense that 'the future is predetermined', while this assumption often appears to be repeated in vulgarized publications on esoteric sciences. The truth is that no diviner can ever predict 'the future', as the future is simply an extrapolation of present thought content, and subconscious thought patterns, as well as emotional patterns.

What the diviner does is in fact scan the content of our unconscious and project this content into some or the other cognitive system that renders it visible and intellectually graspable. Hence, what divination explains is but the status quo of the asker, the person who comes to the diviner, with a particular question or project. While it is true there is a certain probability that the present of consciousness perpetuates itself into the future, by extrapolation of its content on a timeline of events, this is no 'prediction' of the future, simply because the asker can change their content of consciousness *hic et nunc.*

This is why I developed, years ago, the idea of combining astrology and other forms of divination with what I came to call *Creative Prayer* as part of my *Life Authoring* self-coaching technique. The prayer technique is used as an add-on to the astrological consultation in the sense that it helps changing the present content of consciousness, after it has been rendered cognizable by the projective system of astrology. I learnt the technique basically from three books by Dr. Joseph Murphy, *The Power of Your Subconscious Mind*, *The Miracle of Mind Dynamics* and *Think Yourself Rich*.[66]

The solution to the riddle of how divination works is contained in one single phrase of this book. Here it is:

Dr. Joseph Murphy

Remember that because your future is the result of your habitual thinking, it is already in your mind unless you change it through prayer.[67]

Published by Sirius-C Media Galaxy LLC, 2010

What divination does, to repeat is, is to read our habitual and repetitive thought patterns, and extrapolate them on a virtual time line into the future. This is, then, what is colloquially called 'predicting the future'. When you know what it's really about and how it is done, you see that it doesn't make sense, but can be understood as an oversimplification of the truth.

After all, if the future was predestined, as Calvinism assumed, Dr. Murphy and many other new thought authors would not have written their books; and they would not lecture as ministers and spiritual guides. They do it because they have realized that wrong beliefs about life and living are destructive and make for much of the misery we encounter in human lives, and in the world at large. Our mind is fragile in the sense that it can easily be manipulated by the mass media; worse, when fortune tellers, astrologers and diviners come along to pretend they are 'predicting the future', the outcome may even be dangerous as their assumptions may by naive souls be taken as hypnotic spells that then may gain the power to realize as *self-fulfilling prophecies*. The reader may easily imagine where this can lead, and how much strife and turmoil this may produce in the lives of humans around the world.

Murphy has seen it all around himself, and even in his own family, how people fall ill and even die, without having to die, because of the suggestions they receive from others in the form of hypnotic spells wrapped in various forms, and also, unfortunately, in professional divination, when done by unspiritual, greedy and dishonest people.

And it's a fact, only to look at the Internet, what masses of scam artists are around in all those fields called esoteric, new age, mindpower and all the rest of it. When such accumulated power of irresponsible manipulative greed meets the fragile and ignorant mind of the 'man in the street', then we can virtually predict disasters to happen.

This messy situation with people being mislead by both their own beliefs and predictions received from others was one of the reasons that motivated Murphy and before him, Ernest Holmes, to write their books. The science that I have in mind when I put up the dichotomy science vs. divination is the *Science of Mind*, also called *Religious Science*, as it was founded by Ernest Holmes in 1927, and expanded and commercialized in the 1960s by Dr. Joseph Murphy and Catherine Ponder, and others.

I studied the *Science of Mind* thoroughly over the last fifteen years; it clearly emphasizes the priority of mind over matter – spiritual monism – and also the priority of the present over the past and any form of predestination.

It is not known to many that the Bible is against both astrology and any form of fortune telling. For example, Deuteronomy 18: 9-12 affirms:

Deuteronomy 18: 9-12

9 When thou art come into the land which the LORD thy God giveth thee, thou shalt not learn to do after the abominations of those nations.
10 There shall not be found among you any one that maketh his son or his daughter to pass through the fire, or that useth divination, or an observer of times, or an enchanter, or a witch.

Published by Sirius-C Media Galaxy LLC, 2010

11 Or a charmer, or a consulter with familiar spirits, or a wizard, or a necromancer.
12 For all that do these things are an abomination unto the LORD: and because of these abominations the LORD thy God doth drive them out from before thee.

When I came across these Bible quotes in 1991, I was first revolted! I found the Bible forwarded here a form of Christian fundamentalism that was completely against my convictions and spirituality. Yet I wanted to understand what the Bible meant here, what the deeper meaning was behind these admonitions.

Thus I was asking 'how does the Bible relate to divining'? And why does it exhort us to be careful with it? To begin with, let me quote an example from Murphy's book *The Power of Your Subconscious Mind.*

Dr. Joseph Murphy

How Suggestion Killed a Man

A distant relative of mine went to a celebrated crystal gazer in India and asked the woman to read his future. The seer told him that he had a bad heart. She predicted that he would die at the next new moon.

My relative was aghast. He called up everyone in his family and told them about the prediction. He met with his lawyer to make sure his will was up-to-date. When I tried to talk him out of his conviction, he told me that the crystal gazer was known to have amazing occult powers. She could do great good or harm to those she dealt with. He was convinced of the truth of this.

As the new moon approached, he became more and more withdrawn. A month before this man had been happy, healthy, vigorous, and robust. Now he was an invalid. On the / predicted date, he suffered a fatal heart attack. He died not knowing he was the cause of his own death.

How many of us have heard similar stories and

shivered a little at the thought that the world is full of mysterious uncontrollable forces? Yes, the world is full of forces, but they are neither mysterious nor uncontrollable. My relative killed himself, by allowing a powerful suggestion to enter into his subconscious mind. He believed in the crystal gazer's powers, so he accepted her prediction completely.

Let us take another look at what happened, knowing what we do about the way the subconscious mind works. Whatever the conscious, reasoning mind of a person believes, the subconscious mind will accept and act upon. My relative was in a suggestible state when he went to see the fortune teller. She gave him a negative suggestion, and he accepted it. He became terrified. He constantly ruminated on his conviction that he was going to die at the next new moon. He told everyone about it, and he prepared for his end. It was his own fear and expectation of the end, accepted as true by his subconscious mind, that brought about his death.

The woman who predicted his death had no more power than the stones and sticks in the field. Her suggestion in itself had no power to create or bring about the end she suggested. If he had known the laws of his mind, he would have completely rejected the negative suggestion and refused to give her words any attention. He could have gone about the business of living with the secure knowledge that he was governed and controlled by his own thoughts and feelings. The prophecy of the seer would have been like a rubber ball thrown at an armored tank. He could have easily neutralized and dissipated her suggestion with no harm to himself. Instead, through lack of awareness and knowledge, he allowed it to kill him./

In themselves, the suggestions of others have no power over you. Whatever power they have, they gain because you give it to them through your own thoughts. You have to give your mental consent. You have to entertain and accept the thought. At that point it becomes your own thought, and your subconscious works to bring it into experience.

Remember, you have the capacity to choose. Choose life! Choose love! Choose health![68]

Published by Sirius-C Media Galaxy LLC, 2010

There is a difference between *foolishly accepting* a 'prediction' by an astrologer or fortune teller, or to use, for example, the I Ching, for decision making. The same is true regarding serious astrology; it is a question of *professional ethics* to avoid being suggestive in any way. This is equally true for a serious Feng Shui consultant, Tarot expert, and even for paranormals who practice their profession within the rules of the unwritten ethical code set in Antiquity for all *Hermetic Sciences*.

From the side of the client, a certain level of emotional maturity is equally required! How many people die because they receive 'death sentences' from their physicians, taking for granted that the gods in white coats determine their destiny, and for the most part ignorant about the pitfalls, limitations and outright ignorance of Western medical science!

There is a responsibility linked to every new piece of knowledge you learn and digest. This responsibility requires you to use the knowledge not with a foolish, immature and infantile mindset that takes everything for granted when it comes from a so-called 'authority'.

Science and Gestalt

As I have shown in my reviews of some of the lesser known books by Wilhelm Reich, this scientist's conceptual framework had firmly embodied the *Gestalt*. Reich's genius as a scientist was his gift of observation, and his particular talent to see not single elements of a process, but the *whole* of the process. Reich was in this respect really different from the main bunch of his Cartesian-minded professional colleagues.

In our days, Reich would probably be considered as one of the leading-edge scientists. Generally speaking, when we observe living processes, we can either put our focus on single elements, or the substance, or we can focus on the *process*, and the *form*. Both form and substance are present in living systems.

Our culture has created the line as a symbol for evolution. However, the line is an artificial construct, inexistent in nature, a purely mental achievement, while evolution is cyclic. It allows the line only in combination with the circle, so as to say, resulting in the *spiral*.

Merriam-Webster's Dictionary defines the spiral as *relating to the advancement to higher levels through a series of cyclical movements*. The curving movement of the spiral is what it has in common with the circle; the increase or decrease in size of the spiral is a function of its moving upward or downward.

Interestingly enough, the spiral is by far the dominating form to be found in nature, and in all natural processes. It is a symbol or Gestalt for evolution in general. Life is coded in the spiraled double-helix of the DNA molecule. The spiral is the expression of the periodic, systemic and cyclic development that is in accordance with the laws of life.

The progression of the spiral shows that it always carries its root, however transporting it through every cycle onto a higher level or dimension; whereas the line leaves its root forever. All towers of Babel are manifestations of the line; they are linear and are created by linear thought structures.

True growth is always cyclic and spiraled, and nonlinear.

Published by Sirius-C Media Galaxy LLC, 2010

On the subject of bringing in Gestalt thinking in the logic of healing, Manly P. Hall, in his book *The Secret Teachings of All Ages* writes about Paracelsus and states:

Manly P. Hall

Paracelsus discovered that in many cases plants revealed by their shape the particular organs of the human body which they served most effectively. The medical system of Paracelsus was based on the theory that by removing the diseased etheric mumia from the organism of the patient and causing it to be accepted into the nature of some distant and disinterested thing of comparatively little value, it was possible to divert from the patient the flow of the archaeus which had been continually revitalizing and nourishing the malady. Its vehicle of expression being transplanted, the archaeus necessarily accompanied its mumia, and the patient recovered.[69]

It was Gestalt considerations and the insight that nature is basically an assemblage of patterns, and not of randomly arranged matter, that led researchers recently to eventually corroborate the age-old assumption that our universe is *holographic*, and thus programmed in holographic patterns that are all mutually interconnected. Ervin Laszlo writes in his remarkable study *Science and the Akashic Field*:

Ervin Laszlo

In a holographic recording – created by the interference pattern of two light beams – there is no one-to-one correspondence between points on the surface of the object that is recorded and points in the recording itself. Holograms carry information in a distributed form, so all the information that makes up a hologram is present in every part of it. The points that make up the recording of the object's surface are present throughout the interference patterns recorded on the photographic plate: in a way, the image of the object is enfolded throughout the plate. As a result, when any small piece of the plate is illuminated, the full image of the object appears, though it may be fuzzier than the image resulting from illuminating the entire plate.[70]

Before I come back to the role of the shaman as a scientist within shamanic culture, which is the main part of this study, I shall explain why shamans make ample use of entheogens to get connected with their inner god or the quantum field, and what some of the most powerful entheogens are.

Further, I am going to provide a real life example, report of a *shamanic voyage* I made back in 2004, with a Shuar shaman in Ecuador, when ingesting the traditional Ayahuasca brew for the first time in my life. It namely becomes clear through my report that one can well maintain one's scientific approach and mental clarity when being in a state of ecstasy.

I would even say that my cognitive abilities were widened and enhanced by the brew, and nothing interfered with my clear observation.

While my experience with Ayahuasca was not really a positive one as I was unable to move further down the rabbit hole, it was a very important learning experience for me, and I was able to derive considerable conclusions from it. I was namely able to see that the common mechanistic theory that says the trance state is triggered by nothing by the DMT contained in the brew is at least incomplete, if it's not completely wrong. What triggers the trance state and the whole of the experience is the psychic power, and *intention*, of the shaman

Published by Sirius-C Media Galaxy LLC, 2010

when he prepares the brew and during the ceremony itself; it is the power of this intention that as it were regulates the whole of the experience.

CHAPTER FOUR

Shamanism and the Use of Entheogens

Video Presentations

Consciousness and Shamanism

http://vimeo.com/channels/shamanism

Video Description

In these videos, I am reading excerpts from this chapter.

Introduction

When looking for evidence that shamanism is a *science*, I came early in life to realize that it is far from obvious what science really is, or what science *naturally* is – when we look at it from a point of view that is not conditioned by our own cultural bias. In fact, as a boy already, when building a small radio and experimenting with electronic circuits, I became aware of the *primal scientific method*, which is trial and error, and acute observation, and then the inductive logical work of comparing the outcome, changing the settings, and comparing again, and so forth. While I did later not become a scientist, nor, as I had planned, a recording and television technician, I kept a basic intention to look at life not from a magical and purely artistic perspective, but from one that is based upon information, knowledge, observation and methodology.

Since leaving high school and during my extensive law studies, I was deepening my knowledge in psychoanalysis and medical hypnosis as well as perennial sciences both from the West and the East. My law studies and work as a legal researcher taught me that so-called *scientific knowledge* is often a mere formula for manipulating opinion by invoking *established research* or *scientific authorities*. This is unscientific because it means to measure the knew with the parameters of the old. Only much later, and *inter alia* through the writings of Edward de Bono[71] and J. Krishnamurti[72] became I aware that this form of reductionism is inherent in the functioning of the brain and the way perception works. This brought me to the conviction that *exact science* is a myth because all science is relative to the observer's own set of beliefs. Meanwhile, my insights about science philosophy are widely confirmed by quantum physics and *recent consciousness research*, but when I first uttered them, more than twenty years ago, I was encountering but suspicion and estrangement.

What I learnt from this is that instead of accumulating knowledge, we should always check if new knowledge we acquire *resonates within ourselves* and thus confirms our own intuition and sense of truth. For only then it is useful and reinforces our own higher intelligence. There is no knowledge outside of this source and all knowledge is thus *individual* in the sense that not all truth is valid for all. An example for belief hidden behind *methodological* reasoning is the famous meeting of Wilhelm Reich and Albert Einstein in Princeton when Reich demonstrated to Einstein that the temperature difference close to the orgone accumulator contradicts the second law of thermodynamics, the so-called *law of entropy*.

Einstein, reportedly dumbfounded on first sight, handed the matter over to his assistant and Reich got to hear that *for methodological reasons* the experiment could not be reproduced under the same conditions as during the demonstration. In other cases, when Reich showed the bion reactions in burnt sand, he was told that *air germs* were causing the living cell structure in the burnt sand molecules.

The *air germs* theory was a *methodological* trick that did exactly what the Church committee did a few centuries earlier in order to *not* see the Jupiter moons through Galileo Galilei's telescope. This is how mainstream science works, on *beliefs*

Published by Sirius-C Media Galaxy LLC, 2010

that *elegantly are veiled by methodological considerations.* Through further study over more than twenty years, I became convinced that an alternative science and life paradigm always existed, and I retraced it and developed a terminology that I called *Emonics.*

I also found that our science patriarchy had relegated it to the underground, and officially denied it. And I also found amazing parallels between this alternative science and life paradigm deeply hidden within dominator cultures, on one hand, and science and life paradigms of tribal peoples or *scientific* native religions, such as the *huna* religion from Hawaii, on the other.

The present chapter will show that there is no split in the *cognitive assessment of natural healing* practiced within shamanism and by shamans when looked at through the eyes not of mainstream science, but using the millenary cognitive tools of *alternative science*, and particularly our insights into the functioning of hypnosis.

In addition, let me address an important argument people often come up with: they repeat that 'drugs will render you addicted'. While this argument is true for alcohol, for coffee, for tobacco, for most tranquillizers, for some medical drugs, for some of the hard core drugs such as heroin, it is probably not true for Cannabis or Hemp, and it is certainly not true for entheogens. Reputed shamanism researchers such as Michael Harner, Adam Gottlieb, Richard Schultes, Ralph Metzner or the McKenna brothers have repeated it over and over in their publications.

Regarding a synthetic entheogen, LSD, which was was used in psychotherapy, the literature has clearly defended the view that it renders in any way addictive. The discoverer of the substance, the Swiss chemist Albert Hofmann from Sandoz Laboratories, testified for LSD being non-addictive.[73]

Why the substance ultimately was put on the index of forbidden drugs has nothing to do with a problem of addictiveness but has purely political reasons. Terence McKenna notes in *Archaic Revival (1992):*

Terence McKenna

The solution of much of modern malaise, including chemical dependencies and repressed psychoses and neuroses, is direct exposure to the authentic dimensions of risk represented by the experience of psychedelic plants. The pro-psychedelic plant position is clearly an antidrug position. Drug dependencies are the result of habitual, unexamined, and obsessive behavior; these are precisely the tendencies in our psychological makeup that the psychedelics mitigate. The plant hallucinogens dissolve habits and hold motivations up to inspection by a wider, less egocentric, and more grounded point of view within the individual.[74]

Published by Sirius-C Media Galaxy LLC, 2010

What is Ayahuasca?

Ayahuasca, already mentioned as a favorite entheogen brew among shamans, is a brew intended to bring about a trance for self-discovery, healing, contacting spirits or exploring consciousness. The brew consists of two ingredients, *Banisteriopsis Caapi*, also called Ayahuasca liana, and the Chacruna shrub *(Psychotria viridis)*, a Mono Amine Oxidase or MAO inhibitor.[75]

There are many other entheogens used for religious purposes, and they form an integral part of shamanism. Entheogens are plants that contain psychoactive compounds, such as DMT, and others, and that, when taken at appropriate doses, produce a consciousness-altering effect upon our psyche and perception. While there are methods to alter consciousness without plants, modern researchers agree that from a point of view of *effectiveness* there is a large gap between those latter techniques, on one hand and the use of entheogen drugs, on the other. In fact, entheogens are several hundreds of percent more effective than non-plant based methods. Some researchers have seriously tackled the question why this is so, and one of the most persisting on this specific point was the late Terence McKenna.

In his book *The Archaic Revival (1992)*, McKenna affirms that entheogenic plants contain the very essential genetic code, the basic information about the evolution of life on earth, and that for this reason their ingestion, or at least the ingestion of the psychoactive compounds they contain, leads to an *immediate opening of consciousness,* which was something much broader and much more intelligent to experience than mere colorful visions.

In fact, McKenna's visionary and illuminating books would never have had such a powerful impact on the consciousness change of Western society if they only talked about some nice hallucinogenic visions. Anthropologists and shamanism researchers who don't understand the unique phenomenon of shamanism, tend to reduce the entheogenic experience to a mere social game, a distraction or a search for some kind of artificial nirvana; they are of course deeply misled. It is therefore not surprising that most anthropologists, and especially those of them who really do not understand shamanic cultures, tend to employ expressions such as 'hallucinogens', 'narcotic drugs', 'narcotics' or 'psychedelics'. Apart from the fact that these plants are *not narcotics*, because a narcotic drug, such as for example *opium*, renders somnolent but does not alter consciousness, the important thing to know is that entheogens are not understood, in shamanic cultures, as leisure drugs, 'to have a nice time seeing some visions', but really are considered as assets of the religious and numinous experience.

That is why the only expression that comes close to the shamanic mindset is the term *entheogens*: facilitators for getting in touch with the god within. And as such they form part of the religious ritual, and not of party time. This is generally little understood in most Western countries where so-called 'drugs' are usually associated with leisure, distraction, party time and sex. Yet in native cultures the very idea to for ex-

Published by Sirius-C Media Galaxy LLC, 2010

ample relate an entheogenic voyage with sex would offend every shaman if told about it.

It has been equally affirmed that entheogens, apart from their helping us to reach the inner mind, also dissolve nasty and somehow destructive habits such as alcoholism or heroin addiction, and generally help in a process of social deconditioning. Entheogens help us to see behind the veil of the normative behavior code in any given society as they show us options of *different* behavior. What we can thus learn from taking these plants as a sort of *social medicine* is to recognize the pattern of normative behavior we are caught in and that obstructs our creativity and self-realization.

People who are socially oppressed, racial, ethnic, religious or sexual minorities, may want to inquire into the possible dissolution of rigid behavior rules and oppressive normative standards in society. They may thus look for the ultimately most intelligent catalyzer that exists to see all the options reality offers and, as a result, might want to engage in a consciousness-opening voyage.

Another important observation regards mental health. It has often been wrongly stated that indigenous shamanic populations were psychotic or at least pre-psychotic and it comes to mind that usually pedophiles, in our society, once convicted and subjected to *psychiatric expertise*, are labeled in exactly the same way.

When we remember the times of communism in Russia or read books by *Aleksandr Solzhenitsyn* and others, we learn that under that totalitarian regime the same murderous psychiatry with exactly that same vocabulary had been used to eliminate intellectuals who were treated as system enemies because they defended human rights and democratic values. The effective mechanisms to defend a given *standard behavior* paradigm are all founded not upon natural pleasure-seeking behavior but upon adaptive perversity.

Hence, the necessity to look beyond the fence of behavior patterns and inquire into realms that seem apart from it but aren't.

The human soul expresses its originality always in paradoxes and sometimes in extreme behavior and the very attempt to 'classify' human behavior into rigid *standards for all* is in itself an ideology – idiotology. The more a given society puts up general standards, the more it is alienated from life and its creative roots and the more it is subject to decay and perversion. Shamanism is an effective guidepost for reentering the realm of nature's wisdom and true connectedness to *all-that-is*.

As far as I can see, people caught in minority groupings and the social fight involved with minority lobbying hardly ever come up with beyond-the-fence solutions such as studying shamanism and experiencing entheogens, which makes for the extreme poverty of many of those movements, not to say their ultimate system-obedient stupidity. And this unconscious system-obedience can be seen in many a limitation that social activists impose upon themselves and that are, ultimately, still system-prone. As Krishnamurti said, repeating an old wisdom: the social revolt or bloody revolution is still within the same frame of mind as the society it revolts against.

Published by Sirius-C Media Galaxy LLC, 2010

The entheogenic quest is therefore an inner quest, not necessarily a defeatist approach on a social level, but certainly an important add-on to any social activism for any possible social or humanitarian cause. For it opens the mind for the difference, for what Apple, in their publicity, call *Think different* and which is a very handy formula for what I try to convey here. The shaman is not a theorist, not a scientist and not a theologian. He is practical, a pragmatist, a solution-finder and his first rule is effectiveness. He is something like a highly effective Zen manager in his universe of natural laws, and he is a communicator. He communicates with the spirit world, the world of the ancestors and the world of the animal and plant spirits.

A shaman receives his basic education from the entheogenic plant teachers, and only at a minor degree from another, elder, shaman tutor. Shamans around the world, asked why they knew this and that secret about healing, about certain hidden connections, as Fritjof Capra calls *holistic knowledge* or about specific illnesses, answer they knew it directly from the plant spirits. They tend to affirm that they themselves know very little and that they just humbly ask plant spirits every time they can't solve a problem or not find a remedy for a certain illness. And the effectiveness of a shaman, then, is exactly to maximize the response-ability he has for all possible problems he is asked to solve. This involves curing sickness, doing counter-magic, finding the right timing for harvest and even the task to solve political questions regarding tribe relations. The shaman does that by maximizing his unique communication with the invisible world. By the same token, shamans around the world, when asked about *reality* tend to affirm that our visible reality, the one most Westerners think was the only one, is a very minor and rather insignificant part of reality and that the *real* reality is the hidden one, the one that is unveiled during the entheo-genic visionary experience.

If we refuse this bias of a *more or less* in terms of reality assessment, we can still enrich our mindset with the option that there might be *parallel realities* and that all realities, visible or invisible, or visible only through facilitating drugs or other consciousness-altering devices, are equally valid and equally important. An opening of science toward parallel universes, and an acknowledgment of a multitude of possible realities that are intersecting would be a great advance and evolution of Western science.

I shall now first report my Ayahuasca journey, and then evaluate it in a hopefully objective manner, using scientific methodology to evaluate it.

Published by Sirius-C Media Galaxy LLC, 2010

An Ayahuasca Experience

I had two options to experience Ayahuasca on my trip to South America in June 2004. It was either the *Santo Daime* ritual in Brazil, which is a mixup of psychedelic experience and a modified Christian liturgy, or encounter the wistful plant as part of a shamanic ritual in Ecuador. I chose the second, not only because I got the contact with this shaman already before, but also because I felt that none of these churches in Brazil were really welcoming me as a participating visitor. In Ecuador, then, Esteban, the shaman, told me that the original ritual was a pure and solitary experience of encountering the plant teachers, and not a group event. I was thus confirmed in my intuition and was glad to have chosen the more original initiation instead of indulging in a vulgarized and sectarian mix of perhaps incongruent elements.

Some months before my starting for the travel, and when I first exchanged with Jimela, a friend of the shaman who, like him, belongs to the *Shuar Nation* nation of tribal peoples, I got several dreams that I will just paste here from my daily dream journal:

> – I was talking on the phone with some people from Ecuador; they gave me a Spanish lesson and then one of them said I was already speaking Spanish at eighty-four percent. Then I was outside of the house and there was somebody who said I needed to be careful with Columbians. One moment later, all were getting down on

186 | Chapter Four

the earth for protection because they expected an explosion to occur.

– In the next dream I was in a room with several people, one of them a nude woman. All was a bit dark in the room and now I saw that this room was actually a library. For one moment I thought I wanted to ask the woman if she did not feel cold, but immediately changed my intention and went to the other side of the room. One moment later I was outside and held in my hand a book that I wanted to restitute to a library as the book cover was damaged already when I bought the book. I said to somebody: 'This book is only one of many that I want to restitute to the library.' At the end of the dream I hear this sentence in Spanish, and the sentence reminds me of García-Marquez book *Cientos años de soledad*, 'Cientos años de liberdad.'

In the following dream I realized that somebody had stolen me money from my bag.

In the next dream I saw a small boy and the idea was that I was going to drink Ayahuasca together with this boy. A guru or shaman asked me if I eventually had made the decision to drink Ayahuasca? An astrological divination had told him that it would be good for me to drink Ayahuasca. In fact, shortly before I had seen that young boy drawing an astrological chart for me. And this divination seemed to result in a magical outcome: there were suddenly two big water containers and on top of these containers were various things. It seemed to me that all these things were serving the *purga* (inner cleansing). I replied to the shaman that the astrological divination was valid only for the boy, but not for me. Then the shaman put objects close to my body to see how my vital energy would react to them or interact with them. I don't remember what the first object was. The second object was a large oven that had the *form of a square*. The shaman turned the oven in a way that one angle was pointing to my body. Then he said that during six hours I would feel very cold and that the oven could not help me because it was standing in a wrong angle to my body and thus was creating bad Feng Shui. I silently agreed with this reasoning of the shaman because I remembered in the dream that one principle of good Feng Shui was that objects should not pointing sharp corners to a human body.

The morning after my arrival in Misahualli, I met Esteban, the shaman, who was a young man with little shamanic experience. I actually wished to discuss the dreams with him, but he did not seem to be a listener and expert of the soul. I intuitively felt that he had not the maturity of a real shaman. He was a short, strong and healthy native in his thirties who had a rather *pragmatic* approach to life; he and Jimela were working together to promote their native culture to tourists.

Esteban did not hide that he was not a grown shaman, but just a beginner in shamanism and I got an intuition that he might not be able to help me with my fear problem. In addition, the two seemed to be caught in a negative pattern regarding Jimela's husband, a Polish businessman, the person who had invited me. From what they said, he seemed to be a strongly authoritarian character who had not understood their culture and had reacted several times in a rather displeasing way during the Ayahuasca experience.

After breakfast, I went out with Esteban to collect the plants. Close to Esteban's own wooden house, we found a very nice and about 15-year old Ayahuasca liana that Esteban called *la madre*. It was a strong liana of about fifteen centimeters diameter at the bottom and becoming finer and finer higher up. It was spiraling around its host tree and at its top end firmly rooted in the branches of the tree. Esteban first decided not to take this plant but a *daughter* of the mother, but looking around we found only very small and thin ones that Esteban, after careful examination, rejected as not being fit for the brew. So we went back to the mother liana and Esteban climbed up the tree, asked me to hand

Published by Sirius-C Media Galaxy LLC, 2010

him the knife, climbed higher, put the knife in a fork of two branches, and balanced it exactly so that it didn't fall down, climbed still higher and cut a medium-sized liana of about two meters that fell down next to my feet.

I was praying to the plant to allow us taking it, as this is a holy custom of the natives. Not far from the place we found the liana, we also found the second plant for the brew, which is a Mono Amine Oxidase inhibitor, that is called *Chacruna* or *Psychotria viridis*. Esteban used for it another name in Spanish that I did not retain. He put the liana and the leaves carefully in a little bag that he carried on our way back to the hotel.

When we were arriving back, the sun was getting out a bit and immediately the air became hot and sticky and I had a heavy headache that lasted until the afternoon when I could eventually get some sleep. As I am not used to this climate, the two hours walk had exhausted me.

However, after sleeping I felt alert and fine when, at about eight in the evening they called me for the ceremony. Esteban, his wife, and Jimela were sitting around a fireplace, in a the little hut of the hotel that looks over the river and where the floor was covered with a thick layer of beach sand. It was strange that, following my intuition, I went to this place while I did not know that they were there.

Jimela knocked at my room door and left and I did not know where she had gone. Some intuition told me they might be in a hut behind the restaurant that previously I had seen during lunch and that I thought was the private apartment of the hotel owner. Still more alien was the fact that when I got there, I saw three people, two women and one man, sitting in arm chairs, and in the dawn I did *not* recognize them. This never happened to me in my life before, that I saw people I know and did not recognize them. I was looking at them a moment and they did not say anything. Finally I said:

– Can you tell me where I can find Jimela and Esteban?

And Jimela got up and greeted me, and I went to the chair she advised me, and was very confused. The hotel owner, this slim aloof French, was stitching in the fireplace with a fork, trying to light the fire, in vain. He had no regard for me. At the end, he was not even able to light the fire. Only on my angered remark about the misplaced amateurism of that guy, Jimela went there to light the fire again, and that time successfully.

I wonder why all this happened as it happened?

It wasn't dark yet, all was well visible, but in that moment upon entering the spacious hut that is open all around and well lit, I had not recognized them. Had I been in a state of hypnosis already?

By the way, I experienced hypnosis already when I encountered a magic healer from the Philippines in the cabinet of a natural healer in Frankfurt, Germany. He put the audience in a state of hypnosis, suggesting that he was opening the belly of the patient, taking out the intestines, cleaning them and putting them back. I had seen all this with the other participants, the open belly, the blood, the strange sound like a wobbling noise when he put his hands deep into the intestines of the person in order to find strange objects

Published by Sirius-C Media Galaxy LLC, 2010

that he threw out on the floor and that looked like black stones.

The woman who had been treated before me, and who was sitting next to me, complained about strong pain in her belly after the treatment, while I went through the experience unhurt. Whatever the rationale is of magic healing, one thing is sure: this kind of group hypnosis works perfectly well and suggests a scenario to our minds that is more real than reality. You can't but affirm that you have *seen* it while you know of course that no belly was open and no blood was flowing and no black stones were to be found on the floor. And indeed, the first spot I was gazing at when we went back to normal perception was where one of those black stones were formerly to be seen, and even heard, on the floor. Needless to say that there was nothing.

I complimented them about the choice of the place that I found ideally suited. They said it was the first time they did it here, actually for my convenience, and that formerly they had done it in a little hut outside of the hotel. The advantage here was that I could any time go to my room, relax, or go to the toilet if I needed to. And that had been a wise precaution indeed. Esteban was looking very handsome in his traditional dress and I found set and setting really appropriate. I was in a very positive mood when drinking the cup and was surprised about the good natural taste of it. In all the books was written that the taste was horrible and that some people even vomited only because of the *appalling taste*. Nothing of that. It was a plant taste, bitter, but agreeable. I must add that I am quite different from most Westerners as I eat as good as nothing sweet, drink every day bitter herbal tea without sugar and also take now tree times a day a plant medicine that is bitter. And even with wine and cigars, I like the dry and slightly bitter ones.

During this half hour of waiting for the DMT to be absorbed by the blood so that the trance could start, I made up in my mind a hypothesis that I have not found in any book, and it seems that no researcher has hitherto come up with it. My strange experience when entering the hut was giving me the idea: *And what if nothing in that plant really produced the effect, but if all was a hypnotic trance induced by the shaman himself?*

While I knew that it was a daring hypothesis because we cannot deny Ayahuasca containing a considerable amount of DMT, and knowing from much laboratory research about how DMT is affecting human consciousness, it is, I think, something that needs to be addressed.

When we are scientific, we actually *must* address this possibility because it is scientific method to scrutinize *all the causative factors* of a result produced in an experiment. And when we do that, we cannot per se exclude from the start the hypnosis explanation I am forwarding here. While I find it little probable myself in view of the fact that Esteban was a rather young and hardly experienced shaman, I think we would have to disprove the hypnosis scenario in order to more firmly establish the plant teacher theory. And this, nobody has done it yet.

After my experience with the Filipino healer, I know how powerfully suggestive hypnosis can be, and it was exactly for this reason that I drove the two hundred miles to Frankfurt to see for myself what I had found in several German magazines about the spectacular stories about Filipino healers.

The experience started about half an hour later. I was feeling the trance actually from the fingers up, in the lower arms, then climbing into the whole body. I felt very clearly how my arm muscles gradually relaxed and this experience reminded me of the medical hypnosis I experienced as part of the hypnotherapy I did about ten years ago. Only that, with Ayahuasca, the trance was getting much stronger. It was a unique experience. I immediately was aware that the whole thing was not an automatism, but a *directed* voyage. It was directed by an intelligence. This intelligence communicated with me in a subtle way. It was very gentle, loving and caring, and the trance was very gradual, actually as if it had been especially adapted for me and my needs. With my eyes closed, I saw some faint colored forms, like arrangements of flowers or ornamental drawings and thought they were not as colored as in the books, but rather a physical phenomenon of the retina of the eye that was stimulated by the drug. My logical intelligence – that was at no moment veiled or dysfunctional during the whole experience – suggested me that these visions had no meaning in itself and that the experience was something much beyond the mere visual effects.

It then came to my mind that most reports about the Ayahuasca experience only focused upon these visions, and in my view people have taken the symbol for the truth or the finger for the moon. The young French couple who manages the hotel had told me at lunch about their own Ayahuasca experience: *Just visions …, you know, very nice!*

There was a noise developing that I found very interesting. I was clear enough in my mind to detect within this noise the river's sound, the cricket's sounds, the frog's sounds, and yet I found that noise being of another quality than just a mix of all those natural sounds. And it was gradually raising, and after some time, I clearly understood that this sound was something like the *pulsation* of the universe, of life, and that creation was a result of *sound*, that life intrinsically was vibration, was sound.

Then I got clear telepathic messages that I should go beyond that first phase and intuitively knew that there was three or five depth levels of the trance and that I was just in the first. Then the intelligence gave me signals. It was like a gentle knocking at my doors of perception. It was a flickering of about five red squares at the left upper corner of my vision, when my eyes were closed. I intuitively knew that this was the signal to go deeper, and yet I could not. Every time it happened, the fear came up and blocked me to go beyond. I then remembered that it was this fear, from childhood, that had blocked me in realizing myself, that it was this fear that was blocking me to experience hypnosis really deeply and not only slightly, that it was this fear that blocked me to really give my best when I played piano in front of an audience or did anything I really love to do.

The rest of the experience is without importance. It was a terrible struggle between the *call* that I received to go beyond the fear and the fear block itself. I began to vomit and had to call Esteban two times for help. To make it worse, I could not remember Esteban's name, and thus formulated in

Published by Sirius-C Media Galaxy LLC, 2010

my mind a cry for help, and that, to put this cry in language, needed some really hard effort:

– *Chamano, ayudame por favor!*

Esteban then got up from his chair, put his hand on my head, blew cigarette smoke in my face and treated my aura with a special device made from dried leaves, a kind of rattle that is used like a feather to give passes from top to feet, similar to what is done in hypnosis. This treatment helped me wonderfully, but only as long as it lasted. Actually, I opened my eyes several times just to see if the others were still there.

Interestingly, the communication with the intelligence stopped *at once* when I opened my eyes, and in that condition, I felt swindle. Only when closing my eyes again, the communication would gently be taken up by the plant intelligence. Sadly, I felt isolated from the other humans or any other humans, as if I was totally alone on earth, as if there was no soul connection between them and me, while intellectually I since long affirm this connection to be true and part of all life. But emotionally I could not feel it, I could not understand it. *I felt I was locked in my body, cut off from all other life, and cold.*

I saw the two women sleeping in their chairs and Esteban, on my right, was looking strange. The trance view of him was not the normal view of him. When he was talking to me, his face appeared diabolic and grinning and it made me still more afraid. His mouth was not in the middle of his face, but shifted to the right side, and very large, as if it was open, and white inside.

Now I must recall a similar experience. It was when I got a Bach plants treatment from a very experienced natural healer in Germany. This woman had studied with Filipino healers and was using strong hypnotic trance for healing. Holding one flacon after the other of the Bach essences in her left hand, she let me put my left hand into her right hand in order to establish the *flow of the current*. In that moment, I had experienced the same what happened with the Filipino healer and during the Ayahuasca experience.

I suddenly relaxed deeply and my consciousness shifted. When I looked at her face, I saw that she had *three eyes*, one very large third eye on her front, at exactly the location where there is our *third eye* chakra, only that in this state of trance, I saw a real eye there, a large one, much larger than a human eye. I was not afraid to see her and telepathically knew that she was spiritually a very developed soul with true healing capabilities. The strange difference between her and her aura and Esteban and his aura is that this lady's aura had a calming, luminous and beneficial effect on my soul while the aura of Esteban had a frightening and confusing effect.

After each treatment Esteban gave me and despite the fact I vomited out all the contents of my stomach, there was obviously still enough DMT in my blood to continue the experience. Well, in hindsight, this would actually favor the hypnosis theory again because from what I read in books, with most other people, once they had vomited out the contents of their stomach, the trance was broken, while in my case it was not. The intelligence called me still, so gently, so lovingly, so calmly, and I could not get over the block. But it

Published by Sirius-C Media Galaxy LLC, 2010

might also be that the DMT was still in the blood while the actual substance, the Ayahuasca brew, was already evacuated from the stomach.

The most important experience was that this intelligence questions everything. Every thought I had was immediately returned to its contrary and every phrase that I pronounced in my mind was immediately reduced to some more general insight. For example I heard myself thinking:

– I love life …

And it was immediately reduced to:

– I *love* …

Or I heard myself thinking:

– I feel to be alive …

And it was reduced to:

– I *feel* …

And the intelligence seemed to tell me telepathically that I was *locked* in language, that all my experiencing of life was conditioned upon language, and that I hardly ever perceived life *directly*, spontaneously, as an immediate connection.

I knew that this intelligence was connected to all, was connecting all and was all. It simply *was*, and it gently invited me to enter this connection, this wonderful all-encompassing love that it irradiated. Several times, when I wanted to communicate some of the experience to Esteban, the intelligence seemed to hold me back. Once I wanted to say something regarding the intelligence, and I was stopped inmidst of the sentence:

– *Este inteligencia* …, was all that I could say.

The intelligence seemed to wanting to free me from the conditioning I had received through language, and through using language for describing reality. It might be that my fear is in some way connected to language and my fear block a hypertrophy of language or of my left brain. This is how I can try to explain it while all what I write here and can express in words never can come close to the actual experience.

I was constantly communicating to the intelligence as well. I had prayed already before the experience to reveal my life's mission to me and to free me from fear, and I kept addressing Ayahuasca in respectful terms during the experience to assist me in my quest for truth. And as if it was a final test that this intelligence really understood me, I said this, when I felt I wanted to go back to my room and sleep:

> – Dear Ayahuasca, thank you so much for all the insights, please stop calling me now. I know that I was not able to get over my fear block now, and I hope I can find the courage to continue because I know you will reveal me so much more when I get deeper in the trance. But for now I have decided to stop. It is enough.

And immediately the *calling* stopped, the flickering lights did not appear any more and the *alien* noise was calming down and stopped as well. This alien noise, I understood then, was the *language* of that intelligence. It was something like many people, in some distance, speaking all at the same time, and as if many conversations converged to a chaotic no-sense that was but the secret code or pattern of a deeper,

Published by Sirius-C Media Galaxy LLC, 2010

much more unified language of the universe, and of which our human language is only a tiny and almost insignificant part.

Esteban suggested to guide me back to my room and I stumbled back, with his support, because I was hardly able to walk and felt swindle. In my room I spent at least two hours of dreadful vomiting and diarrhea. It was *la purga* as it is described in many personal reports about the Ayahuasca experience.

Eventually I became very quiet and full of gratitude and I said this in my thought, in German:

– *Ich möchte Liebe!* (I want love)

And immediately the response was:

– *Ich möchte lieben …* (I want to love)

And with all my heart and my soul I understood this subtle difference and it was clear to me that in this difference of language was *all the difference*. It was the secret to happiness. I had taken love as a commodity that can be received, instead of understanding that love was not something to receive, but something to build, something like an attitude, and this attitude, it seemed to me then, simply was total openness. And I think this insight transformed something in me.

Shortly thereafter I fell asleep and slept wonderfully, without dreams, until around eight in the morning, and woke up like newborn, feeling light and alert. While usually, after the many strange dreams I normally have, I wake up fearful, anxious and powerless and have to do a whole program of prayer, meditation and exercises every morning, I now felt completely free of any fear and of any past. It was the most exhilarating morning in my life, and perhaps, in some way, the start of a new life.

Published by Sirius-C Media Galaxy LLC, 2010

Hypothesis

My hypothesis is that the consciousness-transforming cognitive experience subsequent to ingesting the ritual brew *Ayahuasca* is not, as it is often suggested, the direct result of plant chemistry, but of the shaman's consciousness reaching the experiencer's consciousness through the medium of plant chemistry as a thought and energy transmitter. This view is not to be understood in a reductionist way. I do not say that all is to be *reduced* to one single root cause, but propose to consider *one more option* in our scientific investigation of para-normal phenomena. To corroborate my hypothesis I shall explain more in detail my mind-opening experience with *Ayahuasca* during a visit to a Shuar native shaman in Ecuador, back in 2004.

There are several facts and events around my Ayahuasca experience that are explainable more soundly when applying my own hypothesis instead of trying to match it with the of-ten voiced theory that psychedelic experiences are caused, as a *linear effect*, by a plant-contained chloride named DMT.

To anticipate a little on my conclusions, my hypothesis provides a *non-linear* explanation of the psychedelic experi-ence. My hypothesis does not deny the existence of the chlo-ride and its possible effects on the human psyche. But I con-tent that the mind-opening effects noticed by the novice after ingesting the brew are a result of the shaman's consciousness impacting upon a *passive perception matrix* that is part of the plant realm and that the shaman uses as a transmitter platform.

I repeat that I do not discuss away any possible other explanation, so much the more as I myself, after the trance began to grip on me, had the impression there was an intelligence in touch with me, an intelligence that was *alien* in a way, and that I attributed to the plant realm. My point is not to invalidate any of the current hypotheses about psychedelic plant substances, but to help finding a valid theory that shows what it is that effectively opens, modifies or expands human consciousness during the Ayahuasca experience.[76]

Before I am going to give some flesh and bones to my assumption, let me report that I found till now at least a couple of references that seem to confirm my point. As the result of a general research on shamanism and entheogens, and particularly Ayahuasca, that I undertook over several years, I must conclude that most of the researchers seem to defend a rather *mechanistic causation theory* that sees the source of all paranormal phenomena in the chemical plant substances. For example Terence McKenna, under the spell of his large knowledge on ethnopharmacology, and his brother Dennis McKenna, an enthnobotanist, never left a doubt in all their writings on the subject of psychedelics that causation of altered states of consciousness is due to psychoactive compounds in plants called *entheogens*.

The question what exactly the role is that the shaman plays in opening greater pathways of consciousness is left open or subject to speculation. Dr. Rick Strassman, a researcher on DMT for many years, has quite the same linear

Published by Sirius-C Media Galaxy LLC, 2010

idea about causation and sees plant chemistry as the activating force, and the title of the book is a typical American euphemism. The book's approach is all but revolutionary, it is hardcore mechanistic and linear, with very little holistic insight.[77] I have found so far only two researchers who express a view that really makes sense. Their research seems to corroborate my own findings. Jeremy Narby, in his book *The Cosmic Serpent (2003)*, puts up the daring hypothesis that causation is due to biophoton emission, not plant chemistry.[78]

He observes that initially his approach to psychedelics research a valid branch of research that however from the middle of the 1970s onward disappeared from the scientific literature. What is seen in psychedelic visions, according to Narby's research, are photons emitted by the DNA. Narby writes in *The Cosmic Serpent (2003)*:

Jeremy Narby

Researchers working in this new field mainly consider biophoton emission as a cellular language or a form of nonsubstantial biocommunication between cells and organisms. Over the last fifteen years, they have conducted enough reproducible experiments to believe that cells use these waves to direct their own internal reactions as well as to communicate among themselves and even between organisms. For instance, photon emission provides a communication mechanism that could explain how billions of individual plankton organisms cooperate in swarms, behaving like super organisms.[79]

Now, succinctly speaking, what Narby wants to show is that what the shamans perceive as *spirits* are in reality biophotons emitted by the cells of the human body:

Jeremy Narby

What if these spirits were none other than the biophotons emitted by all the cells of the world and were
picked up, amplified, and transmitted by shamans'
quartz crystals, Gurvich's quartz screens, and the
quartz containers of biophoton researchers? This
would mean that spirits are beings of pure light – as
has always been claimed.[80]

In fact, Narby's theory does not exclude that causation
might also be due to plant chemistry, but he surely concludes
that what is seen, what is perceived, is not parallel reality but
the reality contained in our own DNA and the superconscious memory surface that is connected to it. A perhaps
more convincing evidence of causation being an effect of the
shaman's own superconscious powers, and not of plant
chemistry, is brought forward, or at least hypothesized by the
American medical anthropologist, shaman and psychologist
Alberto Villoldo. In his book *Shaman, Healer, Sage (2000)*, Dr.
Villoldo introduces the third chapter entitled *The Luminous
Energy Field* with an entry from his journals. Don Eduardo
was one of the powerful Inca shamans Villoldo studied with
for many years:

Antonio Villoldo

I've found that the San Pedro potion does nothing
other than make me sick. (...) I'm convinced that the
altered state I'm in is created by Don Eduardo's singing. And then there is the energy that he claims enters
the ceremonial space when he summons the spirits of
serpent, jaguar, hummingbird, and condor. (...) What I
can't explain is the fact that I'm seeing energy. It only
happens when I sit next to Don Eduardo. When I go
more than a few feet away from him I sense nothing.
It's like he is surrounded by an electric space, where

Published by Sirius-C Media Galaxy LLC, 2010

> the air actually tingles. When I'm inside his space I see everything he sees.[81]

The perhaps most convincing corroboration of my research comes from theosophy and the pulpit of Charles W. Leadbeater. If clairvoyant research is or is not considered as valid scientific research under the present reductionist science paradigm is not *my* problem. It may be a problem for those who, for reasons of psychological defense or emotional distortion, or else for reasons of professional reputation, adhere to a reductionist, limiting, exclusive or sectarian science paradigm as the ultimate hanger for their lacking emotional stability. From a point of view of *real* science as methodically sound, holistic, mentally sane and intelligently communicated observation of nature, clairvoyance *is* science.

As a little excursion, let me quote what a clairvoyant herself has to say about this extraordinary faculty of perception that is clairvoyance. Dora van Gelder writes in her book *The Real World of Fairies (1999)*:

Dora van Gelder

The fact is that there is a real physical basis for clairvoyance, and the faculty is not especially mysterious. The power centers in that tiny organ in the brain called the *pituitary gland*. The kind of vibrations involved are so subtle that no physical opening in the skin is needed to convey them to the pituitary body, but there is a special spot of sensitiveness just between the eyes above the root of the nose which acts as the external opening for the gland within.[82]

In his book, *The Astral Plane (1894)*, Leadbeater very much stresses the fact that we can hardly judge a human being by their acts only; in fact, as thoughts are much more important as an influence upon the world than most of us in the West know, when we go to praise somebody for his achievements and judge him or her 'a good person', we may be completely wrong, because that person may have exerted a ravaging influence on others and the world by their self-talk, by their way of thinking about others, and by their way of judging others harshly over years and years, in their mind. What self-talk namely creates are *elementals* or thought-forms and these thought forms are more or less permanent, and gain permanence over time and depending on the emotional energy we invest in these thoughts.

I think it's good that Leadbeater addresses this point so clearly because most people in our culture are completely ignorant about the impact of thought on the world, on others and on their own karma. Leadbeater writes:

Charles W. Leadbeater

The fact that we are so readily able to influence the elemental kingdoms at once shows us that we have a responsibility towards them for the manner in which we use that influence; indeed, when we consider the conditions under which they exist, it is obvious that the effect produced upon them by the thoughts and desires of all intelligent creatures inhabiting the same world with them must have been calculated upon in the scheme of our system as a factor in their evolution. In spite of the consistent teaching of all the great religions, the mass of mankind is still utterly regardless of its responsibility on the thought-plane; if a man can flatter himself that his words and deeds have been harmless to others, he believes that he has done all that can be required of him, quite oblivious of the fact that he may / for years have been exercising a narrowing and debasing influence on the minds of those about him, and filling surrounding space with the unlovely creations of a sordid mind.[83]

Now, regarding the elementals that are created through thought and intent, and the gestation that is brought about by the repeated fostering of a well-defined thought pattern, Leadbeater explains that these elementals are not autonomous in the sense that they can begin to act on their own and trigger changes; they must be pushed to do so:

Charles W. Leadbeater

But the 'elemental' must never be thought of as itself a prime mover; it is simply a latent force, which needs an external power to set it in motion. It may be noted that although all classes of the essence have the power of reflecting images from the astral light as described above, there are varieties which receive certain impressions much more readily than others - which have, as it were, favourite forms of their own into which upon disturbance they would naturally flow unless absolutely forced into / some other, and such shapes tend to be a trifle less evanescent than usual.[84]

The spirits of nature, shunned by Christian fundamentalism and reborn now in the course of the New Age movement, and the revival of the folk lore of fairies, as it was, for example, rediscovered by Dr. Evans-Wentz in his remarkable study *The Fairy Faith in Celtic Countries (1911)*, and observed by clairvoyant Dora van Gelder in her book *The Real World of Fairies (1999)* have certain well-defined characteristics and they are quite distinct of human beings. Leadbeater explains:

Charles W. Leadbeater

We might almost look upon the nature-spirits as a kind of astral humanity, but for the fact that none of them - not even the highest - possess a permanent reincarnating individuality. Apparently therefore, one point in which their line of evolution differs from ours is that a much greater proportion of intelligence is developed before permanent individualization takes place; but of the stages through which they have passed, and those through which they have yet to pass, we can know little. The life-periods of the different subdivisions vary greatly, some being quite short, others much longer than our human lifetime. We stand so entirely outside such a life as theirs that it is impossible for us to understand much about its conditions; but it appears on / the whole to be a simple, joyous, irresponsible kind of existence, much such as a party of happy children might lead among exceptionally favourable physical surroundings. Though tricky and mischievous, they are rarely malicious unless provoked by some unwarrantable intrusion or annoyance; but as a body they also partake to some extent of the universal feeling of distrust for man, and they generally seem inclined to resent somewhat the first appearances of a neophyte on the astral plane, so that he usually makes their freaks, they soon accept him as a necessary evil and take no further notice of him, while some among them may even after a time become friendly and manifest pleasure on meeting him.[85]

What this means is that by *impacting upon reality* through thought and intent, and through emotional focus, we actually create *elementals*, which are thought-forms that are somehow embodied and individualized. Now, what I conclude from this insight, extrapolating the research of clairvoyant Charles W. Leadbeater to shamanism, is that the shaman, when concocting the traditional Ayahuasca brew, and when focusing on it, actually builds *elementals* by his strong intent and the thought-forms resulting from this focus. These elementals

Published by Sirius-C Media Galaxy LLC, 2010

then, are absorbed by the plant matrix, or the psychoactive compounds in entheogenic plants, and are transmitted to the adept who desires to be initiated by the shaman, through ingesting the traditional Ayahuasca brew.

There is another very recent research that also seems to corroborate in some way my hypothesis, and also Leadbeater's clairvoyant observations. It has this time nothing to do with shamanism but comes from a core physics research.

It is William A. Tiller's highly innovating research on the power of intent involved in the transformation of matter. In his book *Conscious Acts of Creation*, and the DVD with the same title, Dr. Tiller, Stanford University Professor Emeritus, claims that there is ample evidence for the fact that conscious and condensed thought, and intent impact upon matter, and actually change matter.

DVD Back Cover Text

Based upon years of detailed research, Dr. Tiller has amassed convincing experimental data showing that in seemingly the same cognitive space, basic chemical reactions and basic material properties can be strongly altered by human intentions. Essentially, he says, we are all capable of performing what we typically think of as miracles.[86]

If our intent projected upon time and space creates what Leadbeater and others call *elementals* or if it creates *thought-forms* or if it creates a collapsing of the wave function, to use an expression of quantum physics, really does not matter. We might as well call it magic thought power or telepathy. What imports is that we see that there is no magic other than the impact of spirit upon visible and tangible reality.

When I extrapolate this research that is amply documented by additional research in the meantime, then I must conclude that shamanic power is more than the mechanistic ingestion of plant chemistry 'to make things happen'. Then I will see that it's the preparation of the concoction much more than its chemical ingredients that make for the outcome of the experience, and that's ultimately *intent projected into the subtle matrix of plant consciousness* that is the trigger here.

Published by Sirius-C Media Galaxy LLC, 2010

The Consciousness Theory

I shall discuss the theory that I bring forward in this paper using my own experience with Ayahuasca as a point of reference. There are eight arguments for supporting my hypothesis that I will bring forward and illustrate with examples. These arguments are:

▶ 1) Preparation of the brew;

▶ 2) Trance lasting hours after extensive vomiting/diarrhea;

▶ 3) Effect of the shaman's use of cigarette smoke;

▶ 4) The shaman's *focusing his thoughts* on the client;

▶ 5) The *strange* reception;

▶ 6) The *hypnotic view* of the shaman's face;

▶ 7) Similarity with hypnosis used in natural healing;

▶ 8) Similarity with hypnosis used in psychiatry.

Let me now discuss each of these points.

1) The Ayahuasca Preparation

Upon my question about the details of the preparation of the brew, it is highly interesting what the shaman conveyed to me the morning after my arrival in Misahualli, Ecuador. We had been discussing the religious nature of the Ayahuasca experience and Jimela, his assistant, was talking about her husband *Rafal*, a Polish businessman who, from what they said, appeared to be a prototype of a skeptic. For him, the plant simply contained a chemical that brings about certain psychedelic effects in the human brain, and it was all a matter of ingesting that chloride so as to experience the consciousness-altering effects.

Interestingly, from what I was told, things worked out in a way to firmly contradict his positivistic worldview. After several attempts to convince Rafal of proper *set and setting*, and the necessity of careful preparation of the religious experience, Esteban and Jimela said they had given up on him and let him prepare the brew by himself.

The occasion soon was given through the reception of several business friends from Poland that Rafal wished to initiate in the Ayahuasca experience. To everybody's surprise the ceremony ended with a total failure. Rafal complained later to Esteban that his friends *did not feel anything*. And Rafal himself, while having gone through the experience several times successfully when the brew was prepared, with all due care, by Esteban's mother, could not understand that the effect was practically zero this time, plus headache for several hours.

I was intrigued.

– How can that be? I asked Esteban.

– Well, he replied, you see it's not just a matter of cooking that stuff and reducing a large pot of the brew to a tiny cup that you later drink. It's not just concentrating that substance. It's much more.

– What more? I insisted.

– It's *respect*, basically. This respect must be shown in many ways. The tradition for example says that a woman who has her days must not be in the house, and the couple must not engage in intercourse during the time the brew is on the stove; the children must not make noise and preferably remain playing outside. And then, there is one more element. During the whole three to four hours the Ayahuasca is on the stove, the shaman must focus on the pot. His entire psychic energy must be *focused* upon the brew all along it's on the stove. He should not engage in any other thought, he should not leave the house, he should not have any leisure time and he should by all means not touch a woman. He must give his entire respect to the Ayahuasca until the moment it is ready.

– What does this imply practically? I asked.

– It implies that he concentrates his thought energy on the event. He is mentally preparing the event in the sense that he prays to the serpent, the *Boa Constrictor* which is the spirit of the Ayahuasca, to please provide a meaningful experience to the ones drinking the brew so that they may be guided on the right way. That means he will not quit the room where the stove is burning and he will kind of *court the brew* with his whole attention, caring for the fire, the right level of heat, and so on. He may smoke, but his mind must remain focused upon a positive outcome of the later experience. He also must feel a sort of *reverence* toward the Ayahuasca spirit. It means that he really is pure in his intentions, and not doing it for money, for example, but with the intention to help people, either for their spiritual advancement, or for healing, depending which branch of shamanism he is personally adhering to.

While I found that some of the precepts Esteban mentioned were simply destined to assure that the tradition not be spoilt by the ignoramus using it as a showcase kind of thing, I listened very carefully when he spoke about the *focus* the shaman had to give to the brew, about the mantras he had to recite over the brew, about the prayers he had to say and over the energetic input he had to give to the brew in form of *total attention* and a stringent concentration of his thought energy upon the outcome of the experience.

2) The Lasting Trance

The second interesting point in the experience is that after having vomited out completely the contents of my stomach, and very violently so, the effects of the trance lasted for several more hours.

I already felt sick after the first thirty minutes passed agreeably, went on my knees and experienced a rather painful purge because my chest and back muscles began to hurt intensely. It took at least half an hour until I was back in my chair. And even then, the vomiting reflex persisted while nothing was coming out any more, except small quantities of stomach saliva.

Published by Sirius-C Media Galaxy LLC, 2010

And every time I closed my eyes, all would restart as a meticulously working computer program. Closing my eyes was like a code contained in that software. When I closed my eyes, and not before, the intelligence namely asked me telepathically if I wished to restart; affirming silently, I was at once experiencing the *slight visual effects* again and the trance began to get hold of my body once more.

Unfortunately every time instead of being able to let myself get deeper and deeper into that agreeable state of numbness and comfort, something in me terribly resisted and built an annoying level of anxiety that I was absolutely unable to get rid of. The anxiety, then, in turn triggered the vomiting reflex. I was mentally very clear about it all and how it was setup altogether, and told Esteban several times that I understood my resistance to the brew as a psychological defense against self-abandonment; and yet I was unable to cope with that dreadful anxiety.

Esteban however had no advice for me except suggesting me to go back to my room and sleep, and I felt helpless, then, and kind of *guilty*. Why was I resisting?

This went on for several turns. Each turn I would call him for help and he would slowly get up from his chair, put one of his hands on my head, blow cigarette smoke in my face and give strokes to my luminous body with the feather, and this felt *great*. Interestingly, when he blew smoke in my face, the trance immediately vanished, even if I left my eyes closed. How could that be, as the DMT was all the time in my blood?

These few moments filled me with delight, with confidence, with positive feelings and a sense of well-being. And then I would thank him, he would go back to his chair and sink in his own trance, and all would get back to the former desperate condition, and the vomiting would restart.

And eventually I asked to be guided back to my room and even there, about one to two hours after I had vomited out the brew, the process would restart in just the same way as before, just upon closing my eyes as the signal to restart.

Sleeping was only possible in the early morning hours, after more than two hours of a virulent duplex reaction of my body: I got diarrhea and vomiting at the same time. I was once really laughing about it, while I felt so much pain and misery, because I was getting into a *rocking* movement between the toilet and the sink that were luckily quite close to each other. That was my Ayahuasca rock!

Only when I was *completely empty and pure*, I felt I was ready for sleep. And then it was really wonderful, like being in heaven, a wonderful dreamless sleep and a very happy *light* feeling when waking up in the early morning.

3) The Shamanic Treatments

How can it be that the treatments I received from the shaman could completely change my condition for the time they lasted? As I said, they consisted of several elements some of which I have not yet mentioned:

▸   Putting his hand on top of my head for a moment;

Published by Sirius-C Media Galaxy LLC, 2010

▸ Gently blowing cigarette smoke into my face;

▸ Giving me *magnetic strokes* with a feather-like device;

▸ Chanting an *ikaro*, a magic chant.

There was a moment when I became aware that my terrible anxiety had to do with my early childhood, that I had been abandoned as a small baby several times, a fact that my mother, when I was already adult, more or less hesitantly told me once when she was drunk.

I felt a great sense of comfort from the care I got from the shaman and it seemed to me that this care contained the same *love* that a mother bestows on her baby and I was just absorbing that love as if I was a small helpless baby.

But scientifically speaking, if the hypothesis is true that the trance is induced by DMT, how can the effects of the trance completely stop, while the DMT is still in the blood, under the influence of a kind of *psychosomatic energy treatment?* If the trance was *only* induced by the DMT, as a linear kind of causation, there is hardly any even remotely logical reply to find to this question. However, from the perspective of my hypothesis that the plant is only a *passive matrix* absorbing and remitting, and perhaps amplifying as well, the *thought energy* received from the shaman, then what I am saying begins to make sense.

Truly, *all* the effects of the DMT ceased, including the disagreeable swindle that set in every time when I opened my eyes, and I left them open every time I got the treatment as I found it pacifying and nice, and liked to see the shaman in his traditional dress and his serious and caring allure when gently blowing me cigarette smoke in my face.

There was a tenderness in his simple silent movements that greatly contributed to comforting me and alleviate my anxiety. But logically, if the shaman was the *agent* of the trance and not the DMT, or if it was both, if he was the *primary* agent of the trance using plant consciousness as a transmitter of his thought forms, it is clear that he could stop it at will and re-trigger it at any moment in time!

4) Focus and Intent

In our initial talk about the Ayahuasca experience at the morning upon my arrival, Esteban had addressed the issue of *focus and care* in dealing with his clients. I think he wanted to come over as a serious practitioner of the shamanic science as there are today many charlatans, especially in the tourism-plagued areas in Peru and Ecuador. And explaining me more about his work and attitude, he said he would very intensely focus upon the client when beginning the trance. I was intrigued and asked:

– Do you drink the Ayahuasca brew as well, or only the client?

He replied that it depended *on the wish of the client*, and if I wished him to drink the brew together with me, that was okay with him. I affirmed that I wished him to do that and asked him why he focused or concentrated upon the client after the latter had ingested the brew? He said he focused his mind on the client in order to help him better access the *boa*

Published by Sirius-C Media Galaxy LLC, 2010

spirit, the mother spirit of the Ayahuasca, as he himself had been in touch with this spirit from the start, and that this was after all the precondition to being a shaman.

Thus, consciously remembering this conversation with Esteban, I tightly observed him when we started the ceremony, and indeed, after we had drunk the brew from the beautifully carved traditional silver cups, he was sitting down not like me in the beach chair, stretched out and relaxed, but in a fetal forward position that suggested to me he was *concentrating on something*. And he remained in that position for a long time, actually as long as the preparation lasted, the time before the DMT begins to being absorbed by the body and deploy its effect – which is about thirty to forty minutes.

Then, when I felt the trance began, he was stretching out in the chair in a more relaxed position. And I think that even for a third observer this connection of events, this *Gestalt* as it were would suggest the hypothesis that it's the shaman who is the *primary trigger of the experience*, and not just some chlorides in a plant concoction.

5) The Strange Reception

Upon entering the hut, I saw three human figures in beach chairs and did not recognize them. I was *unable* to recognize who they were, as if my thoughts were seized by an unknown power, or as if there was an overlay over my thinking process. I felt I was floating. Please note that at that moment I had just come from my room, refreshed after some sleep, and yet suddenly, upon entering the place where the ritual was going to take place, dominated by an alien force.

I have never had memory lapses or anything even remotely close to a lack of memory for human faces. I have a very good memory for human faces, even after long years to have known a person, and meeting her again. But here, there is no valid reason to be found why I did not recognize the three persons sitting there.

Something like that never happened to me before. I felt as if *under the influence* of something that rendered me incapable of assessing the reality of the simple scene I was facing. It was just after sunset, and it was not yet dark, around eight in the evening. The distance between them and me was not more than five or six meters, not more and perhaps less. The strangeness or queerness of the moment was the fact that I did not *feel* their presence. It was as if there were humans present but that these humans were strangely *disconnected* from me. And they did not invite me to come closer and just remained silent, and I felt like an idiot, not knowing what to do, not knowing what to say. Eventually, feeling really *embarrassed*, I asked if they knew where *Jimela and Esteban* were? And only then the female got up from her chair and approached me and I saw it was Jimela.

When I apologized for not having recognized them upon entering the hut, they said it did not matter, *but to me it mattered a lot.* How could something like that happen? It remains a mystery to me until this day.

I namely evaluate this event as an overlay of consciousness. Already at that moment the shaman tried to get into my

Published by Sirius-C Media Galaxy LLC, 2010

consciousness interface, and he did this by what he said he always did when preparing for accompanying a client on an Ayahuasca trip: he *focused* on the client's consciousness. This *focusing on the client* simply was a native variation of what in Western medical science we call hypnosis. He had *hypnotized* me, even before I had taken in a drop of the Ayahuasca brew.

And this is essentially the core of my hypothesis. And as I am not a newcomer to hypnosis, because otherwise I would probably not have discovered this intriguing explanation of how shamanism works, I erect this now as a theory and ask scientists for evaluating it by either corroborating or falsifying it.

I will honestly relate in this paper my earlier experiences with hypnosis both in the setting of natural healing and in the medical hypnosis setting.

6) The Hypnotic View

Strangely, when I was in trance, I saw the shaman's face differently. I guess this experience would frighten someone with lesser paranormal knowledge and experience. His face seemed distorted, ugly and diabolic. His mouth was shifted to the right, open and white inside. And this *strange mouth* did not move while I heard Esteban talking, and even talking quite fast, so fast that I wanted to tell him to talk more slowly because my understanding of Spanish, while normally quite good, was reduced in trance, but finally I did not get a word out and just remained silent.

I remembered my experiences with hypnosis and especially the one I went through with the quite extraordinary
female German healer that I will report further down.

Needless to add that this *strangeness* about Esteban's face
resulted only in increasing my anxiety and discomfort about
the whole of the Ayahuasca experience.

7) Hypnosis and Natural Healing

It is important to retain here that hypnosis itself was not
at all something strange or alien to me because I have had
quite extensive experience with it in the past, and I think it is
very important that I relate this experience shortly.

To repeat it, without this experience, I do not think I
would have come up with the present hypothesis; I would
probably have accepted my Ayahuasca experience as a *queer*
one and forgotten about it.

Now, let me be very precise what I am talking about. I
am talking here about *medical hypnosis*, not about *stage hypnosis*.
There is almost a world of difference between both forms of
hypnosis. Medical hypnosis is a form of auto-hypnosis in the
sense that it builds upon the full consent of the patient and
his active participation in the progression of the trance, while
in stage hypnosis the willpower of the individual or the entire
audience that is hypnotized is as good as put to zero. Let me
illustrate this a little further. It is recognized in the meantime
that, for example, the acrobatic trick of *sawing the woman* can
be done in two different ways. The traditional way was to
hypnotize the entire audience and *suggest* to them what they

were going to see, as for example the woman cut in two, who yet afterwards leaves the box unhurt.

As stage hypnosis has been *discredited* as a form of abuse and also because the art of stage hypnosis was an orally transmitted knowledge and got more and more lost with the disappearance of the *circus* as an institution, today, in almost all cabaret's, when you see the *sawing the woman* trick, it's a simple trickster effect you are succumbing to. And yet still today a famous popular figure such as *David Copperfield* admits that part of his magic is hypnosis, and not only tricking the audience out with visual and sensory effects, immense speed of action, and uncanny ways to act.

Thus, the two experiences of hypnosis that I wish to relate here, and the further ones that I will report in the next paragraph exclusively deal with medical hypnosis, and not with stage hypnosis. And it goes without saying that I do not suspect Esteban to be a charlatan, but an integer shaman who applies, as exactly as possible, a tradition that he shares with several tribes of Shuar natives, and that he learnt from an old and experienced senior shaman.

The two experiences of hypnosis that I will relate are quite different while both can be qualified as *medical* hypnosis. Their difference is that in the first experience the hypnosis is a person-to-person one, while in the second experience the hypnosis is a person-to-group one and thus a form of collective hypnosis.

The first experience was taking place back in 1997, when I had just returned to Germany from an exhausting two-year business trip to Asia, with my intestinal flora completely down. I had gone through severe diarrhea and intense dehydration over months and was at that time still consulting Western doctors who gave me high amounts of antibiotics to fight the diarrhea.

The result was that the diarrhea continued despite all but my intestinal flora was completely destroyed by the long-term antibiotics treatment. I was suffering from chronic fatigue and felt very lethargic, lacking motivation and appetite. It was for this reason that I consulted a *homeopathic healer* for help. I got several effective treatments and one of them, the one I remember most vividly, was a treatment with the famous *Bach flower essences*.

The female practitioner who told me she studied hypnosis and Reiki with a powerful Filipino healer, first wanted to choose the right Bach flower essence for me and my problem. She explained me very patiently the various methods for finding the energy essence corresponding to my organism's *energy code* and asked me which one I preferred. I chose the most direct one, the one that is done through hypnosis, and the experience was going to be a particularly revealing one for me.

She was sitting at a forty-five degree angle at my right and asked me to put my left hand in her right hand. Then she told me to look in her eyes while she would take one flacon after the other in her left hand to *sense* the effect the vibration of the plant essence had on my organism. Never before was I hypnotized so easily, so effectively and so joyfully. It was a very agreeable condition and I felt very clearly how each of the essences impacted energetically upon me.

Published by Sirius-C Media Galaxy LLC, 2010

She said I was going to feel either joyful, peaceful, positive and happy, which indicated that the essence was right for me, or I was feeling queer, anxious and negative, which was indicating that the essence was *not compatible* with my aura's vibrational structure. Now, what I wish to report about this experience is something really unusual and that I would call *the hypnotic view*.

What I want to say is that from the moment I was hypnotized I realized that *my perception of ordinary reality shifted* and I saw things differently, in a distorted way, or I even saw things that we ordinarily never see. And what I found most amazing was that I saw on the front of my healer a *huge third eye*, as real as it could ever be, a very intelligent-looking large human eye that did not for the least frighten me. In the contrary, in that special hypnotic condition I knew and acknowledged that I saw her *third eye*, her sixth chakra, only that in this special view I could visualize this eye that all the old mythologies abound of telling us.

And I told her at once about my discovery and she was not the least astonished. She said that other patients had seen it as well and that it was quite a normal experience for her, and that according to her Filipino teacher it showed that she had a great innate potential for healing – and this was really true as her treatment was the most effective one can imagine.

It was almost miraculous. I was completely cured within three months, and with only six sessions.

Two years later I started off to another fascinating healing experience, this time directly with a Filipino healer. At that time the magazines in Germany were full of reports about Filipino healers and the photos were absolutely dumb-founding. I saw that the healer had both his hands deep within the belly of a female patient and that there was blood all around, so much blood, and that suddenly he took out something from her intestines, something like a black stone, and threw it on the floor and declared to have found the *evil* in her body. I really thought it was all a dirty trick and wanted to know the truth about it.

Thus I signed up when a befriended natural healer from Hamburg called me and invited me for an audience in a na-turopathy practice in Frankfurt where I was going to meet one of the most powerful Filipino healers.

While I have studied quite a bit of parapsychology over the years, already during my law studies, I was very skeptical regarding this kind of healing. I told to myself I'd be very watchful to find out what was their trick, while I thought to myself that the soundest hypothesis for all this to happen simply was group hypnosis.

However I was struck by the fact that this hypnosis can affect photographic plates as well, and be taken on video. So, after all, can it be explained with group hypnosis? I had seen this in those magazines even before I went to Frankfurt, and thus it was possible to photograph something that is not to explain within our present reality paradigm. It is and remains a miracle until this day.

I plead for the hypnosis theory here simply because I was experienced with medical hypnosis already and knew how it *felt* and how ordinary perception reacts to it. Because it's really that once you have experienced hypnosis, you know for all times *how it feels* when somebody tries to hypnotize you. You are *aware* that you are being hypnotized in a certain moment, and you can fight it if you don't wish to succumb to it, however pleasurable it may feel. Let me add that indeed, generally, it feels very pleasurable if your general anxiety level is not, like mine at that time, higher than average and you build a resistance against it. Now, they seemed to try everything to avoid resistance when we were comfortably installed in that natural healer's practice in Frankfurt.

We were being thoroughly prepared for the experience, for more than two hours, during which the healer was still busy with another group. We did not see him and had to remain in a special meeting room where we were being instructed about the strictly spiritual principles that traditional Filipino healers are bound to. We were really well informed and I got to know the first names of most of the participants that were women in their great majority. The atmosphere in the group was friendly and amicable and I felt really relaxed when we entered the room where I saw the healer on the floor, near to a bed and next to a vessel with burning incense. He seemed to be in a deep state of prayer and meditation. In the room several pictures showing the *Holy Virgin* and *Jesus the Christ* were hanging on the walls, and the audience was visibly collected and in a state of respect and awe upon entering the room.

During the whole experience I did not feel the slightest discomfort or fear and I wish to state this right at the start of this report because it so sharply contrasts with my Ayahuasca experience. Then, all went exactly as I had seen it reported in the magazines. The healer called one of the women to the bed and let her stretch out comfortably, asking her to slightly open the belt of her trousers or skirt so that he could plainly touch her abdomen.

By the way, none of the women felt the slightest discomfort at this demand of the healer. His requesting her to take off her shirt equally did not result in any resistance. Then he *opened* her belly by the navel entry, and virtually penetrated into the navel with the fingers of his right hand, until the navel was an open hole of about the size of the healer's fist.

This opening bled abundantly, and the healer then introduced both of his hands through the navel hole until deep in the woman's intestines. He seemed to search for something in there. Several times, he extruded a part of the large or small intestines, looked at them or rubbed them. He appeared to search for something particular and almost in every case he found it: it was objects like small stones, most of the time of black color, and when he got one he threw it visibly and audibly on the floor.

One of the most interesting details is namely that upon awakening from our collective hypnosis, not only was there no blood anywhere, but there were absolutely no stones to be found on the floor!

Then the woman who was sitting to my left was called. She was a nice young lady who had complained in our previous talks about a problem with digestion that lasted over many years and that she thought could not be cured with Western medicine, as she had *tried everything already*. With her, the treatment took longer than with any other participant

Published by Sirius-C Media Galaxy LLC, 2010

and she seemed to suffer from it, was howling two times very deeply from the depth of her body, like a hurt animal, and gave me a deep regard when she returned, saying:

– I feel *very* hurt, I am suffering great pain!

I was kind of shocked, so much the more as I was called as the next patient. But in my case all went fine and I did not feel anything and avoided to open my eyes, somewhat afraid I could be shocked to see blood, yet did not feel any pain. And upon returning to my seat, I immediately asked her if she had seen me bleeding and she replied that, yes, I had been bleeding abundantly and that the healer had put his two hands in my intestines but that in my case the treatment had been a lot shorter than with most of the other patients, and especially herself. I was asking her then if her pain was less and she said that indeed, the pain was gradually decreasing, giving rise to a feeling of comfort that she had not experienced in many years.

Later we were discussing and exchanging about this daring experience but never got a chance to talk directly to the healer. However, the naturopath and owner of the practice offered a free consultation and revealed to be a very good initiate into this practice that he had learned from several Filipino healers. Thus, we all left that experience with a great feeling of delight and amazement, and I can only say with Goethe that school knowledge will not suffice to explain this extraordinary form of healing.

To avoid a misunderstanding, let me be very clear and to the point: I do *not* say that healing induced by hypnosis, re-

230 | Chapter Four

lated to hypnosis or which may be a veiled form of hypnosis was charlatanism or was not effective. In the contrary!

What I say is that healing which seems *miraculous* to us is in fact a treatment that is so effective that we can't believe it when we compare it to our ordinary, and rather palliative, healing methods. I further say that this *effectiveness* is due to *suggestion* and that suggestion is a command that uses the *power of the word* to impact upon the condition of the body.

The only difference between the normal waking state and the hypnotic state is that in the latter, the body is more suggestible.

Now, what most people don't want to see is that the healing brought about by suggestion is by no means fake healing, but healing that is as good or even better than ordinary healing.

Sometimes we understand certain things when we look at their contrary, or their negative side. So let me give an example that demonstrates what I am saying. It is an old experiment and has been repeated often by *Milton Erickson* to demonstrate the power of hypnosis as a verbal suggestion that directly *impacts* upon the soma. The patient is in deep hypnosis and is told by the hypnotherapist that a sizzling hot iron will be applied to her arm for a few seconds.

Then, the therapist takes the iron which is of course cold, and applies it to the patient's arm. Immediately the body reacts *is if the iron was really burning hot*, and builds a huge watering blister. This blister is still present when the patient wakes up and it takes the same time to heal out as if she was really burnt. It goes without saying that she also experienced

Published by Sirius-C Media Galaxy LLC, 2010

some local pain and had to give her full consent before engaging in that really dumbfounding experiment.

Now imagine that what can be done negatively to the body through verbal suggestion can as well done to it *positively*. This explains that under hypnosis healing can be instantaneous and totally effective!

The most important to report in our context is the similarity in the way the hypnotic trance manifests. You will read about it again in the next paragraph regarding medical hypnosis used in psychiatry. Let me summarize so far that this trance usually begins in the arms with a relaxation of the hand and lower arm muscles and then gradually *mounts upwards* into the body, passing region by region like a gentle embrace, thus the upper arms, the shoulders, the face and head, then the chest and the muscles around the heart.

The fear block I experienced was located in my heart region and this already had been confirmed by the homeopathic healer in Germany who, after an extended treatment, told me that from my earliest childhood I had suffered a problem with being abandoned and that my heart chakra was *closed*, whereupon she opened it by slightly touching my heart region with her finger. This slight touch triggered an amazing amount of tears and I suddenly remembered early childhood feelings and went through a really difficult moment during about one hour, upon which I was left in total peace and serenity, feeling like *newborn*.

To summarize, I reacted very differently during the inducement of the hypnotic trance. In two instances, with the Filipino healer and the German homeopath, it was a very agreeable and smooth experience, while when I did it with psychiatrists, it was rather anxiety-creating. And with the Ayahuasca it was frightening as well.

I conclude that the fact to experience fear or not depends on the hypnotizer and not the particular kind of hypnosis he or she uses. I tend to believe, and I am open to change my opinion if it should reveal as scientifically unsound, that hypnosis requires a high amount of immediate trust between the patient and the hypnotherapist or shaman, or natural healer.

When this trust is lacking, anxiety will interfere with the depth of the hypnosis or make it a rather disagreeable negative experience.

8) Medical Hypnosis

The first time I experienced medical hypnosis was in 1989, in Geneva, with a quite famous transactional and Gestalt therapist, *Dr. Margareta Robinson*.

It was the first time that I ever got in touch with hypnosis, and unfortunately I was *not informed* that it was hypnosis that I was going to experience! When she presented her approach to me, Mrs. Robinson was talking about a combination of transactional and Gestalt therapy that she seemed to have melted into a powerful approach for healing various problems from neuroses to narcissism. I was surprised at the ease of how that hypnosis was brought about. It was by means of using a pet, a *teddy bear*. Indeed, upon holding that magic pet I felt a deep and very *sad* kind of relaxation affecting my body and mind. I suddenly felt extremely tired, and

Published by Sirius-C Media Galaxy LLC, 2010

powerless; that sensation was not at all joyful and agreeable, but an experience that left me very depressive.

I began to feel apathetic and helpless, *very powerless*, like a baby abandoned at the mercy of some or the other untrustworthy caretaker. My muscles became weaker and weaker, as if I was given a sleep potion, until I could not lift my arms up any more.

Then she asked me to get up, handed me a tennis racket and said:

– Here on this bed your mother is stretched out. She is sleeping. Hit her with this racket as much as you like!

And I could not do it and told her about it. She said:

> – Well, this proves only your problem to me. You have internalized all the violence and resistance against your mother and are for the moment unable to exteriorize it. That's why you suffer from depressions. They indicate the deep hatred against your mother and we would have to work on releasing this energy.

I did *not* agree with her and her approach. I felt she was pushing the therapy in a very hurried and jumpy way that was not bringing me relief while I must say that when coming home from each session, I felt I was seeing the world literally with other eyes, so much all seemed to have changed for the better.

I then began a hypnotherapy with an American therapist and this therapy progressed *much more carefully*, while I must admit that we never entered a really deep level of hypnosis because of the fear problem I already described earlier in this report.

Summary

The experience with *Ayahuasca* as I made it with the Shuar shaman back in 2004 is in my view supportive for a non-linear and multi-causative, rather than a linear and single-causative theory of cognition regarding the psychedelic visions and insights subsequent to ingesting the traditional brew.

In addition, in my discussions with the shaman and his assistant, equally a Shuar native, it appeared clear that they themselves rejected the linear and single-causative theory of the kind stating 'it's the DMT that makes for all that Ayahuasca does', explaining that all the art was in the traditional procedure of preparing the cure and the consciousness focus that forms part of it. In fact, the negative experience of the Polish business man with the brew, and the resulting ineffectiveness of it, shows evidently that the single-causative linear theory of cognition regarding the Ayahuasca is flawed.

The cognitive experience with Ayahuasca is probably not a simple direct consequence of the plant-containing DMT, as this has been suggested, for example, by Terence McKenna and his brother, the ethnobotanist Dennis McKenna in their book *The Invisible Landscape (1994)*.

In the eight specific particularities that I have brought forward and commented on in the previous chapter, there appears to be a certain weight of the evidence for a causation

Published by Sirius-C Media Galaxy LLC, 2010

of the cognitive experience by shamanic consciousness acting as a hypnotic agent on the plant matrix that serves as a resilient *transmitter and amplifier of thought energy*.

The specific cognitive elucidation and the insights experienced after ingestion of the brew, that I shall report in more detail in the last chapter, are brought about through a multi-causative impact of the *consciousness imprint* on my own consciousness interface. This impact was brought about through the strong focus of the shaman's thought energies on my perception matrix and reception frequency both during the preparation of the brew and at the onset of the intake ritual.

This concentration of *thought and attention* is known both from parapsychological research and clairvoyant experience, and from medical hypnosis to bring about an energy imprint in form of a *consciousness overlay* on the perception interface of the receiver.

I am talking about a multi-causative effect here because the evidence at stake does not allow to exclude any *proprietary additional impact* of the plant consciousness in the process of triggering the consciousness overlay. In fact, there are details in my report that indicate such an additional impact directly from the side of the plant realm, as a *genuine plant-proprietary consciousness* reaching out into my human consciousness.

The most striking detail in this context was that I had myself the clear intuition of being in touch with a proprietary *plant consciousness* or even an unspecific *universal consciousness* that I was in an ongoing telepathic exchange with as long as the trance lasted, and that I was, strangely enough, in state of turning that telepathic communication on and off by simply closing or opening my eyes.

In addition, the obvious parallels with my previous hypnosis-induced alterations of consciousness demonstrate that the focusing of thought energy that the shaman did as a preparation for the ritual in accordance with traditional native tradition has in some way to do with hypnosis, or brings about an effect or imprint on another's consciousness that is similar to a hypnotic injunction.

The most important detail in this context is well the fact that there is a *plant-specific matrix* involved in this process, and not just a shaman focusing thought energy on myself as his client. This is the specific contextual link with plant consciousness acting as a *matrix receiver* for *intent*, similar as this has been reported for water, by the elucidating research of the Japanese researcher Masaru Emoto.

As Emoto's water research suggests, it is possible to leave imprints in the memory interface of water by positive or negative affirmations, for example in the form of textual labels glued on the water bottles for some time, that produce or not in the water specific crystals. Typically so, the aesthetically appealing crystals are formed by positive and uplifting intent and correlated affirmations rather than by negative and defeating intent and affirmations.

My argument here with regard to the cognitive imprints received in the form of insights during an Ayahuasca trip is on the same lines of reasoning. My idea is that the consciousness interface of plants, at least of plants that are qualified as *entheogens* or as plants containing mind-altering

Published by Sirius-C Media Galaxy LLC, 2010

compounds, serves as a transmitting and amplifying interface for the thought imprint given to it by the shaman's consciousness and thought energy. In how much the plant here participates with its own consciousness-altering compounds, such as DMT, cannot be evaluated from this experience with Ayahuasca, but needs additional, tightly curtailed research. It is namely possible that the plant chemistry, instead of being a unilateral agent of altering human consciousness, serves as a receiver, transmitter and amplifier interface for human intent and thought energy, as this has been reported by Masaru Emoto and others for the *hado*, the specific energy-interface of water.

But even Masaru Emoto has not found, and not even tried to explain the ultimate reason why human intent can have an energetic impact on water, and other substances. The explanation, or one possible explanation is given by Charles Webster Leadbeater in his 1894 booklet *Astral Plane* where he describes the function of elementals in the communication between humans and all realms of nature.[87]

Leadbeater explains that thought is an energetic phenomenon that creates certain vibrations, called thought-forms or *elementals*. These elementals, he says further, gain permanence over time and depending on how much *emotional energy* we invest in those thoughts. And interestingly so, here we encounter the philosophy of the natives who speak about *spirits* when asked what the communicating agents were between humans and plants. And the solution of the riddle is to view the natives' explanation and the theosophical or clairvoyant view together.

The technique consists thus in imprinting intent in the plant matrix by gestating, through the power of thought-energy, certain elementals that function as communicating agents between the human and the plant realm. These elementals, I suppose, are created during the process of collecting the Ayahuasca liana and carefully preparing the brew, and it is these elementals impacting on the plants' psychoactive substances that are becoming active and *communicative* as it were in the initiate's consciousness.

And, to come to an end, I think what the contextual scope of the present Ayahuasca experience well indicates is that a simplistic linear cause-and-effect mechanism between DMT and cognitive insights can safely be discarded as a theory.

Published by Sirius-C Media Galaxy LLC, 2010

The Cognitive Experience

When asked to summarize the insights I got through the ingestion of *Ayahuasca*, I can establish the following catalog:

▶   The intelligence's *alien* noise interface;

▶   The intelligence's *pulsation* as a cosmic energy;

▶   The intelligence's attempts for *calling me* in touch;

▶   Insights about my conditioning through language;

▶   Insights about relationships with others and the world;

▶   Insights about love and life.

These insights did not come up in my consciousness all at once, but rather through little chunks that were repeated several times and with variations, extending virtually until the moment, early in the morning when I feel asleep. Interestingly so, the very last insights I got immediately before falling asleep, at around four in the morning, when stretched out on my bed, after the *purga* eventually had stopped.

It was then that I was suddenly intensely aware of my solitude, my utter lack of relationships, and my general feeling of being disconnected from other people. And it was then that the insights about relationships and about love and life came through.

Another interesting detail is that most insights only came up after I had asked questions to myself or this specific intelligence I was in touch with, while I had not always formulated my questions as questions, but often as affirmations that were then propelled back to me like in a boomerang effect. And what happened was that most of the time the affirmations had been slightly altered in the process. I shall give examples.

Alien Noise and Pulsation

The first phenomenon I noticed about the specific intelligence I felt approaching right at the onset of the trance was its *alien noise*. This noise was clearly distinct from the frog concert and other natural sounds that surrounded me.

My senses were not dulled by the trance but in the contrary sharpened and I could clearly distinguish between the multitude of natural sounds around me in the clear evening air, and the specific alien noise of this intelligence.

When I should put it in words, which is somehow an impossible quest because of the paranormal reality as the contextual background, I would say it's like many, thousands or millions of human voices simultaneously whispering a mantra. I think it is an important detail that it's not just like a machine-noise, or a tone, but that it bears a resemblance to a whispering human voice multiplied by the millions.

I know that Terence McKenna has spoken of the *machine elves* and their *alien sound* as a typical manifestation during the DMT-induced trance. But what I am saying is that in contradistinction with McKenna's perception, the sound was *not*

Published by Sirius-C Media Galaxy LLC, 2010

machine-like but came over to me as *organic*, and somehow related to nature.

I call it *alien* only because there is no sound or anything you could have ever heard in your wake life that bears any similitude to this sound. Actually I prefer the term *noise* over sound because noise is a term used in telecommunications, as something that can either be a background hiss, such as the hiss on vinyl records, or a certain unspecific hum contained in ultra-short wave receivers, or else a term used by graphics designers for the lacking smoothness of an image. All these connotations fit here, in my opinion.

The presence of the *natural intelligence* of the Ayahuasca spirit is related to sound as it manifests not visually in the first place, but audibly, and thus it bears an impact of *resonance*. It has to do with vibration, and with frequency. It has to do with cell resonance and with Sheldrake's notion of *morphic resonance*, as it resonates the mix of nature's frequencies, and comes over as the *Universal Communicator*, and at the same time, the Universal Bearer of all these energies.[88]

This is exactly what I wanted to convey actually about the *organic quality* of the sound: to me it bears a morphological resemblance with organic life, and organic sounds, only that it overlays many or a multitude of such sounds in its audible presence. So, as with ultra-wave communication, the Ayahuasca intelligence comes through on *a certain frequency*, and not on another, and the frequency is tuned by ingesting the brew, and here the DMT may well be active as the attunement agent.

From the onset of the trance, the alien noise was gradually rising in volume, and after some time, I clearly understood that this sound was something like the *pulsation* of the universe, of life, and in that moment I got an intense awareness of the fact that all creation is in fact a result of *sound*, and that life intrinsically is vibration, is sound. This is an insight that I have well today acquired through having studied hermetic and modern literature on sound used for healing, but three years ago, when I went through this experience, I was not yet consciously aware of this fact.

Hence, I can say that this insight was novelty for me. And yet, it sounded completely sound and solid, so to speak, and did not come as a surprise. It is as if the intelligence had not just communicated me something using telepathic touch, but as if it had subtly awakened my intelligence to a novel insight that from that moment could not be unthought any more from my consciousness.

The Five Depth Levels

From the onset of the Ayahuasca trance, I got clear telepathic messages that I should go beyond that first phase during which I saw subtle geometric forms that bore however much of a lesser brilliance and luminosity than those I had seen in some research volumes, such as Pablo Amaringo's *Ayahuasca Visions (1999)*. And I knew in that moment that, contrary to what I had learnt and heard from others about the Ayahuasca trip, the visions were of no importance at all. I simply *knew* this or it was communicated to me in that mo-

Published by Sirius-C Media Galaxy LLC, 2010

ment. At the same time I was called upon by the intelligence to go beyond that first rather insignificant level and explore into the next depth level, but that for getting there I needed to *relax more* and let go some of the fear that I felt was like a congested knot in my heart chakra. The intelligence also communicated to me that there were in total *five depth levels* in the Ayahuasca experience.

I think it is significant to note that I found this in none of the books I had read when doing my research on shamanism and entheogens, and I have not found it subsequent to my experience in any additional books I read about the Ayahuasca quest. What I have well learnt from Michael Harner's seizing account of his own primary Ayahuasca trip, during which he almost died, in his book *Ways of the Shaman (1990)*, was that he had himself experienced, right from the start, the toughest depth level – encountering the primal dragons. But neither Harner nor other researchers and experiencers have given an account of how many depth levels there are, while all seem to agree that there are in fact several levels of intensity to be possibly experienced during the psychedelic trip.

There is something like a *consensus doctorum* in the literature about the ritual use of entheogens that sets a relationship between dose of the substance intake to the intensity of the trip – and here I made a nice and meaningful typo, writing *insensity* instead of intensity. This is true in so far as the experience gets weirder and apparently more insane as a result of our cherished assumptions about reality, and the *sense* that we give to certain experiences being shifted in the course of the strong psychedelic experience.

However, I want to warn here again falling in the trap of single causality and of linear thinking when it goes to evaluate a type of experiences that is intrinsically multi-causal and non-linear in character. In my view, the *depth levels* that Ayahuasca contains are probably not triggered by the dose alone, but also by the *intensity of the focus and intent* bestowed upon the brew from the side of the shaman.

Now, what is also rather uncanny and that I have not found in any description of Ayahuasca trips anywhere in a book is that the intelligence gave me *signals*. It was like a gentle knocking at my doors of perception. I typically saw a flickering of five red squares at the top upper left corner of my vision, when my eyes were closed.

I intuitively knew in these moments that this was the signal to go deeper in the trance, or jump to the next depth level, only that to my sadness I could not follow the invitation because of my fear block. Every time it happened, the fear came up and blocked me to go beyond. And this blockage was not just mental. If it had been mental only I could have overcome it. In fact, I wanted to overcome it, but then the blockage somatized and manifested as vomiting, and later also as strong diarrhea. It was not only my mind that resisted the experience, but also, and perhaps primarily so, my body.

Calling Me in Touch

I was constantly communicating with the intelligence, from the first to the last moment of my Ayahuasca experience. I had prayed already before the onset of the experi-

Published by Sirius-C Media Galaxy LLC, 2010

ence, while still on my bed in my hotel room, that through this wisdom quest I might receive guidance for finding my life's mission and for being freed from my constant anxiety, and I kept addressing the Ayahuasca spirit in respectful terms during the experience to assist me in my quest for truth.

And as if it was a final test that this intelligence really understood me, I said this, when I felt I wanted to go back to my room and sleep:

My Silent Prayer

Dear Ayahuasca Spirit, thank you so much for all the insights, please stop calling me now. I know that I was not able to get over my fear block now, and I hope I can find the courage to continue because I know you will reveal me so much more when I get deeper in the trance. But for now I have decided to stop. It is enough.

And immediately the *calling* stopped, the flickering lights did not appear any more and the *alien* noise was calming down and stopped as well. This alien noise, I understood then, was the *language* of that intelligence. It was something like many people, in some distance, speaking simultaneously, and as if many conversations converged to a chaotic no-sense that was but the secret code or pattern of a deeper, much more unified language of the universe, and of which our human language is only a tiny and almost insignificant part.

Freeing from Conditioning

The plant intelligence I was in touch with seemed to convey to me telepathically that I was *locked* in language, that all my experiencing of life was conditioned upon language, and that I hardly ever perceived life *directly*, spontaneously, as an immediate connection. At the same time, I received the instant confirmation that this intelligence was *universal* in the sense that it was connected to all, and that it was constantly trying to connect all, such as a total communication matrix of the universe, and third that it was and represented all-that-is.

I became keenly aware that this intelligence was the *Logos*, and that it simply *was*, and it gently invited me to enter this connection, this wonderful all-encompassing love that it irradiated and communicated.

Several times, when I intended to tell the shaman some of my cognitive experiences while still in trance, the intelligence seemed to wanting to hold me back. Once I wanted to say something regarding the intelligence, and I was stopped in midst of the sentence:

– *Este inteligencia* …, was all that I could say.

The intelligence seemed to wanting to free me from the conditioning I had received through language, and through using language for describing reality. It might be that my fear is in some way connected to language and my fear block a hypertrophy of language or of my left brain. This is how I can try to explain it while all what I write here and can express in words never can come close to the actual experience.

Published by Sirius-C Media Galaxy LLC, 2010

Love, Life and Relationships

Before I come to talk about the important insights I received about love, life and relationships at the end of my Ayahuasca trip, I would like to expand a little on that peculiar question-and-answer game that was developing between the intelligence and myself. In fact, I was naturally the one who, puzzled, asked the questions, and the intelligence always *instantly* replied. When I say instantly I really mean that the reply did not even take one second to appear, but it was instantly in my mind upon formulating the question. I think this is quite uncanny as a fact while it is probably known to other researchers.

And the reply could have various forms. It was always economical in the sense that it never wasted even one syllable, and when this was possible, it was just turning around my question, or simply *shortened* it, in order to give the answer. This is something I really have never heard of before, and it reminded me of a higher evolution of certain circus jokes or mind games, and it definitely had a note of humor to it.

The most important experience in this context was that my questions were *somehow returned* as questions-that-question-again-my-questions so as to give the answer to any question as a result of the question itself – and not as some kind of outside input. So in a way, the phenomenon suggested that the intelligence I was communicating with was altogether *not* an outside or outward or distinct intelligence, but simply a part of my own higher consciousness. Now, succinctly speaking, this manifested in a way that every thought I had was immediately returned to its contrary and every phrase that I pronounced in my mind was immediately reduced to some more general insight. For example I heard myself thinking:

– I love life …

And it was immediately reduced to:

– I *love* …

Or I heard myself thinking:

– I feel to be alive …

And it was reduced to:

– I *feel* …

And now, when you evaluate the returned, condensed or shortened statements *as answers*, you will notice that they are indeed *highly intelligent answers* to underlying questions. The first statement could be read as an underlying question of the kind: 'What does it mean to love life?'

Now from the returned shortened version, it becomes evident that when I love, I simply love – which means I am in a state of love, and thus as a result I love all-that-is, and thus *also* life. So it's somehow unintelligent to say a sentence like 'I love life' because love cannot be reduced to just a concept like 'love of life' or it is that: a mere concept.

Published by Sirius-C Media Galaxy LLC, 2010

By the same token, to divide love off in concepts such as 'filial love', 'passionate love', 'love for children' or 'love for the elder' does not make sense, as what it produces is splitting the holistic notion of love off in tidy compartments that are *concepts of love*, but not *love* any more. So somehow the intelligence politely corrected what I was saying without correcting me! And I immediately understood the hint, like when you solve a *koan* in Zen.[89]

Eventually I became very quiet and full of gratitude and I said this in my thought, in my German mother tongue:

> – *Ich möchte Liebe!* (I want love)

And immediately the response was, in German:

> – *Ich möchte lieben …* (I want to love)

And with all my heart and soul I understood this subtle difference and it was clear to me that in this tiny difference of syntax there was *all the difference*. It was the secret to happiness. I had taken love as a commodity that can be received, stating that I wanted to be loved, instead of understanding that love was not something to receive, but a state of being to develop into, something like an attitude, and this attitude, it seemed to me, simply was *total openness.*

And at the same moment I became intensely and acutely aware that I was not in that state of love, that I was not giving love to others, and I also knew *why! It was fear that blocked me off to love.* And Krishnamurti's saying came to mind that for the first time I really understood: *where fear is, love cannot be.* And I think this insight transformed something in me. And really, afterwards my relationships changed much for the better.

Published by Sirius-C Media Galaxy LLC, 2010

Literature Review

The first book I read on the subject of shamanism in the 1980s was *Les appeleurs d'âmes* by Sabine Hargous, a French ethnologist. This study published in 1975 with the well-known French publisher *Albin Michel* and that translates in English as *The Soul Callers* is a well-documented thesis on the shamanic universe of the Andine native population. This book clearly centers on one aspect of shamanism only: the healing. The study divides into three main parts *Indigenous Pathogenics*, *Diagnostics* and *Magic Rituals*. However, implicitly, the author expands on virtually all aspects of shamanic spirituality.

After all, it was perhaps a good thing to have begun with this study and not with something that conveys a *felt sense* of indigenous living, as for example the excellent books of Michael Harner, because Sabine Hargous' approach represents decidedly a Western view, with all that this implies. She calls natives *primitives*, as it was the custom in traditional ethnology and that says in one word more than a whole thesis about what quantum physics has taught us about *the observer standpoint*. But for this reason the study is not to be discarded.

I would even say this consciousness split is important for some people who else would never read a similar study because their anxiety to *remain fixated in their own cultural belief system* is greater than their curiosity to explore other, and certainly more direct methods to approaching reality. And as, at this time, I was still working on my international law doctorate and not yet on the daring path to question the Western science approach – while I became more and more critical to it – this was certainly a good book to begin with. As the author was keeping her Cartesian distance to an alternative worldview, she was nonetheless getting deeply immersed in the world of the natives, and her book is all but a dry thesis paper.

The next book that virtually fell in my hands, as I found it on a garage sale, was Michael Harner's bestseller *Ways of the Shaman (1980/1982)*. This book had a different impact upon me than the first one. It was a shock and a revealing new learning!

While I found that Sabine Hargous' study had a rather philosophical touch, Michael Harner's study really moved me into planning myself a voyage, and it was there and then, that I took the decision to engage on the Way of the shaman – or, to be honest, the book rather revealed me that since childhood I was on this way already, without ever knowing it, without ever being able to voice, to describe the special mission I feel is mine.

Before I got to read other of the real power stuff, so to say, because written by empirically minded researchers, and as I had to wait quite a long time to get the books from Amazon USA shipped to France, I ordered two books by Spanish authors in Barcelona, Spain, that I got within two days only and I did not regret to have read them, as they dealt with the philosophical and conceptional issues.

The first was a book from one of the foremost Spanish authorities on the subject of shamanism, Josep M. Fericla,

Published by Sirius-C Media Galaxy LLC, 2010

entitled *Al Trasluz de la Ayahuasca (2002)*. Reading this book, I realized that in Spain there is absolutely no *moralistic bias* against psychedelics such as, for example in France, and in Great Britain or the USA.

But compared to France, the legislation even in the United States is still quite liberal. France is really the worst one can imagine in any of the Western nations; this is after all comprehensible when you look at the Cartesian mindset of French people, their extreme left-brainism and their almost total lack of true spirituality.

In Spain, the exact contrary is true which obviously has nothing to do with left or right-wing governments. In Spain, nobody would get the idea to put Cannabis on an index. It is *as legal as Cuban cigars* and perhaps, definitely, more healthy than those. The second impression was that for this Spanish author, the psychedelic quest was a real parallel way of perception, serious not only for freaks and pioneers in consciousness exploration, but also for philosophers.

Fericla is one of the finest in Spain, and not just for his preoccupation with psychedelics, but in general. Thus, a book from a real authority, and written with a serious mind and true commitment for consciousness exploration. This book confirmed my decision to really take on the voyage I had planned and not just do a theoretical research on the topic of shamanism within the greater project of my research on the *Eight Dynamic Patterns of Living, Audio Book (2010)*.

The literary magazine *El Idiota* that I equally had ordered in Spain contained many interesting contributions one of which I wish to mention here as I find it very important.

This special issue of the magazine entitled *Visionarios*, contains an article about Carlos Castaneda. *Carlos Castaneda: El Enigma del Último Nagual*, an intriguing study by Cristóbal Cobo Quintas that deals with the somewhat mysterious content of Castaneda's well-known spiritual apprenticeship with the Yaqui sorcerer Don Juan. This Spanish author sees the importance of Castaneda's books – be they invented as a part of the media debate about Castaneda pretends, be they real accounts of the practices of one of the last living witnesses of Toltec culture.

It is noteworthy that Castaneda, *inter alia* on his web site, claims to be the only legitimate last descendant of the Toltecs and spokesman for their culture within a largely ignorant world. In fact, what we learn through visions is more than just the visions; this was already clear to me when I read Castaneda myself, more than ten years ago. And what all serious studies and reports about visionary experiences, at least those done with Ayahuasca, other DMT derivates or Peyote converge to is to affirm that these visions *enhance our understanding of nature.* In addition, these studies contribute to helping us understand what is *direct perception* or, as we would call it today, *systems intelligence.*

In my personal view it's not even the visions themselves that have this impact upon our intelligence but some kind of telepathic code written into the visions, but that we are not aware of, and which is transmitted directly into our DNA or, if already contained in it, thus activated or stimulated.

Eventually the books arrived from the USA and the first one I read was Ralph Metzner's excellent book *Ayahuasca:*

Published by Sirius-C Media Galaxy LLC, 2010

Human Consciousness and the Spirits of Nature – which is actually a sampler that he edited and in which many people related personal experiences with Ayahuasca. This book, including Metzner's highly interesting introduction, is through and through a master-piece because it gives so many insights *simultaneously*.

It is all but a dry scientific report, but an exciting adventure to read. In a way, one cannot but feel all those individual experiences on almost a gut level, in order to definitely, and once for all, put aside a Cartesian worldview that tries to split the world off in nice little tartlets called 'science', 'emotions', 'perception', 'experience', etc.

One then begins to understand that awareness is beyond all of this while it encompasses all of this, awareness being the very fact of being aware of being aware. To begin with, it is interesting to see what the primary motivation was for most if not all of the people who contributed to the reader, most of them being involved in natural healing, or otherwise working in social professions. These people all had in common that they expected some tangible results from the experience, for becoming better healers or advisors, or for solving personal problems, and there was none of them that did not at the end of their statements confirm that the experience had been worth it and helped them to reach this goal.

The next book I was reading, and that I found even more mind-boggling was Jeremy Narby's *The Cosmic Serpent (1999)*. This book, written by a Swiss anthropologist, originally written in French, takes a completely different perspective. Narby, questioning the native shaman's conviction that plants really transfer knowledge during the visions, states:

Jeremy Narby

First, hallucinations cannot be the source of real information, because to consider them as such is the definition of psychosis. Western knowledge considers hallucinations to be at best illusions, at worst morbid phenomena. Second, plants do not communicate like human beings. Scientific theories of communication consider that only human beings use abstract symbols like words and pictures and that plants do not relay information in the form of mental images. For science, the human brain is the source of hallucinations, which psychoactive plants trigger by way of the hallucinogenic molecules they contain.[90]

He then puts up the hypothesis that what the plants actually do is to open a *perception channel* to our own DNA's photon vibrations. Photon radiation of the DNA has been confirmed in recent quantum science but physicists did not go as far as saying that this photon radiation's information flow was in any way consciously *readable* for the human mind. This hypothesis has something daring about it and Narby makes his point with quite an amount of writing skill. However, I was again and again considering the premises he based his research upon, and in my opinion, Narby made a paramount mistake in failing to question these premises before he set out to write his book. Here is what I would advance against Narby's argumentation:

– Why should mind visions *not* be the source of real information? The fact that this contradicts modern psychiatry means nothing in terms of perception theory, a field that

Published by Sirius-C Media Galaxy LLC, 2010

psychiatry has nothing to deal with and does not understand anything about.

– Why should it be important how *Western knowledge* considers hallucinations? To call them *morbid phenomena* is definitely not a scientific judgment, but a moralistic opinion and as such irrelevant for the scientific researcher.

– Who says that plants do *not* communicate like human beings? Who has the knowledge to deny this possibility? The natives do not say that plants communicate like human beings; they say that plants communicate with us using a form of telepathy *that is part of consciousness itself* and that makes that communication can be cross-species.

– Scientific theories of communication consider that plants do not relay information in the form of mental images, states Narby. I want to see the treatise of communication where this is written! This sentence is highly unscientific in itself in that theories of communication deal with human communication only and are generally silent about plant communication. The mere silence of this research regarding plant communication cannot logically be interpreted as a denial of the existence of such communication, in general. Here, Narby clearly committed a logical *faux-pas*.

– Finally the last sentence in Narby's hypothesis is equally *suggestive*, and not scientific in that it suggests namely that the human mind was seated in the brain and only in the brain, an assumption that is scientifically overthrown. Neuroscience and consciousness research now coincide in acknowledging that the mind or consciousness, while functioning through the physical brain, is not forcibly physically located

in the brain, but certainly also in the pineal gland, the pituitary gland, and especially the luminous body or aura. Some
go beyond and suggest the mind was located probably everywhere, even in the cells of the skin of the feet, for example,
but also *outside of the body*, as psychic research has confirmed
since long.

But Narby's study certainly has high value in the present
discussion be it as a contradicting resource. One thing was
namely clear to me from most of the German and American
shamanism researchers and their publications: they do not
question the possibility of plants being able to communicate
information to the human mind in what form however this
takes place! They also do not, like Narby, start from a concept of *Western knowledge* but rather take a pioneering, open
and experimental approach while sticking to the facts and
avoiding speculation. And they all seem to take for granted
that the mind and the brain have in common only that the
brain functions like an interface for the mind to operate
within us, and within all. Thus, I found that Narby was, from
the start of his book, much more restrictive and skeptical
than, for example Michael Harner, Adam Gottlieb, Ralph
Metzner or the McKenna brothers in their respective studies.

On a similar line of reasoning while from a totally different perspective is the DMT research of an American doctor,
Rick Strassman, *DMT, The Spirit Molecule (2001)*. Strassman
was all but mystic-minded when he began his study with
hundreds of patients to experience precisely dosed DMT
injections in their veins. The book, or how much I could
stand of it, was quite boring to read and, as a result, my

Published by Sirius-C Media Galaxy LLC, 2010

quotes collection is rather scarce, which is certainly not a mishap of the book itself but more of the reader. I just do not find much interest in this kind of soulless research that goes out to understand life from monkey experiments. Okay, a similar marathon study once conducted in France by two sociologists to disproof astrology, was finally exactly confirming its functionality. But for a serious astrologer, such a study is a circus joke for the ignoramus because since thousands of years initiated individuals know about the cognitive value of astrology and they do not need monkey experiments to confirm this perennial knowledge.

I trust more a critical intelligent and initiated human such as, just to give an example, Michael Harner or Terence McKenna, who have taken DMT and who say with unshaken conviction that they received *knowledge*, real knowledge through the experience, and not just experienced a kaleidoscope of silly flash lights. If shamanism research was so dull and insignificant as most medical doctors and skeptic anthropologists think it was, highly intelligent and trustworthy writers and researchers such as Schultes, Metzner, Harner or McKenna, would have to be called schoolboys. You and me know they were and are not, and that in shamanic voyages we are *not* dealing with kaleidoscopic games.

I was glad, then, to read something from a different mind, and frame of mind. Aldous Huxley, the author of the novel *Brave New World* and other great fiction and non-fiction writings, was one of the foremost witnesses in experiments with perception, altered perception and immediate perception. His book *The Doors of Perception* is an enlightening account of someone who approached the psychedelic experience at first rather as a philosophical curiosity.[91]

Huxley had no or very few preconceptions and his mindset was the one that Zen calls *the beginner's mind*; thus the ideal explorer of an unknown world, at least to our Western mindset. And Huxley's experience was entirely positive. Reading his account, you are thrilled and charmed and at one point or the other seduced to try it yourself. Huxley is very outspoken about the philosophical implications of his experience and he values it positively, so positively that he is cited in every book published on entheogens; he figures almost like an authority while, reading him, one does not have this impression at all. His style is artful, witty and charming, more than in some of his other books, in my opinion.

The book is written from the heart, and there remains no doubt that Huxley loved this mushroom and its hallucinatory compound: *mescaline*. I think it is not a bad idea, for anybody interested in altered consciousness, to read this book as a kind of introduction.

Published by Sirius-C Media Galaxy LLC, 2010

CHAPTER FIVE

A Science of Pattern

Video Presentations

Eight Dynamic Patterns of Living

http://vimeo.com/channels/8patterns

Video Description

In these videos, I am reading excerpts from this book.

Introduction

Shamanism is essentially a science of pattern, and interestingly enough, was so already thousands of years ago while our Western science only know, through systems theory, has begun to see the paramount importance of pattern in the composure of living, and in the functionality of living systems.

I found *Eight Dynamic Patterns of Living* to be present in the lifestyle of most tribal peoples around the world. These eight patterns, *autonomy, ecstasy, energy, language, love, pleasure, self-regulation* and *touch*, are for the most part shunned and belittled as life-fostering values in most dominator cultures, including our postmodern Western civilization.

I am saying that it is precisely because we, as the most economically powerful society of the world do not comply with the eight patterns as basic regulators of life that we are at the border of mass destruction, insanity and ecological disaster.

I began identifying the perennial pro-life patterns in living by firstly invalidating the fake principles that mainstream Western science declares to be the founding concepts of our universe.

To put it more precisely, there was actually nothing to invalidate; I found that these alleged principles were but *intellectual assumptions*, and thus simply invalid as founding principles of life. At the same time, diligent study of the I Ching and the almost daily use of it for divination distilled in me an intuitive understanding of the real and valid patterns that are inherent in all living. I therefore simply call them *patterns of living.*

Let me first of all explain why I use the term *patterns*, deciding to discontinue the use of the term *life principles.*[92] I indeed think that here we are facing a key point that marks the essential difference between *death science* and *life science.*

A pattern is a set of things, a certain arrangement I can make out in the complex scheme of reality, and the main characteristic of this arrangement is that it forms a *relationship* of elements with each other. It is something I can observe. A pattern can be fix or it can be changeable. It can be static or dynamic. By contrast, a principle typically is the start of a down-hierarchy. It's a top-something in a kind of up-to-down order. It is *not* something I can *observe.* Its reality is *merely intellectual:* the outcome of a conclusion I draw in my rational mind *after* observing nature. A principle thus contains my observer point or my judgment about reality.

Death science looks at life through the glasses of principles it has set before it was going to observe. It is essentially blind and proceeds by imposing characteristics upon nature. Western science traditionally has been death science; it gained its conclusions about life by vivisecting cadavers, not by observing the moving changes of living. It is, and remained, a *cadaver science* that is far removed from the changing patterns of reality.

Life science looks at life without any set principles or assumptions and observes the dynamic patterns or changes in the texture of life. It is a science that since its start in China, around five thousand years ago, was interested in life, and thus drew conclusions from life, and not from death. Traditional Chinese science together with most other ancient science traditions of the East is a *life science*, one branch of this very large body of science and philosophy being Feng Shui.

The I Ching is based upon life science, and is perhaps the highest condensation of it. Needless to add that, as such, it is non-judgmental and thus bears no moralistic judgments about human behavior. It looks at human behavior in exactly the same way it looks at all life patterns, and sees the changing nature of it before all. Fritjof Capra in his book *The Web of Life (1996)* explains the importance of pattern when he explores the meaning of *self-organization*, which is a major characteristic pattern of living systems:

Fritjof Capra

To understand the phenomenon of self-organization, we first need to understand the importance of pattern. The idea of a pattern of organization – a configuration of relationships characteristic of a particular system – became the explicit focus of systems thinking in cybernetics and has been a crucial concept ever since. From the systems point of view, the understanding of life begins with the understanding of pattern.[93]

In order to scientifically explore the nature of pattern we need to *change our basic setup of scientific investigation*. Capra explains:

Fritjof Capra

In the study of structure we measure and weigh things. Patterns, however, cannot be measured or weighed; they must be mapped. To understand a pattern we

Published by Sirius-C Media Galaxy LLC, 2010

must map a configuration of relationships. In other words, structure involves quantities, while pattern involves qualities.[94]

This really involves a *radical change* in our scientific thinking because traditionally Cartesian science was quantity-based and measure-oriented, while *systemic science* is quality-based and relationship-oriented, a truth that Capra exemplifies when looking at the properties of pattern:

Fritjof Capra

Systemic properties are properties of pattern. What is destroyed when a living organism is dissected is its pattern. The components are still there, but the configuration of relationships among them – the pattern – is destroyed, and thus the organism dies.[95]

The next important point to understand how nature *thinks* is the cell's metabolism, the network that serves recycling. Capra succinctly elaborates in his book *The Hidden Connections (2002):*

Fritjof Capra

When we take a closer look at the processes of metabolism, we notice that they form a chemical network. This is another fundamental feature of life. As ecosystems are understood in terms of food webs (networks of organisms), so organisms are viewed as networks of cells, organs and organ systems, and cells as networks of molecules. One of the key insights of the systems approach has been the realization that the network is a pattern that is common to all life. Wherever we see life, we see networks. (…) The metabolic network of a cell involves very special dynamics that differ strikingly from the cell's nonliving environment. Taking in nutrients from the outside world, the cell sustains itself by means of a network of chemical reactions that take

> place inside the boundary and produce all of the cell's
> components, including those of the boundary itself.[96]

But the most revolutionary outcome of the systems view is that our usual habit of dissecting parts of a whole for further scrutiny and scientific investigation *does not work* with living systems. Why is this so? Capra pursues in *The Web of Life (1996)*:

Fritjof Capra

Ultimately – as quantum physics showed so dramatically – there are no parts at all. What we call a part if merely a pattern in an inseparable web of relationships. Therefore the shift from the parts to the whole can also be seen as a shift from objects to relationships.[97]

My hypothesis is that Western culture has *never* until now applied the *Eight Dynamic Patterns of Living* and that it *therefore* is *at the border of chaos, destruction or another kind of worldwide catastrophe;* I allege that this culture is suffering from a schizoid mindset, the perversion of love into sadistic hate, rampant violence, the impudent slaughtering of ethnic and cultural minorities, famines that could easily be avoided, and generally a total lack of genuine spirituality which, by itself, already makes for a large part of depression and psychosomatic disorders.

What I say is that the *Eight Dynamic Patterns of Living* have been respected and applied by all major tribal cultures including the North American Indians, and that *therefore* they have lived peacefully. With 'peacefully' I do not mean an artificial *Western* peace concept which is complete nonsense as it is stuck and rigid, but a *dynamic peace continuum* that includes little fights and small wars as required by the dynamics of *yin* and *yang*, but that is so balanced that it will never trigger a major and global destruction.

The fact that Western culture has triggered this destruction in all possible ways, economically, socially, health-wise, militarily and ecologically shows that the *continuum balance* that the *Eight Patterns* provide is completely lacking in Western philosophy, science, military policy, diplomacy, politics and strategy. Western culture has brought about what Wilhelm Reich called *the emotional plague*, symbolized by the atomic bomb's mushroom.

The *Eight Patterns of Living* could be taken as a guide concept for being implemented in a new kind of lifestyle to be worked out as part of our presently evolving postindustrial global culture. That is the basic idea. Besides, I think that the *Eight Patterns of Living* are tremendously useful as a base layer for establishing the ground principles of a new peaceful society, instead of beginning with Adam and Eve and go time and again through all anthropological material. I have actually done this and found that there is no novelty any more in this. The *Eight Patterns* cover all spheres of life and living.

To finalize the present chapter, I shall reprint here an overview over the eight patterns, which is of course incomplete; it is destined to give you an approximation of what each pattern is about. For further study, I refer the reader to both my consciousness guide and my audio book.[98]

1) Autonomy

All peaceful tribal societies have in common that they grant their children an utmost level of autonomy. In *dominator cultures*, that today represent the bulk of large and typically industrialized societies worldwide, the lacking autonomy of the consumer child is a truly pathological phenomenon that often takes the form of co-dependence, which I call *symbiotoholism* or emotional abuse and in general the unhealthy fusionary clinging of members of the family, or as *collective fusion* through the identification with groups, organizations and ideologies. In fact, observing the growth processes in nature, we can see that autonomy is something built in all living, and as such takes part in all growth. In order to realize our personal identity and become whole human beings, we have to be able, still in childhood, to form an original personal identity. This is however impossible if we are reared by narcissistic parents, those namely that are *indifferent* to the unique person of the child they have brought to life.

2) Ecstasy

All peaceful tribal societies have in common that they have a strong *ecstasy pattern* built in their lifestyle which makes them once in a while enjoy group events where the usual rules of conduct are more or less set aside. Usually, these events are characterized by magic rituals, the consumption of mind-altering *entheogens*, that is, psychedelics, and the partial or total disregard for the incest taboo or other sexual taboos.

Published by Sirius-C Media Galaxy LLC, 2010

This principle was wide-spread even among major civilizations; still some decades ago, during the *Carnival in Rio*, it was not uncommon that sexual incest was practiced between parents and children. It is also quite probable that intergenerational sex, while practiced in very few aboriginal cultures, is allowed on a larger scale also in less permissive cultures during *ritual events* that serve to liberate and cultivate individual and group ecstasy.[99]

3) Energy

Life is energy! This is recognized as a vital life pattern in all non-Western societies, and thus the overwhelming part of the world. Oriental cultures were historically the most wistful in *recognizing and applying energy patterns* for healing, good fortune and positive relationships.

The Chinese science of *Feng Shui* is perhaps the oldest distillation of this holistic knowledge into something we today call a *science* while traditionally Orientals tend to speak rather of *philosophy* or of *religion* when they talk about the perennial science of the bioenergy. However, even in the West, alternative scientists from Paracelsus to Reich have acknowledged the existence of the *bioenergetic functionality* not only of the human organism, but also of the weather, the atmosphere and the cosmos as a whole.

While in substance these researchers observed basically the same phenomena, the way they termed the cosmic life energy varied. Paracelsus spoke of *vis vitalis*, Swedenborg of *spirit energy*, Mesmer of *animal magnetism* and Reich of *orgone*.

And since millennia this same energy was called *ch'i* by the Chinese, *ki* by the Japanese, *prana* in India and *mana* with the Kahunas from Hawaii and the Cherokee natives of North America.

Furthermore, parapsychologists universally agree that the motor of all psychic phenomena is to be found in our bioplasmatic and egg-shaped *aura*, an energy body of lesser density that we carry around our physical body and which can be seen as an extension of our bioplasmatic energy, as it is composed of the same bioenergetic charge that we find in the bioplasm.

Emotions are energetic, streaming currents that are a direct outflow of the cell's bioplasm. I speak about emotional flow or, within my Emonics research, of *emonic currents*.[100]

These streamings have their seat not in the brain, as modern psychology wrongly believes, but in the bioplasm and in the aura.

4) Language

Psychoanalysis has revealed the importance of language as a condition for the sublimation of instincts. Furthermore, peace researchers found that a lack of language and thus of communication is at the basis of all forms of violence, inner and outer. This insight has not only psychological but also political consequences. For it clearly indicates that only free speech and democracy, both within the family and the nation, can ensure maintaining peace and regulate our natural

Published by Sirius-C Media Galaxy LLC, 2010

instincts and desires, so that they do not become asocial and violent through denial.

To everyone who says that we have democracy and yet are a violent society, I reply that we do not have true democracy and never had. For violence only comes up when verbal communication is impaired, and the one major reason why communication is impaired about vital issues is *shame*. When we feel ashamed about certain vital events in life, such as sexuality, we do *not freely communicate* about these issues, because we are blocked or inhibited by the nagging feeling of shame that comes up every time we tackle the subject.

Lack of communication straight leads to violence; where the mouth is defended to talk, the body takes over the role of the mouth – and the fist talks! We all know this from history and from private experience, and yet there is little general conscience in our society about the almost holy importance of dialogue, of communication, not only outside, in relationships with others, but first of all inside, in the relationship with ourselves.

Our large civilizations do very little to integrate the *wistful use of language* because they are hardly conscious of the power of the word. Tribal cultures, however, are wiser in this respect and generally dispose of an array of rituals that serve exactly the purpose of what in our civilizations we do within a psychotherapy, putting words on things, events and feelings.

5) Love

All peaceful tribal societies have in common that they follow the *love pattern* and not, as most of the larger nations, the morality principle. The present state of violence within and between our larger civilizations, especially those with *high morality* is in my view the result of despising the love principle and the widespread use, also and especially in politics, of moralism. With other words, it is the disregard of one of nature's highest principles, the principle of *biogenic self-regulation*, that brought about the present state of violence and the lack of love and true care among most of the peoples of the earth.

It is the hypocrite manner of preaching peace and democracy by our false and opportunistic leaders while they tread in the dust natural love like the Biblical serpent.

Wilhelm Reich, in his extensive work on the psychological roots of fascism found that it is the repression of our natural emotions, first of all by prohibiting *our young generations the natural acting-out of their love desires* that brought us at the border of the present abyss of *fundamentalism, persecution, slaughter, genocide, war, civil war and worldwide terrorism.* [101]

6) Pleasure

All peaceful tribal societies have in common that they acknowledge the *pleasure pattern*, for example in the way they educate their children. In planning the child's future, what counts is not the *father's job*, that is the typical dominator position, but the natural inclination and interest of the *child* for

Published by Sirius-C Media Galaxy LLC, 2010

their later profession. By doing this, instead of projecting upon children their parents' wishes and desires, education ensures that every generation provides for children support for what they really are gifted for. The result is both a high level of skill and motivation for profession and career.

It is not surprising that now also in modern nations the *pleasure function* begins to be seen as the main motivating factor for a person's advancement in life. Suffices to read biographies of great and successful men and women to see that all achievement *is a result of desire and persistent acting upon desire* and that there is no better catalyzing agent than biological self-regulation that is based upon pleasure.

7) Self-Regulation

All peaceful tribal societies have in common that they follow patterns of self-regulation or *permissiveness* in the education of their small children, and consequently restrain from inflicting violence in form of physical punishment upon them. The most peaceful of those tribal nations, the *Trobriands* of Papua New Guinea are completely permissive as to children's sex play and free mating games.

8) Touch

All peaceful tribal societies have in common that they are conscious about the *touch pattern* and care for maintaining free body touch among family members, nudity, and abundant tactile nutrition for infants and small children in the form of baby massage, baby-carrying, skin-skin contact, and sleeping naked with children. In dominator cultures, life-denying pediatricians were turning down parents' desire for fondling their children and co-sleeping naked with them.

Now we slowly begin to see the macabre results of the *deprivation of tactile stimulation* in infants. Psychosomatic medicine more and more reveals that our immunity against viruses depends on touch and that lacking touch, especially in childhood, leads to more or less acute *immune deficiency* and as a result to higher vulnerability for certain *lifestyle diseases* such as cancer, heart disease, pneumonia, altzheimer or immune deficiency syndrome.

Furthermore, cross-cultural research has clearly shown that *early tactile deprivation* is one of the major inducing factors for the plague of personal, domestic and structural violence in any given society.[102]

Published by Sirius-C Media Galaxy LLC, 2010

The Autonomy Pattern

Autonomy is fundamental for every being-in-growth. Without autonomy, there is fusion, symbiosis, and dependence. While for certain organisms, such as the human newborn, symbiosis for a certain time is a biological necessity, this symbiosis is time-bound and should gradually give rise to autonomy.

While the natural symbiosis is needed for the first eighteen months of the newborn, it should gradually come to an end after that period. Unfortunately, postmodern international culture is more or less completely dysfunctional regarding this primal movement from fusion to autonomy that should take place, dynamically, in the growth process of the human baby.

What happens is that the necessary biological symbiosis with the mother, *eighteen months from birth*, is neglected for various reasons; babies suffer from a more or less stringent tactile deprivation that will leave scars for their whole lives.

In order to compensate for the lack of care bestowed upon the infant, as a guilt-reaction and for various other reasons, the post-symbiosis condition is not better for the child: instead of growing into autonomy most children in postmodern international culture grow into co-dependence with their parents and caretakers; instead of building a gradually larger extent of autonomy, parents tend to gradually entangle their children in a tight net of stiffening dependencies; this form of *emotional vampirism* is so rampant especially in modern Western societies that I have called it *emotional incest*.

I further argue that the present defamation and persecution of affectionate and nurturing erotic love between children and adults and its manipulatory confusion with child-endangering sexual sadism have their origin in shame and guilt Western society is suffering from because of its deprivatory and dysfunctional childrearing paradigm that endorses and purports emotional and, in a hidden way, sexual incest by holding children dependent, helpless and infantile as long as possible so as to *compensate for the crippled emonic structure of their parents and caretakers.*

This is how an ever new generation of emotional and sexual cripples will raise an ever next generation of dysfunctional water-headed babies that are going to live with a perverted bioenergetic base structure. Such a situation is *shameful* for a society based upon *egalitarian principles* and that pretends to respect the *person of the child;* however, this reality is veiled, because shame tends to bring about defensive and projective reactions. The projections are clearly to be seen in the fact that most mainstream professionals, psychologists, psychiatrists, physicians, psychoanalysts or psychohistorians, without further information pretend that *pedophilia* was something like a *metaphorically incestuous behavior*, and that it *therefore* was offending society's written or unwritten behavior code.

This concern, that has been voiced with particular stress by Lloyd DeMause, founder of *Psychohistory*, appears to be rather of an ideological nature and fails to stand against the very principles it is founded upon.[103] Psychohistory is not a morality codex, but a science that regards world history under psychoanalytical perspective, applying to historical events

Published by Sirius-C Media Galaxy LLC, 2010

and human motivation the psychoanalytical method as it was developed mainly by Sigmund Freud.[104] Yet Lloyd DeMause steps out of scientific objectiveness when he associates pedophile sexuality with incest, concluding that pedophilia was *after all* a *prolongation of incest* beyond the borders of the family. I hold against this argument that it is in itself *incestuous*.

In order to demonstrate what is right and what is wrong here, I have to dig a little deeper and get back to the foundations, not of psychohistory, but of *psychoanalysis*. In fact, Freudian psychoanalysis affirms the sexuality of the child, and newer research has shown that even the fetus is sexual in an auto-erotic manner.

While it is true that Freud took a distance to his student and collaborator Wilhelm Reich because the latter began to fight politically for children's sexual freedom, Freud rejected Reich not because of a divergence in scientific perspective.

Their split was a mere cultural controversy, or, as we would say today, a matter of *political correctness*. Freud was convinced of the child's abundant sexual life, but he thought, as he literally replied to Reich, that *culture primes* and that we had to respect Western society's fundamental denial of children's sexual freedom.

As a result, Freud, and after him the overwhelming part of psychoanalysts, simply blinded out from their professional regard any sexual activity of the child outside of the family.

In fact, when you read psychoanalytic writings, you are quickly overwhelmed by the extreme focus of these people upon *incestuous wishes* of the child or both the child and the parents. The Socratic error here is to assume that this view was *scientific* in any way, while it is truly the consequence of a cultural, and in addition a professional *bias*. The cultural bias is the fact that in patriarchal societies, *natural* sexuality is forbidden for children with the result that *unnatural* sexuality is brought about, mainly in the form of rape-centered pornography, sadomasochism and violent child rape, abduction and murder. The professional bias is the fact that psychoanalysts typically deal with neurotic, and not with sane people.

The final and quite far-reaching results of this fundamental position of mainstream psychoanalysis are the following. The regard upon sexuality is distorted in that incest and incestuous wishes are viewed upon with an exaggerated and unnatural focus that veils the fact that the human being, if raised freely, naturally projects sexual wishes outside of the family. The factual *oedipal touch* of the modern nuclear family and the really widespread problem of incestuous wishes, and factual emotional and sexual incest, is the result of patriarchy's denial of *child-child sex* outside of the family; anthropological research within tribal cultures that give their children full emotional and sexual freedom for copulating with other children corroborates that in these societies incest is absolutely non-existent; interestingly enough, what also is practically non-existent in these cultures is violent crime and sexual dysfunctions, as well as homosexuality.

As modern society says violence is good, and sex is bad, it by the same token says incest is the rule for the child, while sex outside of the family is the exception as it is invariably considered as criminal and allegedly damages and traumatizes the child. This is how patriarchy has put nature upside

Published by Sirius-C Media Galaxy LLC, 2010

down: it focuses upon the sick and dysfunctional and disregards the plain, healthy and natural.

Lloyd DeMause and many of his colleagues in fact seem to suffer from a blind spot as they virtually *project* incest outside of the family, not seeing that they confuse the rule with the exception. Suffices to read, for example, Françoise Dolto, who was a mainstream Freudian psychoanalyst with an outspoken Christian orientation. Dolto was in France and later in life internationally renowned as child therapist, and she expressed in her books the view that the healthy child naturally wishes to find love mates of same or different age outside of the family.

Dolto when asked if she did not agree with Freud that every child had incestuous wishes toward his or her parents, replied this to young psychoanalysts during one of her workshops on child psychoanalysis:

Françoise Dolto

You as psychoanalysts have to deal, in your daily practice, with neurotic children. Of course, neurotic children are incestuously fixated, because the very etiology of neurosis, as we know since Freud, is sexual. So, with this bias in your mind, you wrongly assume that the same was true for the healthy child.[105]

What is true for sexuality is equally true for *autonomy*. A naturally sexual child is *typically more independent* and more autonomous than a neurotic and incestuously fixated child. The frequently observed *clinging behavior* of modern city children, their helpless, infantile and irresponsible behavior, even when approaching puberty, their immaturity in handling sharp or fragile objects such as knives or glasses show well their incestuous fixation, their neurotic blockage and their co-dependent entanglement with their parents, and the *early psychosexual damage* a life-denying and pleasure-hostile education inflicted upon them.

There is a natural striving for autonomy built into every growing life. A child of three years of age needs more autonomy than a child of fifteen months of age. A toddler of eighteen months needs more autonomy than a baby of five months. Many parents ignore that babies, toddlers and pre-schoolers, already before reaching the age of primary school, need to develop autonomy. Many adults believe that children grew through magic shifts, like the one from babyhood to childhood, from childhood to youth and from youth to adulthood. The first shift is believed to take place around seven years of age, the next one around twelve years of age and the final one around eighteen years of age.

Sorry, but this is really speaking about myths. These shifts don't exist in real life as all growth is gradual and smooth. This is why all education should be gradual and smooth. While it is a good thing to have certain initiation rites or ceremonies that mark important steps in the growth of children, these rites are what they are: mark stones that border an otherwise seamless road. I arrive at a mark stone, I see the mark stone, I touch the mark stone, I pass the mark stone, I remember the mark stone. My passing the mark stone is gradual, and smooth in time, and the mark stone itself is of lesser importance than my passing it. What is important is that I constantly grow, that I remain *moving*. We

Published by Sirius-C Media Galaxy LLC, 2010

learn the basic *movement into autonomy* during our *first year of life*, and not later on during adolescence or when we allegedly *turn into* that magic world of adulthood.

I do not want to belittle the important changes that take place in the life of adolescents, and their sometimes passionate focus upon getting more autonomy, nor do I belittle the marking shift from adolescence into final adulthood. But often we observe that especially those adolescents who have rather repressive and possessive parents get onto the obnoxious track and really push it through for every millimeter of increased autonomy. There is a logic in every behavior and adolescents who put high stress on autonomy *have a reason* to do so. The reason is rooted in much earlier years, in the years of babyhood.

There is no alternative to autonomy. To make down or belittle children's need for autonomy is to open the door to emotional and power abuse *large scale*. This form of child-abuse is not perpetrated by the proverbial *stranger*, but by *mothers*, first of all mothers, and more and more also by professionals who are working for, affiliated with, or sponsored by the international *child protection* industry.

The Ecstasy Pattern

What is ecstasy – is it merely a pleasure boost? If it was that, this pattern would have to be annexed to *Pattern Six: Pleasure* and a separate pattern would not be needed.

To avoid a lifeless theoretical discussion on this point, let us proceed empirically and *observe*.

Does a sex orgy bring about ecstasy? I have shortly mentioned in the overview over the eight dynamic patterns above that in fact there were in the past taboo-free events in many cultures, usually one or two days per year where incest, even in direct line, was given a free license. It is historically documented that even in major cultures such as Italy and Brazil this custom reigned still in the Renaissance during the *Carnival in Venice* and the *Carnival in Rio*.

Many tribal cultures have similar rites, but after deeper research on shamanism and having myself gone through a psychedelic experience, I came to the conclusion that ecstasy in the true sense cannot be triggered merely by fulfilling one's sexual wishes, be it wishes that normally remain repressed in our culture. We still are with the pleasure function when we think of orgiastic events, intimately or in public, in group sex orgies; we still have a simple sexual satisfaction, an orgasm, at the end. It may be enhanced through the exotic nature of sex, with children or even one's own children, for example, or even in front of others who do the same, but still, ecstasy as I have found it present in, for example, the Andine *Shuar* culture in Ecuador, is something entirely different than high

Published by Sirius-C Media Galaxy LLC, 2010

sexual fulfillment. While I admit that sexual fulfillment is very important for a healthy life and that we should have as little as possible sexual restrictions, this is not all there is when we look over the fence of mainstream conditioning.

Ecstasy induced by psychedelics is not related to the pleasure function, while in some trips there are actual sexual encounters and strong and satisfying sexual feelings, but this is rather the exception. What I found characterizes the truly ecstatic experience is *religion*, religion in its original and pure sense: a backlink established to our soul and our profound holistic wisdom. It's generally not really a pleasurable experience: it's in most cases an awe-inspiring experience of deep wonder and bliss. And one goes out humbled and wistful as a small child.

Sexual pleasure and orgiastic satisfaction, certainly important for our health cannot really be compared with the wonder and the ecstasy that are the result of a psychedelic experience where *set and setting* are correctly chosen.

Ralph Metzner

It is widely accepted in the field that set and setting are the most important determinants of experiences with psychedelics, while the drug plays the role of a catalyst or trigger. This is in contrast to the psychiatric or psychoactive drugs, including stimulants, depressants and narcotics, where the pharmacological action seems paramount, and set and setting play a minor role.[106]

It seems that the opening of mind is a larger experience generally than the fulfillment of desire in the form of orgiastic pleasure. In many tribal cultures, in fact, both experiences are linked to each other during festivities, and perhaps there is something like one feedbacking upon the other and enhancing the other. Further research is needed here.

All the literature I found relating to ecstasy was exclusively dealing with entheogen-induced ecstasy, or ecstasy as a religious, not a sexual, experience. Therefore, let me report this research more in detail instead of putting up assumptions about a greater vision of ecstasy that I cannot for the moment backup with scientific data.

Terence McKenna, when asked by Jay Levin to define *shamanism*, replied:

Terence McKenna

Shamanism is use of the archaic techniques of ecstasy that were developed independent of any religious philosophy – the empirically validated, experientially operable techniques that produce ecstasy. Ecstasy is the contemplation of wholeness. That's why when you experience ecstasy – when you contemplate wholeness – you come down remade in terms of the political and social arena because you have seen the larger picture.[107]

This is why I engaged my shamanism and entheogens research in the first place: it was for elucidating the *Ecstasy Pattern* within the research project for the present production. But what *is* ecstasy, then? Terence McKenna, in his book *Archaic Revival*, explains:

Terence McKenna

Ecstatic is a word unnecessary to define except operationally: an ecstatic experience is one that one wishes to have over and over again.[108]

Published by Sirius-C Media Galaxy LLC, 2010

McKenna is right in not complicating what anyway can hardly be verbalized. Ecstatic joy is different from sensual satisfaction because it is a *learning experience*. We learn each time about ourselves and the universe. That is why ecstatic experiences are often also called *mind-opening* experiences. But there is more.

When you remember your childhood you may recall a form of excitement that was not related to sexuality, but to some form of wonder, a joy that you could not explain where it was coming from and what it was induced by. I have called it in one of my consciousness guides *feeling good without reason*.

You may remember that as a child at times you felt like flying right to heaven, so happy and excited you were, and most of the time without any apparent reason. Well, I succeeded in reawakening this basic innocence in myself and thus I do experience moments of full and unhampered happiness and ecstasy when they come; and they come spontaneously, without being asked for and without being triggered by any drug. But I know that not many adults have kept their childhood innocence or have achieved reawakening it during adulthood.

Now, let me address a very fundamental argument that most people come up with more or less spontaneously regarding psychedelic experience: they tend to argue that drugs will render you *addictive*. This is true for certain drugs, but not for all drugs. It is true for alcohol, for coffee, for tobacco, for most tranquilizers, for some medical drugs, for some of the hard core drugs such as heroin, but it is *not* true for Cannabis and what we call psychedelics or entheogens.

Shamanism researchers like Michael Harner, Adam Gottlieb, Ralph Metzner, Richard Schultes, Alberto Villoldo or the McKenna brothers have repeated often in their publications that *psychedelics do not render addictive;* regarding synthetic drugs, we should remember that once LSD was a scientific substance, used in psychotherapy, and that the one who discovered it, the Swiss chemist Albert Hofmann, asserted that LSD was not rendering addictive in any way.[109]

Why the drug ultimately was put on the index of forbidden drugs has nothing to do with a dependency problem, but simply and clearly has political reasons. Terence McKenna explains:

Terence McKenna

The solution of much of modern malaise, including chemical dependencies and repressed psychoses and neuroses, is direct exposure to the authentic dimensions of risk represented by the experience of psychedelic plants. The pro-psychedelic plant position is clearly an anti-drug position. Drug dependencies are the result of habitual, unexamined, and obsessive behavior; these are precisely the tendencies in our psychological makeup that the psychedelics mitigate. The plant hallucinogens dissolve habits and hold motivations up to inspection by a wider, less egocentric, and more grounded point of view within the individual.[110]

Most experts on shamanism agree with Mircea Eliade who says that the role of the shaman in shamanic cultures is *to be a manipulator of the sacred, whose main function is to induce ecstasy in a society where ecstasy is the prime religious experience.*[111] The subtitle of Mircea Eliade's study *Shamanism: Archaic Techniques of Ecstasy*; the book is really *the* classic of all classics on shamanism and clearly suggests that shamanism is *primarily a system of techniques destined to bring about and maintain ecstasy*, individually and collectively, as a repeated experience so vital for the tribe and tribal life, and for peaceful social togetherness within the tribe and between tribes, that its importance cannot be overestimated.[112]

Why should ecstasy, then, be so important for peaceful living, for social relations, for peace, for a wistful attitude in living? What is so special about ecstasy, and what is different in ecstasy compared with the satisfaction of desire, or sensual and sexual pleasure? Terence McKenna has answered this question when he said, as I quoted it above, that ecstasy is the *contemplation of wholeness*.

Thus, in simple terms, when we talk about ecstasy, we talk about *religion*, not religion in its perverted function as a system of indoctrination or morality catalogue, but religion in its original sense as a *gnosis*, a gain of knowledge about life, an instantaneous holistic revelation about the *deeper meaning* of our lives, a regard that implies the gain of *direct intuitive knowledge* about the sense of living, the meaning of our lives, and the essential nature of *all-that-is*.

In this sense, there is a synchronistic link between shamanism and the experience of ecstasy as the contemplation of wholeness, on one hand, and paranormal realities such as the fairy world, the religious visions and miracles experienced by sages and saints, and generally, the discovery of *soul* in our daily life, on the other.[113]

Thomas Moore, in his bestselling book *Care of the Soul (1994)*, gives practical guidelines about how to bring about authentic ecstasy within our non-shamanic Western culture; as such the book is of unprecedented value, a precious gift that goes beyond all research reports about shamanism because it gives us, within the framework of our own culture, ways of bringing about ecstasy in daily life and in our multiple relationships with self, other and the whole universe.[114]

A similarly precious guide for realizing the authenticity of our own soul reality, while coming from a quite different point of departure, is Karen Kingston's book *Creating Sacred Space With Feng Shui (1997)*. Karen Kingston's unique talent is an inborn and absolutely stunning inborn sense for the higher dimensions of existence, for all that is invisible to our physical eyes and undetectable by our five senses. The novice reader may be astonished about the authority that this text reveals and the power of the author's approach to Feng Shui that is the pragmatic and direct approach of an experienced practitioner. This book is not simply one of those poetic writings that elaborate a *magic* view of life. It is that also, but it is much more. Behind the beautiful appearance and the refined language is hidden a *hard core* manual that is truly scientific – in the sense of a higher and holistic form of science.

Of course, the representatives of reductionist modern mechanism would question most of Karen Kingston's scientific concepts. But this argument is true for almost all publications about Feng Shui, which is not a collection of ancient myths, or superstition, but a real science.[115]

What Karen Kingston does is exactly to go beyond the limits of a Cartesian science that is based upon today admitted wrong premises about life.

Published by Sirius-C Media Galaxy LLC, 2010

To call Kingston's approach to life *animistic*, an argument that has been put forth also against most of Goethe's scientific writings, and first of all against his *Color Theory*[116], would disregard the deep and intuitive truth that is at the basis of this holistic life philosophy. Karen Kingston who is married with a Balinese and lives several months every year in Bali, gives pertinent information in her book about the ways that Balinese use Feng Shui or *Space Clearing*.

There are in Bali actually two levels of handling spiritual wisdom, a professional level – if I may say so – and a popular or intuitive level. The professional level is since many generations in the hands of the first caste, and especially the *Pedandas*. Here, we encounter a highly sophisticated and informed way of handling spiritual information that is so complex and so deep that most Westerners would only shake their heads when they heard about it.

On the other hand there are the people in the street who, in Bali, it seems, are also wiser as anywhere else in the world. For they, too, have this knowledge, only in a more intuitive and less literary form. Having lived and worked Bali for several years, I understand Kingston's natural affinity with Bali and the Balinese. I could not imagine where else somebody like her could live. It all sounds like a miracle but I am convinced that we will fully understand it once we know more about the complex influences that sound and vibrations have on our aura, on all our seven bodies.

From what we learn from such an experience, it appears that neither intellectual brilliance nor extraordinary talent or knowledge, nor else a specific sense for paranormal realities, but simply total *surrender* is needed to experience the transformation entheogens bring about. McKenna states in his book *Food of the Gods (1992)*:

Terence McKenna

Shamanic ecstasy is an act of surrender that authenticates both the individual self and that which is surrendered to, the mystery of being. Because our maps of reality are determined by our present circumstances, we tend to lose awareness of the larger patterns of time and space.[117]

That the *ecstasy pattern* has never been integrated in patriarchal civilizations is kind of logical. This denial is the inevitable outcome of the power structures in *Oedipal Culture* as I call the patriarchal shift that happened around five thousand years ago.[118] The knowledge taboo is inherent in patriarchal or dominator society and it finds its explanation simply in the fact that a *fundamentally undemocratic oligarchy* always focuses upon *manipulation*, not information, indoctrination, not natural knowledge to keep the masses at stake and itself at power.

It has been established since the beginnings of the 20th century, foremost by the Bulgarian researcher Dr. Edmond Bordeaux-Szekely, and his discovery, in the Vatican library, of the *Dead Sea Scrolls*, the original gnosis of the Essenes, in which sordid ways the Church falsified the teaching of Jesus of Nazareth and the whole of the text of the Bible.[119]

Do we need further proof for the fact that, as a culture, we have gone astray from the original and immediate knowledge that direct perception conveys? Direct perception is the original mode of knowledge gathering used by most tribal

Published by Sirius-C Media Galaxy LLC, 2010

cultures that live in accordance with the wisdom of nature.[120] Among these cultures, most use use entheogens such as *Ayahuasca* to get in touch with the perennial wisdom of plants and mushrooms, beings by far older than our human race.

Of course, in our dominator societies, this knowledge has been largely forgotten because of the knowledge repression terror of the Church's Inquisition for more than one millennium.[121]

Yet, direct perception is the angular stone of K's unique teaching that revived this ancient knowledge for our times, and it has been found by famous think tanks such as Edward de Bono and learning innovators such as Dr. Georgi Lozanov, originator of *Superlearning®*, that direct perception is the *primary mode of functioning* of our brain; it was also asserted that it is our original, unspoiled, and highly effective mode of learning.[122]

After the analysis, let us have a look at how we can integrate our lost Gaia knowledge if we want to. The late Terence McKenna, eminent expert on the matter, was rather skeptical because access to entheogens has become tiresome and difficult with modern society's strict taboo on mind-altering substances:

Terence McKenna

The impact of hallucinogens in the diet has been more than psychological; hallucinogenic plants may have been the catalysts for everything about us that distinguishes us from other higher primates, for all the mental functions that we associate with humanness. Our society more than others will find this theory difficult to accept, because we have made pharmacologically

> obtained ecstasy a taboo. Like sexuality, altered states
> of consciousness are taboo because they are con-
> sciously or unconsciously sensed to be entwined with
> the mysteries of our origin - with where we came from
> and how we got to be the way we are. Such experi-
> ences dissolve boundaries and threaten the order of
> the reigning patriarchy and the domination of society
> by the unreflecting expression of ego.[123]

On the other hand, international culture transformed the world into a village and we can travel to cultures with a still largely shamanic world view, at least in their more remote parts, such as, for example Ecuador. In fact, in 2004, invited by a Polish businessman married with a *Shuar* native, I went to Misahualli, Ecuador for experiencing the religious ceremony of taking the Ayahuasca brew, in a context where set and setting were carefully chosen and not left to hazard or tourism-friendly amateurism.

After this experience with Ayahuasca, that I am going to report further down in this guide, I am convinced that we can *directly access nature's original wisdom* and receive guidance from its universal intelligence, as well as benefit from a power of deconditioning that has no parallel in human history and experience.

Published by Sirius-C Media Galaxy LLC, 2010

The Energy Pattern

All native cultures have an innate sense for the energy-nature of life, all life, not only human life.[124] This fact has long been veiled because of the metaphoric language used by most tribal cultures. For example the amazing knowledge that the *Dogon* in Mali have about the cosmos and even details about certain stars, such as Sirius, knowledge that in our culture only astronomers have, is all coded in a beautiful metaphoric language.[125]

The same has been observed by Terence McKenna regarding the terminology used by tribal peoples to describe phenomena pertaining to the human energy field. In *Archaic Revival*, and with regard to the bioenergetic charge contained in plant substances used for religious purposes, McKenna writes:

Terence McKenna

They are the true phenomenologists of this world; they know plant chemistry, yet they call these energy fields *spirits*.[126]

It has often been objected to the amazing knowledge of tribal populations that shamanic reality was a pre-psychotic or *primitive* worldview that only peoples with an archaic mindset could uphold, but that, by contrast, the Western citizen had a far higher developed consciousness. This kind of ignorant judgments that used to be uttered by anthropologists and psychologists suffering from the usual Christian and Colonial conditioning, fortunately have largely been silenced

by advanced and unbiased research on shamanism and parallel realities; what *strongly corroborates* the evidence of plants
being universal radios of life-related knowledge is the fact
that, contrary to prejudice, Western people *who have had exposure to plant teachers* react to this experience in much the same
way as natives do, and report pretty much the same phenomena as natives report and reported over millennia.

What namely recurs in these *visions* or *insightful journeys*, or
contemplations of wholeness or however one may want to name
these explorations in the plant realm, is the fact that nature is
basically coded in *energy patterns*, these patterns being observable, recognizable and subject to conscious manipulation for
beneficial or for detrimental purposes.

Black magic, sorcery, is explainable as a conscious manipulation of energy patterns for the sake of hurting others,
and from this point of view and after scrutiny it is evident
that sorcery is a powerful school of wisdom, as has been
shown, *inter alia*, by Carlos Castaneda's explorative journeys
with a *Yaqui* sorcerer from Mexico.[127] I will now quote three
references from Ralph Metzner's sampler *Ayahuasca*.[128] Kate
S., a woman artist in her forties, reports:

Kate S.

I had the thought that the reason certain cultural or
ethnic art forms appear is because of the *planetary energy* in the location of the origin of that form, and that
the music and the art were intricately connected and
reflective of the energy of the planetary location of
their origin and the energies which exist there.[129]

I.M. Lovetree, an educator in his fifties, states:

I.M. Lovetree

After the ayahuasca sessions, I feel cleansed within, throughout and all about. I have a sense of having been healed at all levels, especially the physical. The ayahuasca medicine seems to have a special affinity for the gastrointestinal system: it snakes its way through the body, seeking out and eliminating obstructions to *life energy flow.* I sometimes think of it as a form of kundalini, a Liquid Plum'r for the soul. For cleansing and healing, for reconnecting with the vegetable kingdom, ayahuasca is definitely my medicine of choice.[130]

Ralph Metzner, in concluding the reader, summarizes:

Ralph Metzner

The fundamental reality of the universe is a continuum, a *unitive field or fabric, of both energy and consciousness*, that is beyond time, space and all forms, and yet somehow mysteriously within them, simultaneously transcendent and immanent. In traditional Asian religions, this unitive field is variously referred to as Tao, or Atman-Brahman or Tantra (the 'web' or 'fabric' or the 'jewelled net of Indra'). Some Native North Americans refer to it as Wakan-Tanka, the all-pervading Creator Spirit. In the traditional Anglo-Saxon religions of the British isles it was called the wyrd, an invisible network of magical forces. In theistic religions like Christianity, this oneness corresponds to what is called the Godhead, i.e., beyond the personal deity. In the systems language of post-modern science it is seen as an infinitely complex system of interrelationships, or 'web of life'. At the level of the planet Earth, this integrated whole is referred to as Gaia – the name of the ancient Greek Earth Goddess that has become the name of the whole Earth considered as a purposive intelligent living super-organism.[131]

Since we are part of the unified system of interdependence, just like every other being, we can never

> actually be outside of it, as a detached, objective observer. *But since the unified field is energy, we are energetically connected to every other form and being in the universe.* And since the field is also consciousness, this enables us, as human beings, to attune with, identify with, and communicate with any and every other life-form, object or being in the universe, from the macrocosmic to the microscopic.[132]

William A. Carey, a medical doctor who did an amazing research on psychedelic substances[133] to find the missing link between *psyche* and *soma*, equally states that the human body, according to Edgar Cayce's prophetic visions, basically reveals to be an energy structure:

William A. Carey

Cayce said, for instance, that this body we find ourselves in the moment is an *energy structure* and will respond positively or negatively to other energies; that the atoms and cells that make up the body are pure energy and have consciousness of their own.[134]

Published by Sirius-C Media Galaxy LLC, 2010

The Language Pattern

In my own experience with *Ayahuasca*, as I report it in *Consciousness and Shamanism*, further down [135], I made an amazing discovery about language:

> And the intelligence seemed to tell me telepathically that I was *locked in language*, that all my experiencing of life was *conditioned upon language*, and that I hardly ever perceived life directly, spontaneously, as an immediate connection.

I knew that this intelligence was connected to all, was connecting all and was all. It simply was, and it gently invited me to enter this connection, this all-encompassing love that it irradiated.

> The intelligence seemed to wanting to free me from the conditioning I had received through language, and through using language for describing reality.

What this extraordinary experience taught me was that language is indeed, as Freudian psychoanalysis strongly emphasizes, a *part of culture*, so that we can say that there cannot be culture without language. However, what this intelligence conveyed to me was, so to say, the other side of the medal. Language is part of culture, but not forcibly part of nature's original setup. I might express it metaphorically the way that nature's language rather sounds like when, before the symphony starts, the musicians test and tune their instruments by playing the A unison. Culture's language, it goes without saying, is the symphony itself, a written score, well defined, well put in print, well *black on white* – while the tuning of the instruments sounds like a form of chaotic improvisation and is volatile, and up to the moment, not fixated, not written down, not put in print.

Language serves social conditioning, or the other way round: conditioning is in our mainstream civilization done by language, by *verbal* language first of all, and by emphasizing the early acquisition of language skills by the young child. I certainly was conditioned that way, perhaps more than others; my mother told me that I already began to speak at the age of seven months and by the age of twelve months my verbal language capacities were as good as complete.

My mother had been a radio speaker and her language training was very astute – which is a great advantage for me today as a writer. But it also has strongly conditioned my perception of reality. I seem to perceive reality more than most others through language, and less through visual input, which is however the way most people function in our terribly visual culture.

I found that nature-abiding cultures have a much more general and encompassing notion of language. Language is understood primarily not as verbal language in the sense we use it in Western culture but as a notion that encompasses the telepathic interchange with other language realms such as the animal and the plant realms. Language, in this sense, is communication in every possible way, be it by sound, be it by thought, be it by gestures or movements. It is clear that such a holistic notion of language as it seems to be common to

Published by Sirius-C Media Galaxy LLC, 2010

most tribal cultures leads to a more subtle understanding of nature because it is *cross-lingual*, so to say.

When I dispose only of human language, and spoke I twelve languages, I will still not be able to communicate with an animal, or with a plant. The native may also want to master the languages of other natives tribes, but he will for that reason not neglect to master what I may call *the universal language* which is a form of sensitivity that includes telepathy and that is founded upon a deep interest in other realms of living.

Thus, I must ask, what does culture in fact mean, if it is only a system of restrictions, if it seems to be a *stupidity-factor in human development* when I compare it to the much deeper wisdom of nature? Culture that defines itself as distinct from nature will use a distorted language. This kind of culture, which is ours, is created by perverted language, conditioned language and restricted language. In native cultures, which are *true cultures* in the sense that they integrate nature, I observe, the taboo is a restriction of action, but not a restriction of language.

When I see that, I understand that taboos that prohibit language are not only impeding communication, but destroy culture. The prohibition of talk creates human beings that are mute, people who have to use other forms of communication than verbal expression. They use violence. This is the status quo of modern international consumer culture where worldwide terrorism had to come up in order to show even the most ignorant members of society what degree of collective violence we have attained. And we have attained it not by talking but by being and remaining mute. Tribal cultures, in their natural wisdom, know all this and care for the importance of language and the fact to *put words on things*. Within native cultures, taboos such as the incest taboo, as well as taboos that concern the world of the spirits are *do*-taboos. They are not talk-taboos. What is prohibited by the taboo is the *deed*, not the verbal expression of the tabooed behavior. Furthermore, in those cultures there are special festivities that are focused upon humanizing the taboo through its verbalization. Language humanizes the taboo and integrates it into the individual and the collective unconscious.

The spirituality of man is in first line its capability to humanize asocial behavior through verbalization and, doing this, to *create culture*. However, a taboo that represents a mere *non-dit* will not preserve culture, but destroy it, for the tabooed behavior cannot be comprehended as long as it is not humanized by language. Any possibility to communicate it is cut-off by the talk-taboo. Talk-taboos are the result of hypocrisy and undermine the taboo since what is not talked about is done secretly while it is denunciated publicly. Talk-taboos are thus against democracy since democracy implies free speech.

It is for this reason that language education and the support of the young to express their emotions and all what can be said, is so overwhelmingly important for the formation and preservation of culture. This was historically the case in humanistic education which was based on the study of the old languages and the Hellenic tradition. These antique cultures namely had less talk taboos than our today's mass cul-

Published by Sirius-C Media Galaxy LLC, 2010

tures. After the end of the Hellenic era and the beginning of the *moralistic epoch of mankind*, namely under the influence of post-platonic and Christian thought, language was more and more tabooed and the possibility of complete dialog more and more narrowed.

With good reason, psychotherapist Robert M. Stein calls Judeo-Christian tradition *primitive* in the right sense of the word, because it's undeveloped on a level of what I might call 'cultural cognition':

Robert M. Stein

Creative psychological development, individuation, is dependent on spiritual freedom. When we say, for example, a man has a free spirit, do we mean that he freely or necessarily transgresses the imposed manners, mores and taboos of his culture? I think not. But it does mean the freedom to do anything or go any place he desires in the imaginal realm. He is a man who has clearly distinguished the sacral, timeless world from the secular, historical world. He knows he can move with unashamed dignity among the gods and demons which belong to the mundane world. Such freedom cannot occur with a primitive form of consciousness in which inner and outer reality are governed by the same laws and values. In this sense, our Judeo-Christian tradition is primitive in that our thoughts and desires are subject to the same dogma, the same regulation, as our deeds. Spiritual freedom requires a break with biblical tradition and the development of a new form of consciousness – a consciousness which promotes the cultivation of imaginal freedom.[136]

Cultures that prohibit language, such as ours, are in truth no cultures as they are in their very root against freedom and against culture. They are barbarous, authoritarian and tyrannical. Their well-sounding democratic setups and constitutions do not alter this fact and contribute rather to veil this truth. The way to personal freedom and creativity, to autonomy and the detachment from collective categories is only possible through individually building a *culture of language*, the civilization of personal expression. Only on this very personal and individual basis culture can be created for the community, for language is every kind of expression, everything a creative mind may come up with, not only speech and writing, but also every form of art, of expression that leads to communication.

Our asocial instincts and perversions can only be humanized through language; there is no other way. Without language the process that Freud called *sublimation of the instincts* is impossible. In case that a given society lacks language to express asocial behavior, repression takes place. What is not integrated, is disintegrated, repressed and projected. All collective tragedies of humanity were and are accompanied by collective psychosis, a process that breaks through the fragile balance of repression and brings to the surface the primary archaic patterns of behavior. It is the same on the individual level. An ego without language is a psychotic ego and to create collective muteness means to prepare individual and collective psychosis.[137]

The Tao of psychic health, then, not only for the individual, but also for entire cultures is the way of language, of communication and the active use of speech and writing. To realize this, it is not enough to write well-sounding guarantees in national constitutions and otherwise shut up as a *politically correct* citizen. An eminent expert on the importance of

Published by Sirius-C Media Galaxy LLC, 2010

observing the formation of language in the human baby was the late Dr. Françoise Dolto, the famous child therapist from Paris, France, whom I have known personally. I visited her *Maison Verte*[138] in Paris in 1986 and interviewed Dolto after the visit in her apartment. An interesting correspondence followed up to our meeting.

Dolto was a pioneer of *child psychoanalysis*, made unique discoveries in this science and was known for her remarkable if not miraculous healing successes with psychotic children. In my observation, her success was due to a unique and almost *shamanic* understanding of language, which of course includes the direct telepathic language of our subconscious mind when it bonds in dialogue with another human's subconscious. In the daily practice of the *Maison Verte*, a center for parents and children, the difference it marks regarding conventional parent counseling becomes obvious when, for example, the child is greeted first when s/he arrives with their parents.

Children bring their parents! Dolto used to say. The fact that the parents remain anonymous has been a very important detail in the good functioning of the place. They are identified as Jacques' mom or Helene's dad which was a very elegant way to protect the parents' anonymity while creating a language system that is child-focused. Without anonymity most parents would not have come to the group in order to discuss family problems.

Dolto's insistence upon language and truthful communication may sound strange in the ears of people who have never heard about psychoanalysis. To remedy this lack of knowledge, there is no better way than to read Dolto's extensive publications and to learn from her extraordinary penetration of the matter and her wealth of experience.[139]

In the *Maison Verte* it is first of all the child that is spoken to, however not in the usual way people speak to children, but as one would speak to an adult. Babies are addressed as *Monsieur X* or *Madame Y* while they are shown around the facilities when they come for the first time. Françoise Dolto revived an old tradition from the European aristocracy. In fact, more than two hundred years ago it was common in aristocratic families to address even small children with the same forms of politeness one addressed adults. Consequently children *are not kissed or fondled or found 'cute'* when greeted by the receptionists. Tenderness, such is the philosophy, has its place only in relations between parents and child. The verbal communication with children is painstakingly truthful and serious, and no lies are told to children. When a child, for example, says it had no father, because the father left the mother, the center assistant would insist:

> – But of course you have a father, although you do not know him. Both your parents have created you in their loving embrace, your mother and your father. Your father has participated in your procreation with his body. You see that you certainly have a father.

No truths of life are hidden in front of children. By the same token, assistants are specially trained to avoid allusions that are unclear and subject to misunderstandings. Facts of life are named and explicated. Even though parents are discussing intimate matters, children are not complimented out of the room if they want to stay with their parents. Nothing that concerns the child is discussed behind his or her back, or over their head.

In many cases nothing but the respectful, rational, pragmatic and fearless way of handling relationships in the *Maison Verte* leads to solutions in behavioral problems with parents and/or children; fixed roles are thus easily dissolved and more flexible and creative forms of behavior can be learned.

The press often spoke of miraculous changes with people who attended the house over a certain period of time.

One of the main goals of the *Maison Verte* was to prepare toddlers and small children for day care. Françoise Dolto, before she opened the center, had come to the insight that an immense number of children are *traumatized* by prematurely entering day care. Not only the child needs to be prepared, Dolto found, but also the mother. The sudden separation from the mother four long hours of day care can represent a great shock for a small child.

Contrary to most day care institutions where the transition of the child into day care is facilitated by a gradual extension of the time span the child passes in care, the *Maison Verte* practices a different approach. Not the adaptation for separation is the decisive factor, but the *maturity of mother and child* for the separation. Often, Dolto found, the children were ready for it, but the mothers not. The result was that the mothers gave to their children highly contradictory messages which were triggering emotional disturbances.

For example a mother would affirm to the child that day care and consequently a separation was needed, but telepathically suggest to the child that she herself was not ready to stand the separation without immense emotional suffering. On the other hand children were found to be not mature yet for the separation when the mother had firmly decided to give them in daycare.

The *Maison Verte* insures that both, mother and child, are ready for the separation. This is of special importance within the current nuclear family structure where children are almost exclusively fixated upon their parents.

Published by Sirius-C Media Galaxy LLC, 2010

The Love Pattern

Culture and Pleasure

The present state of violence came about through wrong relationships, the sacrifice of love and the upsurge of 'moralism', that is, compulsive morality and collectively regulated sexual behavior[140], and first of all by a deficient or totally lacking relationship of the inner parts of the psyche to each other.[141]

We do not just have violence. The problem is that violence multiplies because there are multiple causes that trigger violence, and they are adding up to each other. Succinctly speaking, if our problems were originating just from one single source, we would have since long solved them. As a race, humans are presently learning that most effects have more than one cause; the idea of a *straight line linking one particular cause to one particular effect is infantile.* Yet our mainstream science establishment is today still based upon this myopic view. Only training our systemic and whole-brain thinking capacities can help us further.

Pleasure-Denial and Violence

Violence is the result of a *power vacuum* that comes about through an inner fixation or complex within the *lower self* that acts as a compensation to suffering early in life. People who are in touch with their inner truth and who are liberated of culturally created fear blockages are able to realize constructive personal and collective happiness and they tend to build meaningful relationships.

Our love choices depend not on what our crime laws stipulate but what is ethically sound and viable. More and more, it will be possible to make responsible love choices for relations that are unusual or even tabooed by former moral laws that belonged to the cultural heritage of the Pisces era.

In that era of our collective past, happiness was smashed by multiple nonsensical ideological doctrines. It is these doctrines and the fierce dogmatism they violently enforced that created the almost chaotic state of violence that we face today in the world.

One of the main objectives that flow from this insight is the urgent need to redefine *as natural* all erotic longings and sexuality for all ages through information and social reform. Part of this endeavor is to publish the well-hidden facts about the roots of violence and the current subtle manipulation of the credulous masses into a collective *hyper-violence paradigm* that will result in a more gigantic destruction that was ever seen before on the globe.

Compulsory Sex Morality

Among the main reasons for violence being the repression of natural body pleasure and free love between people of all ages in general, and the child's free sexual life in particular, my task was to retrace the wrong turn that humanity has taken since prehistory and to embed this truth in a cross-

Published by Sirius-C Media Galaxy LLC, 2010

cultural perspective that is focused upon the importance of love as a major factor of human evolution.

Compulsory morality or moralism has been the *major killer app for love* during all ages; it is moralism that brought about most of the current violence and destruction all over the globe. The pioneering work that Wilhelm Reich has done in this field of research is of paramount importance.

The essential truth gained from years of research on the functional processes of life is that all parts of the psyche must be given a voice so that a constructive inner dialog can be set up.

Abuse is ill-defined in our culture. It only considers the victim and not the abuser. However, the abuser is a victim in as much as the person he or she has victimized. In fact, any other than nonviolent and consenting love and sex interaction between two people, regardless of their age, simply is a lack of information and still more a lack of physical love experience. In addition, it is true that nobody can be victimized who has not previously *chosen to act as a victim* in a given situation. The abuser is trapped by the victim's paradigm in as much as the victim is trapped by the abuser's power problem. Both attract each other and there is no abuse without mutual implicit consent about acting out the two sides of abuse, the active and the passive one. Fighting against abuse is therefore not a moral cause but must start from a rational and two-sided view of the problem as an *entanglement situation* that is karmic and inherent in both parties' life matrixes.

Moral wars, by contrast, only lead to more confusion, more destruction and more abuse. For they do not tackle the roots of abuse which are the same roots as the roots of violence, but only are concerned with the reflections that such shortcomings produce on the surface of society. They are for that reason entirely ineffective and superficial.

A true remedy can only come from *tedious study* and observation of all the factors involved in abuse and those factors *are for the most part unconscious elements of consciousness*, entanglements that are hidden in the psyches of both abusers and abused, energetic blockages that have cut off the stream of life in one or the other way so that parasitic patterns came about.

There is thus an urgent need to change the reigning paradigm regarding love and abuse so as to reduce violence and to bring about positive change for constructive new relationships that are based upon the *golden rule of conduct* as it is taught by sages since times immemorial. It is to be seen in what horrendous ways both clerical and politically fascist movements and leaders have since centuries tried to veil this essential truth and thus spread the *emotional plague* all over the globe.

After extensive research on mythology, particularly the writings of Joseph Campbell[142], I saw how the present love-killing paradigm has come about from ancient times and how it was possible that the former love-based world order was overthrown and violently eradicated by a world order that replaced love by morality and natural care by obligatory and institutionalized forms of family relations.[143] Historically, the transition in human prehistory from peaceful and life-affirming matriarchal fishing-farming cultures to violent and

Published by Sirius-C Media Galaxy LLC, 2010

life-denying hunting-killing patriarchal cultures is of particular importance for the understanding of the present macho and hero paradigm with its strong Puritanism and all its life-denying and sex-repressing dimension.

There are important humanitarian consequences of this insight. Special care must be bestowed upon children who can be reformed and healed from biopathic deformations and character armors. There are millions of orphans in state institutions all over the world; if a small percentage could be taken care of in collaborating with responsible institutions that understand the importance of the love paradigm, humanity would be served as a whole and true evolution would be possible. The spiritual or religious impact of this project is obvious. It would lead us back to our divine origins, through reconnecting to our higher self. This is true for all involved, the children, the educators and all those who help building this new worldwide educational system.

There were a great number of publications appearing in those years of the *hippie generation* that reported about experiments with freely raised children. What now seems highly disturbing in the *New Age of Fascism* that we are presently entering is the fact that those children were astonishingly mature and ready to assume appropriate responsibilities, highly flexible and adaptable to sudden changes in their life milieu, intelligent and independent.[144]

Political restoration has spread its jovial and patriarchal wings over good old science, wiping under the carpet what does not fit in the reigning restrictive and somewhat paranoid worldview. *Sexual permissiveness*, once a big word, was and is a myth in Western cultures![145]

Anthropological Evidence

In his book *The Invasion of Compulsory Sex Morality (1971)*, Wilhelm Reich referenced and discussed Malinowski's field studies on the Trobriand Islands in Northwest Melanesia, one of the few still existing matriarchal cultures.[146] In fact, as early as in 1929, Malinowski published his report on the sexual life of the Trobriands in which he draws the reader's attention particularly to the *sexual life of children and adolescents.*[147] Malinowski found, not without surprise, a *high sexual permissiveness toward children's free sexual play*. More generally, he noted the total absence of a morality that condemns sexuality in children.[148] Instead, he observed, that children engage in free sexual play from early age. Initiatory rites were absent with the Trobriands since children were initiated from about three years onwards, generally by older children, in all forms of sexual play.

This play is nonviolent and encompasses, with the older children, complete coitus. The most interesting finding for Malinowski was that in this culture *violence was as good as nonexisting* and that there were as good as *no sexual dysfunctions*. Trobriands were found to be almost ideal marriage partners and divorce was a rare exception. Violent crimes were nonexistent and incest was strongly tabooed and inhibited by social norms.

Published by Sirius-C Media Galaxy LLC, 2010

Other researchers found similar phenomena with the *Muria tribe* in South India where children stay until their maturity in so-called *ghotuls* where they live their sexuality freely and in utter promiscuity. Older children initiate younger ones progressively into sexual play. These researchers found that after a phase of promiscuity, the children, from the age of sexual maturity, form strong bonds and durable partnerships based not on sex, but on *love*. They further found that these first steady relationships form the basis for later marriages that, regularly, last life-long.[149]

Some researchers and sociologists allege nowadays that these findings had 'no significant meaning' for our culture since they could not be extrapolated from their original cultural setting. However, such arguments assume that the human, depending on his or her cultural conditioning, was basically different from one culture to the other. This is questionable, for the *biological foundations are with all human beings the same*, regardless of cultural or social conditioning. If all anthropological or psychological insights were valid only in a given culture, how could psychoanalysis which was founded by Sigmund Freud in Austria be successfully applied in Italy or France, in the United States or even in India or South America? The truth is that those critics hide their own emotional blockages and blind spots behind pseudoscientific arguments. One cannot disregard the extensive field studies of experienced anthropologists such as Malinowski or Margaret Mead or wipe them under the carpet with half-truths and moralistic philippics as it now is the trend, especially in the United States.

Historically, the love pattern was integrated and lived by the majority of tribal cultures, but never was accepted by any of the larger dominator cultures that today form the core of our industrialized nations on the globe.

My thesis is that the destructiveness of civilization *is the result of the repression of the natural emotions of the child and the building of moralistic behavior structures* that have gradually replaced the primary self-regulatory processes that nature has coded for the growth of all living.

Love Osmosis

Violence and destruction that characterize human history have their roots not in a biological or genetic error, nor otherwise in the human setup, but generally in the failure of man to keep in touch with nature's wisdom, and in particular by replacing love with compulsive morality and perverting children into obedient robots that have repressed their feelings in order to survive and be accepted. Examples to the contrary, as already mentioned, are the pre-patriarchal high cultures of Antiquity, the Trobriand and other rather remote island cultures, and some more well-known cultures such as the Balinese culture, where people are generally emotionally balanced, happy and productive, loyal and intelligent.[150]

Crime rates in those cultures, if we take only the Balinese culture as an example, are extremely low. Violent crimes such as murder and rape, or the rape and killing of children, are as good as non-existing. Marriages are generally long-lasting and divorce rates are considerably lower than in modern so-

Published by Sirius-C Media Galaxy LLC, 2010

ciety. These cultures are *more matriarchal in character* than the highly violent modern civilizations which are predominantly patriarchal. History reveals that already the first highly developed civilizations, such as Sumer, Babylon or the Inca culture were early patriarchal systems. With patriarchy began the oppression of women and children and the reduction of sexuality toward certain acts that were allowed and certain others that were prohibited. With the increase of power for the patriarchal system, repression, denunciation, intolerance and violence began to reign where before freedom, peace and tolerance were blooming.

An important factor within this process that keeps worsening until today is the repression of the child's emotional life.

Love versus Morality

As already mentioned, *Bronislaw Malinowski*, a renowned anthropologist, found with his field research on the *Trobriand Islands* in Papua-New-Guinea that this tribal culture has created social institutions that support the *free development of the child's intrinsic sexuality.* This is through maintaining special houses for children and adolescents. From age three, children stay and sleep in these houses, together with other children, and live their child sexuality freely and in total promiscuity. The older children gradually initiate the younger ones into sexual relations, until coitus.

The Trobriands think that the child must live his or her inborn sexual drive in promiscuity in order to be able, after puberty, to form steady and stable relationships with a partner for marriage.[151] Malinowski was astonished to see that with the Trobriands marriages were indeed stable, the divorce rate being below five percent.[152] Regarding violence and crime, it was virtually non-existing on Trobriand.

The way back to love can only start from the destruction of *any and every concept of morality*, be it justified religiously, ethically, culturally or scientifically. If we are to regain psychosomatic health and sanity, we have to stop thinking negatively about nature and accept our humanity without flinching, and also without the senseless zeal *to improve what is as it is.* Without morality concepts, we again have to face the crude reality of *suchness* in everyday life. Instead of escaping in our *comfortable realms of artificial duality*, we have to practice acceptance when we realize that nothing is *per se* black or white, bad or good, unless thought declares it so. Where there is morality, love cannot be. Moralism and love are mutually exclusive as violence and pleasure are mutually exclusive.

Moralism is associated with violence, love is associated with pleasure. Love and pleasure are the original ingredients of life, whereas morality and violence are decay products of the sacrifice of love to the gods of respectability and materialism.

The price that civilized humanity pays for the destruction of love is high, as high as the price of life. Life on earth depends on our turning back to the wholeness, humility and permissiveness of love, abandoning the fragmentation, arrogance and rigidity of morality-based thinking and acting. As all research shows, love brings about natural growth, prosper-

Published by Sirius-C Media Galaxy LLC, 2010

ity and happiness without effort and without strife. Most if *not all our current worldwide problems* could be solved peacefully and without conflict if we abandoned moralistic attitudes in politics and based our strategies on love and cooperation.

Instead of rigidly adhering to static concepts, we would understand the *unendingly dynamic and flexible* way by which nature acts and just like nature, we could bring about positive and constructive changes in every moment.

What I call the *Love Pattern* is by no means an intellectual concept, nor is it an invention of mine. What I propose is looking at life with the eyes of life, instead of looking at it through the myopic glasses of morality. Anthropological evidence and cross-cultural research has clearly demonstrated that contrary to psychoanalytical doctrine, the Oedipal confusion is not a universal psychosexual necessity in the development of human sexuality, but the inevitable result of love-denial in early childhood and the unnatural erotic fixation of children upon their parents.

What modern-day *child protection* does is advocating emotional incest, and thus what it gives to our children is entanglement, not freedom. Children who enjoy their sexuality by having sex play with other children and adult mates are not confused about their emotional attractions and sexual preferences.

Much to the contrary, these children have a mindset that, compared with their virgin comrades, is based upon what Freud called the *reality principle*. Under the definition of modern child psychology, they would have to be considered more adult than most of our current adults that are raised like infantile idiots. They are in fact more responsible, more considerate, more socially minded and less selfish than the prototype of the castrated consumer child.

In my discussions of this topic, I regularly get the question why, if this was so, society was so hard on admitting and accepting the existence of *pedophilia* as a natural counterpart to children's own love wishes? My answer is that this society is *so deeply ingrained in incest in every possible form* that admitting children's emotional and sexual freedom would jeopardize the expectation of most parents to be the foremost and exclusive love mates of their children. There are many people in our culture who think that allowing the child *to be sexual* would naturally imply children having sex with their own parents. When I object that children, if let free to choose their sex partners would in most cases not opt for their parents but for peers or adults other than their parents, people seem to be puzzled in a really interesting way. I namely see in those moments that they are more puzzled and disturbed by the idea that children have sexual relations with strangers than with their own parents.

They are in fact more sympathetic to the idea that children have incestuous relations rather than free choice relations outside of the family web! And this is, sorry, really a perverse idea and it shows the deeply perverse base setup of our patriarchal society that considers children first of all as pleasure toys for their parents and extended family, and only *afterwards* as people like you and me who have the right to choose their partners instead of complying with the unwrit-

Published by Sirius-C Media Galaxy LLC, 2010

ten law to serve as kiss cuties for consoling sexually frustrated elders that happen to be their parents.

After all, let us ask the pertinent question: Why should parenthood grant sexual privileges? My guess is that when one day we have enough discussed, publicized and mourned about rampant incest within the modern nuclear family, and all the cards are on the table, the taboo on pedophilia will be lifted, because it will be seen that it's after all more natural to have children *copulating with choice partners* than to tie them as sex dummies to their parents.

It is noteworthy to observe that people who reject children's right to be sexual and to live choice relations rather than being night pillows for their neurotic parents *are often sexual virgins* and as such sexually as little experienced as most consumer children. In addition, by actively defending incest, such individuals may unconsciously strengthen the repressive *child-abuse paradigm* of the majority and defend parental interests in controlling and manipulating children, instead of serving the *true interests of children* for autonomy and self-regulation in love and sexual matters.

Abandoning morality and returning to love will not result in upholding incest, except in the rare cases that the sexual interaction between parents and children is mutually agreed upon and shared as a conscious bonus in the parent-child relationship. In the regular case, incest serves but the parental interest for affirming and re-affirming control and dominance, and leaves the child little space for yes-or-no decisions. Whatever one may think regarding this subject, nature has given us *millions of potential love mates*, and the moment we choose sex *within the family* instead of *sex within the world*, we show that family life has an undue dominance over us and that we have not made the step from the cradle into the world – which is a world of free choice and not a world of freedom within a lion's cage.

Interestingly, outside of the Western world, I never met a child who upon my question affirmed of wanting to engage in, or having engaged in any sexual affair with a parent or close relative. To most of these children, the very idea of incestuous sex is clearly negative or not even occurred in their mind because they were busy and satisfied with love and sex relations outside of the family.

With prostitute children in Asia I found that these children clearly distinguish between the relationship they maintain with their father, on one hand, and that they wish to be platonic, and the erotic relationships they maintain with their male sex clients, on the other hand.

Incest is nothing bad in my view, nor do I believe that it's per se immoral. And yet it is *not a viable option* when millions of other love partners are available. The deep unhealthy co-dependent attachment that *emotional incest* typically results in is clearly counter-productive to building autonomy and personal strength in our youth. Instead of advocating more incest, we should advocate more erotic love options for our children *outside of the family*.

Published by Sirius-C Media Galaxy LLC, 2010

Rebuilding Trust

Abandoning morality and returning to love means that we begin to trust the dynamic self-regulatory processes that are built into all our life functions, and our natural biological relationships. As most Westerners have desperately lost their continuum and with it their innate trust in the goodness of nature and natural living; this is the crucial point for most of them. As I was just like most of them some two decades ago, I know what I am talking about.

But I also know that this trust can be regained, even if we experienced a deeply negative and humiliating childhood and youth. Without trusting nature, we do not know what it means to trust another human, and without this basic trust, love between humans is impossible. It is as simple as that. All our senseless rhetoric about *child abuse* shows that we do not trust nature and therefore mistrust the natural self-regulatory wisdom of love.

Despite abundant research showing that one of the results of this mistrust is violence, we continue to put our trust in authority and the nonsense of tradition instead of trusting what is in front of our eyes. All our sages told us that draconian laws only show that the natural law of love has been lost, but we continue to put every year more people in jail than we liberate from our prisons thus discarding out a steadily growing part of our societies, thereby diminishing or even destroying their humanity and our society's need for human potential in freedom.

How can we continue to declare more and more people socially inept while we as a society engage worldwide in wars and massacres that defy any of the atrocities that we blame our prisoners with? Building trust is the first step when we are serious to regain love as our foremost regulatory life principle. Here, our scientists and our poets for one time agree that this move is a good one and that nothing can defeat us if we put our trust in the natural intelligence of love.

Published by Sirius-C Media Galaxy LLC, 2010

The Pleasure Pattern

An intelligent society tolerates to a certain extent, and socially codes, perverse behavior in the sense that it considers *violence as the only true perversion*, while it can comprehend that unnatural sex ultimately is human. That is why most of the ancient wistful civilizations that were not perverted by the plague of moralism had social outlets for pedophile, incestuous and other marginal sexual longings.

The human nature only knows *one* taboo, that is, doing harm to another. And violence is what most of the time causes harm, not perverse sex. This is simply so if present-day moralists and world puritans accept it or not. Culture is not always in accordance with nature; and the human being has the unique ability to forge culture that obeys to its own *cultural laws* instead of abiding by the laws of nature. I do not say that this should be so, and I do not judge what ought to be or not be in terms of human behavior.

What I do say is to observe and describe what we see; and what we see is that culture can *well be different from nature's wisdom*. Our patriarchal culture is against nature and yet it has survived until this day. But it will ultimately probably not survive because *it's not intelligent to build a culture that turns natural laws completely upside down*.

Yet I think that our admittedly flawed patriarchal culture, had it not a definite hangup with fascism and fundamentalism, and had it integrated the power of the female and tolerated sexual perversion as the *little crime* in its various forms as unnatural and incestuous sex that serves to defend the *big crime*, that is structural, political and domestic violence, could survive because of its ability to flexibly adapt to new situations and environments. Françoise Dolto was noted to say that the fundamental difference between perversion and neurosis is that the former one day is dropped by satiation, as the energy in perverse behavior is to be naturally exhausted one day, so that the perverse habit is being dropped, whereas neurosis only worsens over time if no therapy is engaged.

This is so because in perverse behavior there is still a good portion of nature's flexibility, which is why the behavior is not rigid and can change. The same is not true for neurotic behavior. The ultimate issue, then, with patriarchy is not that it's perverse and against nature, *but that it's neurotic!*

In fact, there is a natural logic in every pleasure-seeking behavior. For example, men who seek complete love relations with little girls or boys, while mainstream society considers their sexual longings as perverse, can one or the other day outgrow their immature sexual attractions and *grow up* sexually, and as a result turn to older partners. Or they may not, considering their admittedly childish sexuality as a form of poetic lifestyle. And it is. You cannot copulate with a 2-year old the same way you can copulate with a 20-year old. Your sexual behavior *will thus be naturally restricted by the smaller size of the child's genitals,* and if you want to bear up with this restriction, you are either a fool or you accept this limitation for the charm that you derive from mating with sexually immature mates.

Published by Sirius-C Media Galaxy LLC, 2010

This is the reason why pedophilia, while its etiology may be known and understood, is generally not considered an *abomination* by conscious pedophiles themselves, while society well considers it that way. This is because the pedophile will not define his or her behavior with its sexual part only, or based on their sexual longings, but within the greater context of their *emotional attraction* to children, and the poetic world that they build around their relationships with children. And for good reason! Our society has namely destroyed much of this poetic world that not only pedophiles love and cherish, but generally sensitive men and women, and particularly our gifted poets, writers, musicians and geniuses who most love in life what is genuine, innocent, young and exuberant!

Sexually engaging with a person who is under a certain age defined by the law is not a divergence from the behavior code that is so fundamental that it should be demonized socially and considered criminal behavior. Succinctly speaking, societies, like present-day international consumer culture that developed into a barbarous horde of judgmental persecutory terminators will pay a high price for their lack of *erotic intelligence;* they will destroy themselves in the long run by breeding such a *high level of structural and domestic violence* that they will one day suffocate in it.

There is only *one* perverted form of pleasure: it's violence. Violence is the only true perversion in the human setup, while all sexual longings, however they may be considered by a particular cultural setup, are still connected with the healthy base layer of living, because they are not violent per se. These desires only *become* violent through their individual and societal repression because repression distorts and perverts the strong *élan vital* contained in these desires that, when unrepressed, is partially sublimated into art and poetry.

However, as I said already earlier, sublimation only works when consciousness allows the acceptance of the desire, and not when consciousness represses the longing by judging the particular behavior bad and socially destructive. To repeat it, pedophilia is not socially destructive but its very repression is. While pedophile desires in their sexualized form may admittedly not be the result of a healthy upbringing, and while they may be considered forms of pathological sexual wishes, *this does not mean that they have to be criminalized*, and that they have to be repressed. As long as they do not turn into violent rape desires and are not acted out as child abduction, rape and murder, they should be socially tolerated and coded as acceptable, while marginal forms of erotic social conduct.

Our society does not follow the precepts of *erotic intelligence*, that's the problem here. I would say it creeps along the lines of emotional stupidity, and this especially since the last three hundred years of so-called *enlightenment*, during which it has mechanized and robotized pretty much everything that was filled with soul and a certain poetic content – such as, for example, relationships with children.

To be true, today these relations, in mainstream culture, are neurotic, standardized and idiotic, hypocrite through and through. And they are dominated by a denial of the body, by a denial of touch, and by a denial of truth, and of truthful language. This denial is based upon amnesia, the amnesia namely of our own childhood, with its insecurities, its dan-

Published by Sirius-C Media Galaxy LLC, 2010

gers, and its secret erotic adventures. Most of us, in a deeply neurotic society, have forgotten that our bodies were the first and certainly the most natural source of pleasure. As a result, most adults in industrialized societies live lives that are not their own. Alienated from our bodies, we compensate for the lost paradise of *Being* through *Having*, possessing, consuming, to paraphrase Erich Fromm.[153] The things we are attached to *give us a fade ersatz* of the joy we could have if we had kept true to what is freely given by nature. This dilemma begins in early childhood.

The luxury of civilization costs a high price. We pay for it with our bodies that we gradually destroy. A body that is not connected to a soul is a dead body. If the soul is still in that body, death occurs while we live, in the form of alcoholism, drug addiction, heart disease, rheumatism, leukemia, cancer, hiv or other so-called 'lifestyle diseases'. The process of alienation that leads to this gradual decay of the human body is an integral part of the conditioning for the consumer society. Consumption that makes the prosperity of this form of human social life, functions only if people actually consume – which means that need to be alienated from their bodies virtually from babyhood.

This is not a communist view or opinion, but a socio-economic and verifiable fact. Modern society is organized that way, if we personally agree or not. One of the most important conditioning devices are children toys. The toy industry sustains the entire structure of the consumer society. Without the early conditioning toward toys *as a body pleasure ersatz* people would not accept the later ersatz satisfactions that they receive for the sacrifice of primary body pleasure.

What is primary body pleasure? It is the pleasure that already the small child derives from playing with the body.

This pleasure is an essentially sexual pleasure. The moral and societal prohibition of child sexuality is the primary condition for the functioning of a civilization of *ersatz satisfactions*. This fact explains the research results that prove a direct correlation between the civilization standard of a given society and the severity of its child sex taboo.

However, most scientists who have arrived at these conclusions have failed to see the impact of this insight upon human civilization and culture. Freud thought that man develops culture through sublimating primary body pleasure, and thus not through the satisfaction of our innate desires; he assumed that culture was the outcome of a transformation of original libido energy into a form of creative energy that serves cultural purposes. Is this thesis true?

I think it is true, but not in the sense Freud thought it was. It is true insofar as the prohibition and transformation of instincts leads in fact to a form of culture, an *ersatz culture* for the original culture that would have been created through living our original instincts. But it is not true in the sense that sublimation led to a *true and original culture*. After deeper analysis, we can observe that the denial of living our original desires leads to a denial of living our original creativity.

The result is exactly the *fake culture* that today Western consumer culture represents.

Published by Sirius-C Media Galaxy LLC, 2010

By contrast, if we look, for example, at *Minoan Civilization* that did not repress sexual pleasure, neither in children nor in adults, we see how high human civilization can grow on the basis of an originally permissive attitude toward our primary instincts. In many ways Cretan culture, which was not yet a patriarchal, but an *egalitarian culture*, was superior to our modern culture, more developed, more refined and, first of all, more peaceful and more sensitive toward our real human needs.[154]

The sudden and brutal end of this immensely creative culture through the invasion of the patriarchal Phoenicians was one of the turning points in human history. It was this turn into patriarchy and violence that put humanity on the track of pseudo-culture; it was from this time that the artificial and hypocrite, the stupid and doctrinaire, the false and arrogant, together with violence, war and destruction began to dominate the natural and naturally intelligent original cultures that preceded them.

All our great religions have absolved and fueled this turn into a *manipulative and undemocratic pseudo-culture* that still represents present-day mainstream culture. Religions played the role of a catalyzer in our conditioning for war and destruction – although they generally paid lip service for the contrary.

It is significant that tribal cultures who put the human body and body sensitivity in the foreground of cultural, artistic and social life and where love is not faked, do not need to 'preach love', do not need to discuss love and do not need to heal love – simply because they *practice* love. Their religion is not the integrity of pseudo-moralistic values, but the integrity of *love*. Religion, in these cultures, is not a power factor and does not feed on power nor does it exert power over individuals. Religions, in these cultures, do what they should do in accordance with the true sense of the word *religio*, that is, they give guidance to those who are searching for the truth about coming and going, transcendence of suffering, care for the sick, the needy, the marginal and the dying. The North American Indians, for example, have preserved forms of this original and most pure religion that was once universal for all human beings.

Herbert James Campbell, an English neurologist, found in twenty-five years of research a universal principle which dominates our brain: *the pleasure principle*. This sounds like Freud, but it has little to do with psychoanalysis or psychology. What we are facing here are facts proven by natural science, by neurology.

Campbell' book *The Pleasure Areas (1973)* represents a summery of years of neurological research.[155] Campbell succeeded in demonstrating that our entire thinking and living is primarily motivated by pleasure, not only as tactile or, sexual or sensual pleasure, but also as intellectual or spiritual pleasure. With these findings, the old theoretical controversy if man was primarily a biological or a spiritual being, became obsolete. For it is in the first place our striving for pleasure that induces certain interests in us, that drives us to certain actions and that lets us choose certain ways.

During childhood and depending on the outside stimuli we are exposed to, certain *preferred pathways* are traced in our brain, which means that *specific neural connections* are established that serve the information flow. The number of those connections is namely an indicator for intelligence. The more of those preferred pathways exist in the brain of a person, the more lively appears that person, the more interested she will be in different things, and the quicker she will achieve integrating new knowledge into existing memory.

High memorization, Campbell found, is namely depending on how easily new information can be *added-on* to existing pathways of information. Logically, the more of those pathways exist, the better! Many preferred pathways make for high flexibility and the capacity to adapt easily to new circumstances.

Campbell's research indicates that the repression of pleasure that is part of our Judeo-Christian culture, has negatively infringed upon human evolution and impaired the integrity of psychosomatic health. This is exactly what Wilhelm Reich found – without having had at his disposition Campbell's later neurological findings.

Not only neurologists such as Campbell have thought about the basic functions of life and living, but also people who were formerly active in totally different fields of science. The American scientists Ashley Montagu and James W. Prescott came from different research positions and perspectives. Montagu wanted to know why in animal experiments small rhesus apes died when they were deprived from their mother while they survived when a simple felt mat was put in the cage as surrogate of motherly tactile affection.

Prescott researched on the roots of violence. He did from the start not acknowledge the age-old myth that man was *per se* a violent creature even though human history, or what historians saw of it, seemed to prove it. Both scientists basically came to the same result, that is, *tactile stimulation of the infant* is a main source of early pleasure gratification and a condition for human health, for harmony, and for peace. Ashley Montagu's research developed a specific focus on the importance of the human skin as a primary pleasure provider.

Grant's Method of Anatomy defines the skin as the most extended and the most varied of all our sensory organs.[156]

Ashley Montagu's study *Touching: The Human Significance of the Skin* was the final result of thirty years of skin research, not only Montagu's, but of others whose research Montagu evaluates.[157] Montagu's research is interesting with regard to tactile stimulation in early childhood. Montagu's specific focus during his research was upon the mammal mothers' licking the young. He found astonishing unity in zoologists' opinions as to the importance of motherly licking for the survival of the offspring. He discovered that it was in the first place the *perineal zone of the young* that the mother preferably licks. Experiments in which mammal mothers were impeded from licking this bodily zone of the young resulted in functional disturbances or even chronic sickness of the genitourinary tract of the young animals.

Montagu concluded from this research that the licking did not serve hygienic purposes only, but was intended *to provide a tactile stimulation for the organs that were underlying the part of the skin that was licked.*[158] However, he further concluded, lick-

Published by Sirius-C Media Galaxy LLC, 2010

ing hardly ever occurs in the mother-child relationship with primates or humans.[159] With one exception: an Eskimo tribe, the *Ingalik*, was found where the mother licks the face and hands of the baby in order to clean them, until the baby is old enough to sit on the bench.[160]

Most researchers found that with progressing evolution, licking was gradually replaced by eye or skin contact between mother and child. The tactile needs of the small child seem to correspond to the desire of the parents to express love through *tactile affection* such as kissing or fondling, or pressing the child's naked body against one's own during sleep or rest, which is very common with Eskimos or Indian tribes.

In the run of industrial civilization, however, this has changed fundamentally, to this very day. Modern pediatrics or child psychologists recommend parents to put their children in separate rooms and beds so that parents and children are physically separated.

This is the main reason why the *civilized child* gets much less of tactile stimulation in early childhood than children in tribal cultures. A direct relationship was discovered between early tactile stimulation and the immune system of the child.

This relationship was corroborated by France's first and foremost obstetricians, Frederick Leboyer and Michel Odent.[161] Michel Odent writes in his book *La Santé Primale* *(1986):*

Michel Odent

It is not yet completely understood that sensorial perceptions at the beginning of life can be a way to stimulate the 'primary brain', at a time when the 'system of

> primary adaptation' is not yet grown to maturity. More
> specifically, this signifies for example that, if one fon-
> dles a human baby or an animal baby, one also stimu-
> lates his immune system.[162]

Montagu states that love was once defined as the *harmony of two souls and the contact of two epidermises.* In this sense the *peau à peau* that is nowadays is again recommended by pediatricians is a primal condition for the healthy growth of children, the good functioning of their immune system and the early creation of *preferred pathways* in their brains.

Skin-skin contact thus can be said to promote high intelligence! This is really something our society has never understood, and that even most scientists do not acknowledge, despite the abundance of research that clearly corroborates this truth.

If this was really understood, our need for tactile pleasure, shared nakedness and sex, however sex is acted out and however it is like, could never be subject of criminal law!

In his research with rhesus, Montagu reached astonishing findings. When he deprived newborn rhesus of their mother and let them in the *naked* cage, they died. When he did the same, but put a kind of felt mat in the cage, they survived, although they carried away some brain damage from the deprivation of the mother. However, it was a fact that the *felt mat* assured their survival.

How could that be? Montagu went one step further. He replaced the mother through a *felt mother* that was hung in the cage. Now the young did not only survive but they also had almost no more brain damage.

Published by Sirius-C Media Galaxy LLC, 2010

It was especially that the young survived simply by a felt mat being put in the cage that intrigued Montagu. Further observations led him to realize that the young rhesus used the *fur carpet to give to their bodies tactile stimulation*, which obviously served as a compensate for the tactile stimulation they normally got from their mother in form of licking. The interesting fact about this experiment is that it was not the milk of the mother nor her care that was essential for the young's survival, but exclusively some form of *tactile pleasure*. The felt of the carpet was similar to the mother's fur and therefore acceptable for the young as a mother surrogate.

This research amply demonstrates how important tactile stimulation is with mammals, and so much the more with humans where primary symbiosis is even more prolonged.

James W. Prescott's research particularly focused on the consequences of early tactile deprivation. In his article *Body Pleasure and the Origins of Violence (1975)* Prescott uses R.B. Textor's supra-cultural statistics [163] in order to scientifically prove his collective and political conclusions.

Already in the 1930s, Wilhelm Reich disproved the very widespread misconception that sadistic and destructive tendencies were part of human nature. Reich strongly opposed Freud and his theory of a 'death instinct', stating that those destructive instincts were but secondary drives, *a direct consequence of the cultural repression of the natural sexual instinct* which brought about a *collective neurosis in the human animal*. Reich's findings, at the time violently opposed by the majority of his scientific colleagues, are now clearly confirmed by Prescott's research which brings statistic evidence as to the malleability of the human individual through his early tactile experiences or the absence of such experiences:

James W. Prescott

Recent research supports the point of view that the deprivation of physical pleasure is a major ingredient in the expression of physical violence. The common association of sex with violence provides a clue to understanding physical violence in terms of deprivation of physical pleasure. (...) Although physical pleasure and physical violence seem worlds apart, there seems to be a subtle and intimate connection between the two. Until the relationship between pleasure and violence is understood, violence will continue to escalate.[164]

It is interesting what Prescott wrote in the introduction to his study:

James W. Prescott

Unless the causes of violence are isolated and treated, we will continue to live in a world of fear and apprehension. Unfortunately, violence is often offered as a solution to violence. Many law enforcement officials advocate 'get tough' policies as the best method to reduce crime. Imprisoning people, our usual way of dealing with crime, will not solve the problem, because the causes of violence lie in our basic values and the way in which we bring up our children and youth. Physical punishment, violent films and TV programs teach our children that physical violence is normal.[165]

Prescott thus fully confirms Reich's research and corroborates his socio-economic and sex-economic findings. More specifically, James W. Prescott found a noteworthy relationship between pleasure and violence. Referring to laboratory experiments with animals, he could detect a sort of *recip-*

Published by Sirius-C Media Galaxy LLC, 2010

rocal relationship between pleasure and violence, that is the presence of pleasure inhibits violence – and *vice versa*. Prescott states:

James W. Prescott

A raging, violent animal will abruptly calm down when electrodes stimulate the pleasure centers of its brain. Likewise, stimulating the violence centers in the brain can terminate the animal's sensual pleasure and peaceful behavior. When the brain's pleasure circuits are 'on,' the violence circuits are 'off,' and vice versa. Among human beings, a pleasure-prone personality rarely displays violence or aggressive behaviors, and a violent personality has little ability to tolerate, experience, or enjoy sensuously pleasing activities. As either violence or pleasure goes up, the other goes down.[166]

Further, Prescott found a *direct relationship* between child-rearing methods of a given culture, and the degree of violence that reigns in that culture. In detail, he found that societies that tend to rear children in a rather Spartan way, hostile to pleasure and with little or no tactile stimulation, cherish in their value system various forms of violence, do warfare, torture their enemies, practice slavery and progeny and concede to women and children a rather low social status; these societies also exhibit a high crime rate.[167]

Another violence-indicating parameter in a society, Prescott found, is *physical violence towards children in form of corporal punishment.*[168] Further, repression or tolerance of children's sexual life plays a decisive role in the assessment if a society has a high or low violence potential:

James W. Prescott

Thus, we seem to have a firmly based principle: Physically affectionate human societies are highly unlikely to be physically violent. Accordingly, when physical affection and pleasure during adolescence as well as infancy are related to measures of violence, we find direct evidence of a significant relationship between the punishment of premarital sex behaviors and various measures of crime and violence.[169]

As a result of his research, Prescott advocates the *abolishment of corporal punishment of children*, a rise of the social status of women, the reinstitution of the extended family, the reintegration of the elder and a more active participation of men with *child-rearing and the bestowal of physical affection on children* in their role as fathers or educators.[170]

Published by Sirius-C Media Galaxy LLC, 2010

The Self-Regulation Pattern

Oedipus, in Sophocles' famous tragedy, killed his father and married his mother. He didn't do this consciously but was led by the invisible threads of fate or destiny. The whole tragedy was triggered by an oracle that told Oedipus' father he would die through the hand of his own son. Wanting to escape the impending fate, the father, after his wife gave birth to a baby boy, let bind the infant, pierce his feet, and told a slave to abandon him on top of a mountain. The slave, however, did not follow the order and, being moved by the suffering of the baby, took the child home from where, through a chain of coincidences, the baby eventually was being taken to the *King of Corinth* and his wife, who formally adopted the child and called him *Oedipus*, a name that means *swollen foot*.

When Oedipus was grown up, he left in order to find out about his destiny, as some people had made remarks that he did not for the least resemble his parents. Without however knowing the true story about his origins, as his adoptive parents never had revealed it to him, Oedipus traveled to Delphi to ask the oracle and god Apollo warned him he would kill his father and marry his mother.

Oedipus thought he could easily escape the prediction simply by not returning to Corinth, to his supposed parents.

Yet his escape was illusory; it was through Oedipus' solving the Sphinx's riddle that he was going to marry his own mother. The marriage as well as the patricide that preceded it were acts that Oedipus committed innocently, not knowing the truth. He was thus not aware of what he was doing. Or, to put it in other words, he was completely blind regarding the truth of his fate, of his life. However, as the story continues, it becomes clear that, despite his innocence, Oedipus was *not* considered free of guilt.

This strange story became the basis of Freud's theory of the *Oedipus Complex* which is an angular point of his theory of infantile sexuality. Furthermore, anthropological studies have shown that the Oedipal problem is universal in all cultures that repress the free sexuality of the child, but *only in those*.

Please note that this is the result of my own research on the matter, while mainstream Freudian psychoanalysis claims to this day the *Oedipus Complex* was universal, and thus a part of psychosexual growth in all cultures around the world. This view is simply wrong, and has since long been refuted by anthropology/ethnology.

Published by Sirius-C Media Galaxy LLC, 2010

Needless to mention that on Trobriand there is no Oedipal problem or complex to be found in the psychosexual development of children. Equally universal are considered the taboos of incest and patricide/matricide. However, cultures differ in the social mechanisms that regulate child care and the emotional and sexual development of children.

Cultures, as the before-mentioned matriarchal ones, tend to raise the child within the child's own natural continuum whereas all other cultures tend to condition the child to certain cultural or ideological values and a rigid morality codex.

This is *primarily done through indoctrination* and, secondly, through gradually alienating children from their bodies. The most effective way to indoctrinate children within any given culture is to implant in their mind *a deeply rooted doubt about who they are*. This doubt which creates a vacuum will then be filled with magic formulas such as *Be not what you are!*

The next step is to force the child to play roles in order to *please their parents*. The main role in this drama which is the *Drama of the Gifted Child*, as Alice Miller called it, is the role of the child parenting his own parents.[171] This education that I call *rearing narcissistic comedians*, is very common if not the prototype education in modern consumer culture, which, for this reason, I came to call *Oedipal Culture*.

This is why *narcissism* is rampant in Western nations, especially in the United States. However, there are few researchers who see that the main etiology for narcissism is to be found in our child rearing paradigm. Those who do and did, such as Alice Miller, Lloyd DeMause, Alexander S. Neill or Alexander Lowen were and are not representing mainstream psychology, despite the brilliance of their work. They have, *inter alia*, found that education that typically leads to narcissism is rich in inventing and executing several other magic formulas that are given to the child in the form of hypnotic injunctions.[172] Some of these are:

▸   Be adaptable and flexible until self-alienation;

▸   Never be yourself in front of your parents;

▸   Be not child-like, but adult-like;

▸   Be mature in immaturity;

▸   Understand what your parents don't understand;

▸   Be logical and uncomplicated;

▸   Respect your parents while disrespecting yourself;

▸   Mistrust your intuition;

▸   Follow authority without questioning.

Many parents who believe they are modern and generous to their children *are in reality tyrants* because they consume-train their children and mold them into co-dependent *ersatz* partners. In most cases such parents are not conscious about the fact that they act as the long arm of political systems and ideologies subtly hypnotizing their children with the concepts they have themselves been fed with.

Published by Sirius-C Media Galaxy LLC, 2010

It is for this reason still considered as revolutionary, if not some sort of subversive activity to rear children in truth and autonomy, for such kind of education is not compatible with the oedipal-paranoid worldview that mainstream industrial culture is founded upon.

To raise children responsibly does not mean to charge them with a burden of premature responsibility. However, to infantilize children and deny them by so-called 'child protection laws' any even slight responsibility is surely worse, and the latter practice is the strategy of Oedipal education which arrogantly calls itself *child-protective*. The Oedipal confusion brings about over-adapted and deeply disloyal citizens!

Oedipal culture is a community of secret anarchists that obediently say their credo, but silently sabotage the very content of it.

Education toward autonomy is based upon the *unique truth of every single child*, also and especially if this individual truth is contradicting the reigning socio-political ideology. It is especially disturbing for the industrial culture that the child be a complete sexual being from birth, and that, as a result, children have a birthright to have their emotions and sexual feelings respected.

Françoise Dolto, in her book *La Cause des Enfants (1985)*, wrote that it scandalizes most adults that a child be their equal and that, therefore, most parents raise their children as formerly princes ruled their kingdoms.[173] The sociopolitical reasons why this is so are obvious: a body-oriented child is not an easy consumer of toys and a thousand needs and devices artificially created by the industrial consumer culture.

For those who object this view, I recall that the repression of the child's sexuality has exactly started with the onset of the Western industrial bourgeoisies, at the end of the 17th century.[174] And this was surely not a mere coincidence. Françoise Dolto writes:

Françoise Dolto

Until he was six years old, adults behaved with the prince in a perverse way: they played with his penis, allowed him to play with their genitals and to sleep with them and play *the little devil* with them. All this was allowed. But suddenly when he was six years old, they dressed him like an adult and he had to follow the royal etiquette [citing Ariès, L'Enfant et la vie familiale sous l'Ancien Régime, p 145]. Despite the trauma that could follow, he had nonetheless kept something essential since, during the first years of his life, he could live his sexuality with other adults than his parents. He had here more chance in spite of the precocious adult clothing they put him in. His example is only valid for the rich classes. However, in other levels of society, how could a child of that time repress his incestuous desire and sublimate it?[175]

Most researchers came to agree that this is the true reason for the repression of child sexuality; while patriarchy has well repressed the sexuality of the girl child long before, this was not true for the boy child.

Historical studies on child rearing practices in Europe stress the fact that still during Renaissance the sexuality of the child was not interfered with, and that, back in the Middle-Ages, apart from Christian circles, it was completely free. The *paradigm of child repression*, as I came to call it, clearly started out with industrial culture, and as a matter of fact, it treats both sexes equally in the denial of child sexuality, and thus cannot be attributed to patriarchy, as some researchers wrongly believe, but is a *modern phenomenon, and related to consumerism and materialism.*

Consumerist industrialization brought the societal replacement of body pleasure (*to be*) by *ersatz* body pleasure (*to have*), to refer to Erich Fromm's psychoanalytical research on this matter. Ersatz body pleasure is the pleasure that replaces original body pleasure; thus first of all *the toy.* Not the self-made toy that still has some connection with the body, but the industrially produced toy that is completely alien to the child's body. Typically this toy – which is produced by a gigantic worldwide industry – consists of materials that are not akin to the human body, that is, *plastic* and *metal.* Both materials have in common that they are cold and rigid while the body is warm and pliable.

Unconsciously children are conditioned upon the characteristics of the toys they is playing with.

–*Be plastic!* Be without feelings, artificial.

–*Be metal!* Be hard and mechanical.

– These are the characteristics of the culture you are growing into. *So mold yourself accordingly!*

Techniques of confusion are all the educational methods that *gradually alienate the child from their own truth* – which is their body continuum. *The child namely thinks from the body toward the spirit (inductively) while the conditioned adult thinks from the spirit towards the body (deductively).* This means that the child's truth is defined and experienced as the truth of their own body.

Every fact of life that disregards their body or tries to set it aside will not be regarded by the child as truth. It is for this reason that children cannot comprehend moralistic educational concepts since those concepts deny the body and hypertrophy the intellect. The result are lifelong giant babies in the form of adults who never made the cut with their childhood and remain psychosexually immature: *true virgins*. While life has not made us to remain virgins, but to leave virginity and grow into loving copulation, for otherwise life could not continue.

The fundamental conditioning of man is done within his first seven years of life. What comes later is only polish. *Oedipal confusion cheats about this truth*. It creates a confused spirit within an immature and rigid *nonsexual* body that has lost its natural erotic intelligence.

Oedipal confusion plays the game of eternal maternity until the baby is far older than thirty! It loves naive mother dependence and mistrusts children who are precociously mature. It blows up the child care industry to a gigantic worldwide *business* with children as their products! *Children who resist oedipal castration and maintain their natural capacity for sexual love and sensual pleasure and curiosity are put in the corner and labeled as sexualized and delinquent.* If they still dare to play their own game, the child psychiatrist is ready to interfere and to issue a certificate which will mark the social death blow: *schizophrenic* or *epileptic*.

Summerhill School was founded in 1921 in a village near London, England. It was a *free school* which means that there was no moralistic education and no punishments. What is

Published by Sirius-C Media Galaxy LLC, 2010

the difference between the free school concept and other alternative educational concepts? Unlike in concepts such as *Montessori* or *Steiner*, in free schools no repression is used in educating the child. True, Montessori looks very pragmatic, very rational and focused upon the necessities of daily life. Children learn to iron shirts, to do gardening, to cook. They are put in sensitization classes to stimulate their sensorial perception – but this enchanting rhetoric is deeply false. The child does not need to be stimulated sensually since it is naturally sensual. But of course, the moralistic and in this case Catholic background of Montessori education does not allow the children to live out their emotions and display their *sensual and tactile* needs. So they are first emotionally killed and then artificially induced into fake-feelings in so-called sensitization classes – a ridiculous and hypocrite idea after all. With Steiner, it's the stressing of the *soul values* of the child, as if the child himself or herself was not able to realize their soul level, which is why I came to think that Waldorf education is just another make-believe that is unable to see the reality of children because they simply do not allow children the free development of their their natural emotions and desires.

Of course, behind these different approaches are different root philosophies regarding the role of the child in society. Neill, the founder of Summerhill, was against Montessori that he considered as a milder and 'more intellectual' but nonetheless intrinsically authoritarian form of modern education.

There is no doubt that Maria Montessori who was a believing and practicing Christian wanted to revolutionize education and her contribution for more humanity and respect towards children was certainly remarkable. Her teachings brought about amazing change, not only in her own schools.

One of her inventions was the child-oriented furniture and small toilets that we know from modern day care centers. But was it really an approach that served children to be more able to realize their own intrinsic nature?

Montessori's point of departure was the observation that the child's brain, not unlike a sponge, absorbs the intrinsic atmosphere of his or her environment. In her book *The Absorbent Mind (1973/1995)*, she cites *psychological research* that proves that children learn in the first three years of their life more than adults in sixty years of hard study. Although Montessori was in the beginning against all educational programs, she designed a specific educational program for her schools that focused in first line on the *intellectual training* of the child.

By means of a sophisticated system of different games, puzzles and assembler games that by the way are much more complicated than what usually can be bought in toy stores, the child's mind is well prepared to handle both *intellectual and practical* situations in daily life.

As such, the theory really makes sense and actually was very appealing to me at first. But what about the soul level of the child and their communication abilities? My experience of this approach when actually visiting Montessori Schools in various countries was rather negative. First of all, it was extremely difficult to get a permission for visit at all. I had to justify my wish as if I wanted to visit a secret terrain of the armed forces. The permission was conditioned upon my be-

Published by Sirius-C Media Galaxy LLC, 2010

ing very short and my restraining from any communication with the children! The children seemed to be robots, pale, dull, insensitive, without life. But they worked, and how!

Their way to work though the various tasks seemed obsessive, almost neurotic, while they were bombarded with a full-cry Beethoven symphony from a portable stereo.

These schools really were places without soul and without humor. When the children had their pause, they sat down on a bench, opened their lunch boxes and eat silently, without talking to each other. They seemed to have no contact to each other, isolated in their intellectual cages, without all what usually makes the somewhat noisy and happy society of natural carefree children. These children were little adults that had lost their souls and stared in the air with dull eyes and moody faces. I was more than irritated when I left those places. I found them even worse than the violent orphanages I had seen in some third-world countries. This was child dressage of its best, but one that was no circus because here not even clowns were allowed.

I never got a chance to visit Summerhill. I doubt that it is as permissive as it pretended to be, and this simply because it is located in *England*. Can one imagine a more repressive culture as to the expression of natural human emotions? Of course, it is much more humane than most traditional schools and, in particular, *corporal punishment* is absolutely taboo. The education is permissive regarding the healthy emotional and sexual development of the child. Masturbation is not repressed, sexual play only for the purpose to avoid procreation.

That is at least what we learn from Neill's books, but it must seriously be doubted if in practice, the free sexuality of children and youth would ever be tolerated in a school. Today, more than half a century later, it is *not tolerated*. Thus we have to take Neill's statements, *cum grano salis*. Nonetheless, Summerhill has followed up to a great tradition. Neill's educational approach can be seen on a line with famous historical educational methods, like *Jean-Jacques Rousseau*'s educational ideas, or *John Locke*'s, in that it was founded upon the conviction that nature is generally good.[176] Summerhill was thus in flagrant contradiction with Christian, and in particular Calvinistic child rearing which uses harsh punishments and emotional frustrations in order *to better the human soul* that it considers as basically bad and rotten.

The goal of Summerhill was not to bring about conformists, but adults with high self-esteem, strong intuition and sensitivity, humor and respect for life in all its forms.[177]

One of Neill's main motivations was to produce adults with a positive and constructive mindset, people who are free of hatred and the repressed anger that is part of traditional education; people also whose emotional life is intact and lively.

When Neill opened Summerhill, he was already fifty-one years old. He had spent not years but decades with studying human history and education and was deeply concerned that human history was marked by hate and violence, human destructiveness, intolerance, war and slavery. Neill really got to see behind the trail of lies behind traditional education and our modern civilization. He saw this perverting hatred again

Published by Sirius-C Media Galaxy LLC, 2010

and again in the children who arrived from traditional institutions where they had been declared *uneducable*; he saw that the whole concept of the *difficult child* was wrong in that these children were not more destructive than others, but unhappy, emotionally blocked, frustrated and lonely. Their destructiveness and violence was but a symptom of the underlying reasons that were deeply rooted in the hypocrite, paranoid and violent societal system they were raised in.[178]

In addition, most of them had been neglected or even abused by their parents or by educators in the homes they were coming from. These children had lost confidence in adults. They had been deceived and felt the cold pressure that comes from authoritarian ways of child-rearing. They doubted the existence of love and how it feels to being loved. In this sense, Neill acknowledged, all children who are raised in an *authoritarian and repressive system* will be difficult once they are freed from the pressure they are subjected to. This difficult behavior, Neill found, actually was an inner healing process that established a new value system in their mindset. But first they exploded, of course.

With adults it is the same, as we all know. Violent crimes, war, slavery, torture and terrorism are the results of the hypocrite make-believe that we call civilization and that has in truth nothing to do with being civilized. All these consequences of the authoritarian system show that this system is based upon wrong premises, that it is not human and not made for humans since it disregards human dignity in the most flagrant way.

Neill knew that only an education that was based upon love could finally overcome the violence inherent in a society that is full of hatred. For the only way out of violence is fighting its roots: lack of love and respect, lack of positive encouragement, dehumanizing treatments and a belief system that is based on the idiotic and arrogant idea that nature is fundamentally bad and has to be 'improved' or 'reformed'.

Corroborating Wilhelm Reich's controversial research on human emotions, Neill found that moralistic education not only negatively impacts upon the mindset of the children, but also infringes upon the soma, especially the energetic and muscular balance in the body of the child.[179] Also in accordance with Wilhelm Reich, Neill applied in his school the principle of *self-regulation*. Every attempt to impact upon children in a purely intellectual or moralistic way was leading to failure and was soon abandoned by Neill. Instead Neill believed that the children will comply with what they really subscribe to and understand so that they do it because they believe that it has to be done. This attitude of course requires that the educator himself believes that children are beings born with reason and that they will use this reason whatever their age is.

That is exactly where moralistic educators have doubts. They use to dig a ravine between adults and children, as if the child lived in another world with different natural laws. While they acknowledge the necessity of reason, they deny that children possess this reason; some go as far as considering children on one level with animals. Free child-rearing is unthinkable without conceding children the freedom of their

Published by Sirius-C Media Galaxy LLC, 2010

natural erotic and sexual feelings and, not to forget it, the freedom of speech regarding these feelings.[180]

The latter is as important as the former because children who are allowed to do it without being allowed to talk about what they have done will never really believe that this freedom has been given to them. Instead they will experience guilt and shame. Only a child who is sexually free and erotically satiated will develop their full potential of interest and work energy. It is further necessary that children really *feel accepted* as persons in their own right, in their natural wholeness that encompasses an undistorted emosexual setup that wants to be developed and experienced, for all life asks for growth, and it is therefore total nonsense when mainstream psychologists pretend the child had to grow with a *sleeping* sexuality that would *awake* at the end of puberty. Fairy tales.

We are facing this challenge for a more truthful education also on a *collective and global level* when we develop more tolerance, understanding and compassion for others. It is by changing the basic foundation of our educational and pedagogical values. The Summerhill concept realized a first step to more humanity here on earth, by raising more *humane humans*. Interestingly, the overwhelming majority of Summerhill graduates were later seen to score very well on the social ladder while leading healthy and balanced lives and experiencing positive and highly rewarding relationships.

Neill stated that his own measure of success was the capability to work with joy and to live positively.[181] After forty years of experience, Neill was able to conclude that, applying this definition of success, 'most of the Summerhill graduates became successful people'.[182]

The Touch Pattern

In our observation of the *Love Pattern* and *Pleasure Pattern* and of the nonviolent and sexually very permissive Trobriand culture, we saw that originally love and sex were not separated, but were linked in a functional way. For developing our love capacity, we need to live as children our sexual curiosity to its fullest. In order to be able to remain with a partner over many years, we need to have had promiscuous sex previously, and preferably during childhood.

Instead, what we can derive from the anthropological field research is that our Western theory of love and sex being *separate* experiences is fundamentally schizoid, and contradicts the nonlinear logic of life and insights into the connectivity of mind and soma. It seems to be an *elaboration of pornographic clerical fantasies* rather than the worldview of sane and integrated individuals. The word *sex* itself, that interestingly is missing in the vocabulary of many a sexually permissive culture, implies, by the very fact that such word exists, the possibility of a split between sexual function from love so that love remains a kind of reductionist concept of pure caring and affection which, of course, does not exist in real life.

This split between love and sex makes for a lot of damage in our striving for unity and harmony. One of the typical problems that many Westerners face is that they *unconsciously split off* their love feelings and their sexual longings and project them on different partners. The marriage partner receives love, care and affection and all sexual longings are split off and projected upon prostitutes or prostitute-ersatz figures such as secretaries, servants, mistresses or female children.

Many a reader may object that it often happens that sexuality is lived in its cold form, deprived of love, as mere satisfaction and rather brutally, and that some people even experience more intense sexual feelings when they can encounter sex without being obliged to fake tenderness or caring. This may be true but research has also shown that sex which is experienced connected with love engenders a higher level of lasting feelings of happiness and joy than sex that is acted out in form of an ego trip.

Whatever our personal opinion in this respect may be like, there is no doubt about the fact that *sexuality in every form is focused upon our skin as the main sexual organ*. It seems, however, that sexology has rather hesitantly taken the turn to consider the importance of skin contact in the give and take of body pleasure. Without the skin we would be like barrels without ground. Our bodies are water-sleeves containing more than ninety percent of water. Besides, the skin is our temperature regulator, similar to the astronaut's spacesuit. Besides this protective function, our skin plays an important role in our wellbeing and health. When we experience to be caressed or massaged, when we are touched with love, we feel good. *It feels good.*

Ashley Montagu spent more than thirty years researching about the overwhelming importance of children's tactile stimulation. His book *Touching: The Human Significance of the Skin (1978)* is a guide not only for his own research results, but also for the complete range of literature on skin research

Published by Sirius-C Media Galaxy LLC, 2010

collected over twenty years, in this study.[183] Ashley Montagu gives many examples that demonstrate how destructive the *touch taboo* in our culture is, especially for children. He found that the deprivation of tactile pleasure creates a misbalance in the psychosomatic setup of the child.

As I reported it already, Montagu, starting out with animal research, inquired about the biological reasons for the motherly licking is a primary condition for the survival of the young.[184]

It is noteworthy to observe that the mammal mother extensively licks the perineal zone, the region between anus and genitals of the young. Experiments that prohibited test animals from licking and being licked led to grave urinogenital infections of the young animals or even their death.[185] Further, Montagu found that licking is normally *not* part of human child care. An exception was found with an Eskimo tribe, the *Ingalik*. The Ingalik mother would lick the face and the hands of the newborn in order to clean them. She would continue this licking until the baby is old enough to sit on the bench.[186] Montagu found that humans generally touch children more with their hands, that is, by touching, stroking, caressing, than, for example, with their tongues, and thus by licking, and that, in addition, eye-to-eye contact plays an important role in nurturant child care.

Different researchers found that tactile stimulation of the child is of primary importance for building and maintaining a strong immune system with the child's organism.[187] Montagu remarks in this context that love has been defined as *the harmony of two souls and the contact of two epidermises.*

In this sense body pleasure of mother and child is the most basic, the most natural and the most beneficial form of human sexuality. Needless to add that this is a form of *pedophile* sexuality. And this is one of the examples that clearly show that pedophile erotic and sexual sensations are in some way part of nature, and not, as it is wrongly believed, a perversion.

The skin is our primary sexual organ. All stimulation of the sexual organs is effected through the stimulation of the skin that surrounds them. Sigmund Freud has defined sexuality as *any behavior that has a physical connotation to the sexual organs and that is focused on receiving pleasure.* We could ask if body pleasure really must be concentrated on the sexual organs so that we can qualify it as sexual? It is certainly also a form of body pleasure to drink a fresh beer or to eat one's favorite dish. This kind of pleasure could be called oral or culinary pleasure. Hardly anyone would go as far as qualifying it as *sexual*. But how is it with caressing one's chest or bottom? Is it sexual or not? Does it depend on the way we caress that it is sexual or merely affectionate, or does it depend on the intention? Or is the decisive factor which body zone is caressed? Or does it depend on the fact that the one who caresses is sexually aroused by the activity – or not?

Still during the Renaissance it was common in Europe that all members of the family co-slept naked, as today it still is practiced with the Eskimo and many native populations.

The bodily touch or casual caresses that happened during the night were generally *not* considered as sexual or sexually intended.[188] Today, in our culture, many people would

Published by Sirius-C Media Galaxy LLC, 2010

find it unusual to let sleep their children naked in one bed or that adults, be it the parents, would sleep naked together with their children. This is astonishing as the majority of psychologists are now outspoken about our need for direct body contact, warmth, togetherness, tenderness, and nudity – and this independently of age or gender. Several scientists have researched on the consequences of a deprivation of love nutrition, that is the lack of tactile pleasure, and got alarming results. Unfortunately, the greatest part of this research was done with apes, although genetically, as we all know, the human race is very close to apes. But nonetheless it would seem more convincing to have done direct research on human beings.

Pediatrics and modern child psychology have done great work during the last twenty years. After the publication of volumes if not entire libraries of results of this research, now almost all specialists of child care agree that *children who are raised in a milieu deprived of love, tenderness and caring body touch face greater adaptation problems later in life*, frequently show learning difficulties and tend to be more rigid in experiencing joy and pleasure than children who grew up with love and abundant body touch. The first group of children exhibit symptoms such as restlessness or hyperactivity; in school they often have drawbacks because of their low attention span and concentration ability. In the group, they are seen as rather isolationist and uncooperative. They are easily pushed aside as *difficult*, and once this has happened, the above symptoms aggravate, sometimes dramatically. It is only a step from there to autism.

Many of the children who are in institutions for so-called delinquent youth were formerly affectively and tactually deprived children; yet life circumstances and often an intolerant and punitive attitude from the side of the environment made them turn away from sociability and into marginality.

What is the specifically pathological in their behavior and in the circumstances that have contributed to form it? How does it impact on children if in their family tenderness and care were replaced by violence and brutality? Research into domestic violence has shown that healthy forms of body touch and body pleasure do virtually not exist in such families. If there is touch at all, it is one that hurts, violates, humiliates and degrades.

There is almost unanimity among scientists and professionals that for the small child *tactile stimulation is essential* for healthy psychosomatic growth. It has been shown that *close and long-term body contact* between the child and their mother, father or other tactually nutritive persons *decisively strengthens the child's immune system and improves health.* One could conclude that these findings are not only valid for small children but also children between the age of six until puberty, and even adolescents, for what could be called *skin erotics* seems to be a life-enhancing and health-strengthening factor in all living.

In fact, in India, as Frederick Leboyer reports, where it is a common tradition to massage babies with warm oil, there are many mothers who continue massaging their children, which always includes gently massaging their genitals, until adolescence.[189] It is believed that massaging children's genitals will enhance their procreative ability, sexual potency and

Published by Sirius-C Media Galaxy LLC, 2010

resistance against illness. Such tactile forms of childcare are however by no means associated in India with incest or pedophilia nor is it even considered as sexual. It is regarded as a natural and necessary attribute to nurturant parental care.

Much research has been done on the roots of human violence, but there is hardly anything comparable to the findings of the American sociologist *James W. Prescott*. His publications are unique in that they found child care methods, sexual attitudes and violence level in a given society forming part of a subtle feedback system.[190]

As already pointed out, the quintessence of this research is the thesis that cultures that continue to be highly repressive regarding the emotional and tactile needs of small children and that, in addition, prohibit premarital sex, will end in an inevitable chaos of violence and destruction that has not seen an equal in human history.

Violence, Prescott states authoritatively, and with an abundant amount of evidence, is a compensation reaction of the human brain for the deprivation of tactile pleasure!

These research results amazingly match with the findings of the British neurophysiologist Herbert James Campbell, who summarized forty years of neurological research saying that the motivation of every kind of human activity is nothing else but pleasure.[191] In the case pleasure is prohibited, the human brain automatically compensates for this lack by activating the violence areas in the brain. The pleasure areas and the violence areas function, neurologically speaking, in a way that one inhibits the other; their activity levels are thus mutually exclusive. For example, the more the pleasure areas are stimulated, the more the violence areas will be inhibited and deactivated. The other way around is equally true: the more the human brain runs on violence, the more its capacity of experiencing pleasure is impaired.

Thus, to summarize Prescott's findings provocatively, one could say that it is up to us, and within our individual responsibility, if we want to run our brains on violence or on pleasure. Both is *not* possible. We have to decide what we want.

American society seems to have made this decision long ago, Prescott found. It has unequivocally decided for violence and against pleasure. The historical roots? Prescott identifies them in the Biblical and Jewish traditions that are the foundation of American or, more generally, Western society. The logical conclusions of this interconnectedness are:

- The more a person has received tactile nutrition during her early years, the more she has known body pleasure from childhood, the more their pleasure areas will be activated and, as a result, the more their violence areas will be inhibited;

- The more a person was deprived of tactile pleasure during childhood, the more their desire for body pleasure is repressed or body pleasure experienced as a guilt-producing activity;

- As a result, the more a person's pleasure areas will be inhibited, the more their violence areas will be active.

Interestingly, Prescott found that *tactile deprivation* in early childhood does not automatically lead to a violent character. There are namely factors that compensate for early tactile

Published by Sirius-C Media Galaxy LLC, 2010

deprivation, the most decisive of those factors being pre-marital sex.[192] In this point resides the specific appeal for a future world politics that is peace-centered. This appeal is to refrain from repressing children's emotions, their sensuality, and their growing sexuality, to let it grow freely and without moralistic or other inhibitions, and to recognize this freedom by law.

Unfortunately Prescott's research *never really penetrated into the mass media* that seem to discard out from public forums this kind of research and is known only to a relatively small circle of scientists and intellectuals. The problem of the present public discussion about sex is that it is still rigid and mytho-logical, impregnated with fears, inhibitions and taboos.

The *stranger* haunts American talk-shows and the general hysteria about abusive or non-abusive sex with children is all-pervasive in the American media culture. This general atti-tude is far from being supportive for the tactile needs of present-day Western children. Its obsessive focus on protect-ing children can easily be revealed as either a lip-service, or a new money-making device, or else another way to enslave children into a *doctrinaire system* of industrial or post-industrial values.

To partly remedy the present almost hopeless situation would be, for example, to implement *baby massage* on a large scale as it was proposed by Frederick Leboyer. This would implement at least one possible form of tactile stimulation and body pleasure for the growing next generations. It would ensure that next generation societies will be a lot more peace-ful and a lot less violent than the ones preceding them in history. We *spontaneously communicate love* through body touch and skin contact. Parents fondle and kiss their baby. Lovers embrace each other. Children like to cuddle into their parents' bed and siblings naturally share one bed, at least until a certain age. Attitudes in this respect vary from one society and continent to the other.

Historically body touch was natural and not reflected about. Sexual contacts in certain institutionalized forms are to be traced back into the beginnings of human life and history.[193] Moreover, as the French child therapist and book author Françoise Dolto reports in her book *La Cause des Enfants (1985),* still in the 17th century love and sex games between adult women and small boys were not repressed and happened quite often among the aristocracy. They may have been considered by some people as funny or scurrilous, but not as immoral and still less as dangerous.[194] The separation of love and eroticism into different age groups and the sexual mathematics that results from this modern kind of regimenting sexual relations has brought about much confusion and, as a result, much public discussion.

After all, it seems to be entirely unnatural and artificial, to say the least. In fact, a generally schizoid and punitive attitude toward love, sex and tenderness can be traced back to the Assyrians and, in particular, the *Code of Hammurabi,* the first legal code in history that contained sex laws. In addition, the so-called *holy books* of all our important religions are filled with poisonous negativism regarding body pleasure. Tolerance, and as an essential ingredient *sexual tolerance,* has never been practiced by humanity. As a matter of fact, historically,

Published by Sirius-C Media Galaxy LLC, 2010

the times of repression and persecution of sexual minorities by far prevail in human history. After all, modern science cannot compensate for the darkness and stubborn rigidity of a human mind that has been conditioned to violence since centuries.

Only love can bring a change. The problem is not that people are not enough informed or not interested in science, but that most individuals are relying too much upon moral judgments that they have more or less blindly taken over from higher authorities, instead of relying on their own intuitions and the inherent intelligence of their bodies and their skins.

It does not make sense to engage in witch hunts in order to eradicate witch hunts. It is more intelligent to leave those how are armored in their armor and to take care of those who have not yet built such armor against life and against pleasure. Our skin can show us the way and gives the signals.

In order to find back to our natural continuum, we must relearn to reason in a way that includes touching and feeling and to allow, in a real and a metaphoric sense, our skin to remain open for the osmosis of love.[195]

CHAPTER SIX

The Matriarchal Science

Introduction

Shamanism could be called a *matriarchal* science. While the expression seems awkward on first sight, let us look at it from the other side of the moon. If one thinks that the *science of shamanism* could not be qualified as matriarchal, then one must first refute the argument that our Western science tradition was and is, entirely, patriarchal! That a science may be qualified as 'matriarchal' or 'patriarchal' seems to mess up the scientific methodology and bring in an ideological criterium for measuring what science is. I agree with this criticism. However, this is not what I am saying when I speak of the 'matriarchal science', when I speak about shamanism.

What I am saying is not that science is per se ideological, but that the humans who do science, are well *more* than the science they engage in. In other words, scientists cannot as humans be defined by the science they are doing, but rather, they define, as observers, what kind of science they are going to create.

Quantum physics has left no doubt about the fact that the observer is always entangled with the object of observation and that there is no such thing as 'detached' observation. Hence, because of this entanglement, the science we humans create cannot be neutral in the sense to be detached from us, from our humanity. On the contrary, it is *impregnated* by our humanity, and if we follow certain ideologies, our science will reflect it in one way or the other. *This is not a matter of ideology, but a matter of consciousness.* The transition from matriarchy to patriarchy, and the correlating change in symbolism, was not just a historical or psychohistorical event, but something that has affected human consciousness as a whole. This change or transition deeply impacted upon wake consciousness, while visionary and psychedelic experiences, as well as experiences with hypnosis have given repeated evidence to the fact that the matriarchal symbolism is still deeply rooted in our sub-conscious mind and its pictorial vocabulary.

These visions are not just visual effects that show the en-ergetic impact that encountering cosmic spiritual entities has on those who go through such transformative experiences. In fact, the amazing and almost miraculous healing that most have experienced who encountered *mythic creatures in the trance state* that is induced by entheogens can only be explained when we assume a *direct infusion of bioenergy* from these sources – whatever the explanation one may give as to their origin.

What I wish to demonstrate in the present chapter is that the large rhetoric about the matriarchy-patriarchy dichotomy is but an *intellectual problem,* as on the soul level or the level of super-consciousness, such a dichotomy simply does not exist.

The patriarchal gods may have taken their place in our churches, but they were fortunately not able to penetrate in our hearts and the larger parts of our consciousness that are accessible for the spiritually awakened individual. Here, we have the whole of the Olympus, so to say, and not only the official part of it. And for everybody who has once entered that dimension, I do not need to mention the fact that the matriarchal symbolism is the only one that really stands out on that level of consciousness, while the patriarchal gods may

Published by Sirius-C Media Galaxy LLC, 2010

well exist also on that level, but have a minor importance. The psychological reason for this fact was clearly acknowledged by Joseph Campbell. He writes:

Joseph Campbell

[A]s all schools of psychology agree, the image of the mother and the female affects the psyche differently from that of the father and the male. Sentiments of identity are associated most immediately with the mother; those of dissociation, with the father. Hence, where the mother image preponderates, even the dualism of life and death dissolve in the rapture of her solace; the worlds of nature and the spirit are not separated; the plastic arts flourish eloquently of themselves, without need of discursive elucidation, allegory, or moral tag; and there prevails an implicit confidence in the spontaneity of nature, both in its negative, killing, sacrificial aspect (lion and double ax), and in its productive and reproductive (bull and tree).[196]

In a society with mainly patriarchal values, the soul will keep the counterplayer inside, hidden, and in the dark. Thus in a solar culture such as ours, the *counterplayer* is the lunar principle, or, as it was called in antiquity, the *Lunar Bull*. If we want to become whole so as to embrace *our genuine soul values*, we have to heal the phylogenetic split between patriarchal and matriarchal culture. This split was a historical fact and it has left imprints in our soul and our psyche.[197]

But while this may have been so as a matter of history, while there have been matriarchal cultures first and patriarchal societies thereafter, this is not how the soul has experienced these matters. Recent research has corroborated that things are really not as clear-cut as historians for centuries thought they were.

When questioned about *patriarchy* and *matriarchy*, many people, and among them even researchers of repute, tend to jump to quick conclusions. They take either-or positions or they question the whole dichotomy calling it a historic bluff.

And there are those who try to find a way out of that hide-and-seek game that leaves important questions open by declaring them as obsolete. There is one author who stands out, Riane Eisler.[198] She has not declared the dichotomy as a historic bluff, but showed with a lot of evidence that both of these concepts never have existed in a pure form, but that in a way they are complementary. However, in a second step, that was perhaps more important than the first, she has looked at the basic ingredients of each of these cultural opposites and found remarkable, if not striking, differences.

There is one main difference that she peels out and that, once you know about it, cannot be unthought. It is the discovery that, deep down, the two concepts differ by the way they look not at one gender, but by the way they look *at both*.

More precisely, Riane Eisler found that matriarchy is predominantly a paradigm that favors *partnership relations* between the two sexes and generally between all members in a given society, while patriarchy favors dominance and oppression, male over female, and above over below, in the sense of strict obedience-based hierarchies, which means in clear text powerful over powerless. Without knowing more, here already, with this kind of rudimentary knowledge as a quintessence of Eisler's in-depth research, we see that there is something of an automatism built in patriarchy. *It's the automatism of abuse.* It's as if all was setup for it to occur. It's as if the cultural and social framework was exactly drafted for abuse to happen, while abuse is of course eloquently fought, in patriarchal terms, as a sin and an abject behavior. While matriarchy tolerates it and has built rape right in most of its cultural myths. But patriarchy has *institutionalized rape*, in all its forms, sexual, social, racial, ethnic, military and commercial. That is the difference.

Riane Eisler's amazing research has brought to daylight that maintaining the age-old dichotomy of matriarchal versus patriarchal is only accurate when we describe their psychological content, but not when we describe evolutionary changes in the human setup. In reality, Eisler points out, we are dealing with a *partnership-oriented paradigm versus a dominator paradigm*, the first coming close to the idea of matriarchy, the latter more or less synonymous with patriarchy. The merit of Eisler's approach is that we can get away from extreme positions: because there never was a really pure matriarchy or a really pure patriarchy in human history.

When we look, for example, at the mythology of highly patriarchal tribes, such as the ancient Hebrews, we find matriarchal elements, and therefore must conclude that we got a mix rather than a pure soup. In that mix, to rest with the example of the Hebrews, are predominantly patriarchal elements and a few matriarchal elements, as in *yang* is a small portion of *yin*, and *vice versa*. And as Johann Jakob Bachofen found in his classical treatise on matriarchy, even in highly matriarchal cultures there are to be found a few elements of patriarchy.[199] Therefore, when we use the dichotomy matriarchal-patriarchal, we are arguing not from a real-life perspective, but rather from our ideological understanding of *patriarchy* or *matriarchy*. What counts for us within the purpose of this book is the *spiritual significance* of matriarchy as a psychological and archetypal complex in the collective unconscious of humanity not the historical or psychohistorical evolution of humanity.[200] Regarding the evolutionary aspect, Joseph Campbell writes in *Occidental Mythology (1973):*

Joseph Campbell

For it is now perfectly clear that before the violent entry of the late Bronze and early Iron Age nomadic Aryan cattle-herders from the north and Semitic sheep and goat herders from the south into the old cult sites of the ancient world, there had prevailed in that world an essentially organic, vegetal, non-heroic view of the nature and necessities of life that was completely repugnant to those lion hearts for whom not the patient toil of earth but the battle spear and its plunder were the source of both wealth and joy. In the older mother myths and rites the light and darker aspects of the mixed thing that is life had been honored equally and together, whereas in the later, male-oriented, patriarchal myths, all that is good and noble was attributed to the new, heroic master gods, leaving to the native nature powers the character only of darkness - to which, also, a negative moral judgment now was added. For, as a great body of evidence shows, the social as well as mythic orders of the two contrasting ways of life were opposed.[201]

It is not difficult for us today to see that the symbolism of mythology bears a specific psychological scripting. Particularly under the perspective of psychoanalysis, and even more so, of *psychosynthesis*, there is little doubt that the old sagas are of the nature of dream, or that dreams are symptomatic of the dynamics of the psyche. Sigmund Freud, Carl Jung, Joseph Campbell, Otto Rank, Karl Abraham, Géza Róheim, and many others have within the last century developed a vastly documented modern lore of dream and myth interpretation.

With our modern-day discovery that the holistic patterns of fairy tale and myth correspond to those of dream, the long discredited ideas of 'archaic man' have returned dramatically to the foreground of consciousness. One of those archaic symbols or archetypes is that of the *Lunar Bull*, for there is a direct relationship between mythology and astrology. It can be said that astrology uses mythology to a large extent in order to make spiritual energies more visually comprehensive.

When explaining the nature of the planetary energy of Moon, for example, astrology will use certain metaphors. These metaphors are embodied in symbols, and the symbols, as such, build a necessary vocabulary for anybody to study who wants to practice and explain astrology. For example, the main symbols traditionally associated with the Moon energy are: *Cancer, Bull, Female, Shell, House, Black, Water, Shadow.*

As the mythic bull's characteristics are associated with the *lunar* energies, it was called, in Antiquity, the *Lunar Bull*. This expression is not a fancy, even today, because the *bull fighting tradition* that dates from patriarchy, has put the whole bull mythology completely upside down.

The killing of the bull that was once a ritual sacrifice for the Goddess as the tutelary divinity of the bull was transformed into a sport in which the stabbing of the bull is a *symbolic rape* expressing the subordination of the female under the male's sexual dominion. Thus, by analogy, the modern bull, the bull that is stabbed and killed by the Matador within the traditional bull fighting has quite little or nothing to do with the matriarchal mythic or lunar bull. The lunar bull was the object of worship prevalent in the age when our sun was passing through the sign of Taurus.

What was preserved from that time were the mysteries of *Mithras*. The horns of the bull were generally a symbol of fertility and bountiful riches in many cultures for thousands of years. The constellation Taurus may also allude to the Greek story of *Europa and the Bull*.

Europa was daughter of King Agenor. One fine spring day, accompanied by her hand maidens, *Princess Europa* went to the seashore to gather flowers. Zeus, who had fallen in love with Europa, seized the opportunity. Zeus transformed himself into a magnificent white bull, and as such he joined King Agenor's grazing herd. Europa noticed the wonderful white beast, who gazed at them all with such a mild manner that they were not frightened. Europa wove wreathes of flowers for the beast, and wrapped them around his horns. She led him around the meadow, and he was as docile as a lamb.

Then, as he trotted down to the seashore, she jumped onto his shoulders. Suddenly, to her surprise and fright, he plunged into the sea and carried the princess to Crete. As they reached the Cretan shore, Zeus then turned into an eagle and ravaged Europa.

She bore three sons, the first of which was Minos, who is said to have introduced the bull cult to the Cretans. He had Daedalus build a labyrinth in the depths of his palace at

Published by Sirius-C Media Galaxy LLC, 2010

Knossos, which became the home of the Minotaur, the off-spring of Minos' wife Pasiphae, and a bull. Seven young boys and seven maidens were ritually sacrificed to the Minotaur every year, until Theseus killed the monster.

The Lunar Bull

> The mythic lunar bull, lord of the rhythm of the universe, to whose song all mortality is dancing in around of birth, death, and / new birth, was called to mind by the sounds of the drum, strings, and reed flutes of the temple orchestras, and those attending were set in accord thereby with the aspect of being that never dies. The beatific, yet impassive, enigmatic Mona Lisa features of the bull slain by the lion-bird suggest the mode of being known to initiates of the wisdom beyond death, beyond changing time. Through his death, which is no death, he is giving life to the creatures of the earth, even while indicating, with his lifted forefoot, the leftward horn of the mythic symbol.
>
> – JOSEPH CAMPBELL, OCCIDENTAL MYTHOLOGY

What does this myth tell us? Which psychological truth does it reveal? Let us have a deeper look at this intriguing story. We got a seducer here, we have an abduction, a rape, and then, as a result, a child-eating monster that eventually is killed. And we have a bull. What does this bull stand for, psychologically?

Experts of mythology and psychiatrists agree that the bull, despite of his phallic horns is a *symbol for matriarchy*, and this because the bull cannot be seen isolated from the God-

Published by Sirius-C Media Galaxy LLC, 2010

dess that, metaphorically and from the visual depictions, stands on the shoulders of the bull. This is a metaphor because we would not be interested in that bull if it had only a historical meaning for us. We are interested in that bull because we have its energy within us. Joseph Campbell affirms that all the gods are *within us.*

Hence, the bull, as a *sort of matriarchal god,* also is within our own unconscious, a part of our male love instinct that can enjoy to conquer and rape, abduct and possess, enclose and abuse. We are used today to a psychological language that suggests all these longings were abysmal and abject and we tend to project them, as a result of our blinding them out, onto others that we call *the monsters, perpetrators, rapists or sex offenders* and that our morning papers abound of. And yet, all this psychological hide-and-seek is useless: we are facing but parts of ourselves when dealing with these well-hidden issues that often are wrapped into the folder of our best-kept family secrets.

When in *The Power of Myth (1988),* which is actually a wonderful example of human dialogue, Bill Moyers asked Joseph Campbell about the serpent as the seducer in the biblical story of the genesis, Campbell replied:

Joseph Campbell

That amounts to a refusal to affirm life. In the biblical tradition we have inherited, life is corrupt, and every natural impulse is sinful unless it has been circumcised or baptized. The serpent was the one who brought sin into the world. And the woman was the one who handed the apple to man. This identification of the woman with sin, of the serpent with sin, and thus of

> life with sin, is the twist hat has been given to the
> whole story in the biblical myth and doctrine of the
> Fall.[202]

Joseph Campbell basically affirms that patriarchy is but a form of life-denial, a collective neurosis, not a lifestyle, philosophy, or *Weltanschauung*. It's a disease, a *twist* given to life that perverts its very nature. And ultimately, therefore, it's a refusal of humanity. Campbell develops the theme further by alluding to the *Star Wars* plot:

Joseph Campbell

Darth Vader has not developed his own humanity. He's a robot. He's a bureaucrat, living not in terms of himself but in terms of an imposed system. This is the threat to our lives that we all face today. Is the system going to flatten you out and deny you your humanity, or are you going to be able to make use of the system to the attainment of human purposes? How do you relate to the system so that you are not compulsively serving it?[203]

Patriarchy, with its craving for obedience to the father, is a sort of compulsion neurosis. Not only are individuals *flattened out* by systems that are eternal replacements of *real fathers*, those that have typically abandoned their roles as true caretakers, having become troublemakers. These authority-craving individuals have *flattened out* their better halves, their right brains, so as to serve the system even more.

The bull story tells us that rape desires as part of sexual longings are not destructive *per se*, but become destructive when they are enclosed, incarcerated, tightly controlled and discarded out of life by strict moralistic rules rules. The Mi-

Published by Sirius-C Media Galaxy LLC, 2010

notaur became a *child-eating monster* because it was enclosed in a tower, because **King Minos** was afraid for his reputation and wanted to hide the monster from the populace. And this may historically have been the first time when child protection thinking was to be noted in human history, and when the results were obviously as devastating as they are today.

We have the symbolism written into the Tarot where *The Tower*, the 16th Arcane[204], is a symbol of something that is too tightly controlled, to a point to explode, with all that usually accompanies those explosions – that most of us have gone through, once in our lives, in one or the other way, be it a scandal, a public outrage, the revelation of a family secret, an abuse story, or criminal conviction as a *sex offender*.

And then, we ask *'Why has this happened?'*, and we are again regressing in childhood longings for autonomy that were thwarted by over-controlling parents or educators, and we face our *rage* – eventually. The public outrage we encountered was but a projection of our *own inner rage* that we had repressed. We had forgotten about the library with the books that can talk, and the wizard, and magic houses that endure.

And back where we came from, we can eventually ask what we really want when we want to rape, to possess, to abduct, to ravish. And we gradually, very gradually, find out that, then, we want to find unity with our soul, and *make the split undone* that was forced onto us by patriarchal life-denial, by moralism, by a schizoid education that we suffered, individually and collectively. After all, to copulate means to *link!* And then we might finally ask the pertinent question: 'How has patriarchy come about – and what was before?' It all started with a murder. The murder of the Goddess. Which is ultimately a matricide, and implicitly a mother-rape. And it became the foundation of what is called *a culture*. It became the foundation of what is called 'religion'. Joseph Campbell explains:

Joseph Campbell

[I]n biblical times, when the Hebrews came in, they really wiped out the Goddess. The term for the Canaanite goddess that's used in the Old Testament is *the Abomination*. Apparently, throughout the period represented in the Book of Kings, for example, there was a back and forth between the two cults. Many of the Hebrew kings were condemned in the Old Testament for having worshiped on the mountaintops. Those mountains were symbols of the Goddess. And there was a very strong accent against the Goddess in the Hebrew, which you do not find the Indo-European mythologies. Here you have Zeus marrying the Goddess, and then the two play together. So it's an extreme case that we have in the Bible, and our own Western subjugation of the female is a function of biblical thinking.[205]

It seems that when man began to preach *high morality* and confessed to strive 'for goodness', he began to really become diabolic. Campbell writes that *the vandalism involved in the destruction of the pagan temples of antiquity is hardly matched in world history*.[206]

Again we may reflect on the teaching of the Lunar Bull. Zeus married the Goddess by raping her, and that rape ultimately was union and creation. While patriarchy, with its strong emphasis about the *abomination* (sic!) of what it labels *sexual crime* is exactly embodying the perversion that it so

Published by Sirius-C Media Galaxy LLC, 2010

strongly projects upon matriarchal cults, and, today, upon *matriarchal people*.

I mean the so-called *pedophiles*, who are put today in the shoes of that Goddess that early patriarchy wiped out. And as their ancient precursors did, the Hammurabis of old, the *darth vaders* of the modern state are out to euthanize those they don't understand and who *disturb their circles*, by questioning their culturally sanctified paranoia.

The *true abomination* is not matriarchy or Goddess cults, and not rape, but *a cult or religion or cultural paradigm that perverts nature into a total repression of the living impulse* and that puts a single male god as the creator principle, thereby annihilating the eternal *balance of polarities*, manifesting as *yin* and *yang*, female and male, Moon and Sun, red and blue, cool and hot, dry and moist, and that restricts life to a *dead morality*. Campbell explains:

Joseph Campbell

The patriarchal point of view is distinguished from the earlier archaic view by its setting apart of all pairs-of-opposites - male and female, life and death, true and false, good and evil - as though they were absolutes in themselves and not merely aspects of the larger entity of life. This we may liken to a solar, as opposed to lunar, mythic view, since darkness flees from the sun as its opposite, but in the moon dark and light interact in the one sphere.[207]

Many people, even in our days of feminism, women rights and open criticism of patriarchal tradition and values do not really grasp the implication of patriarchy upon our sexual mores and sexual laws. Or they are simply afraid to question the reigning system as deeply as that, scratching the surface with their research.

I have been in touch with several researchers from Germany and the United States who openly unveil and criticize the trap of patriarchy and who also defend the sexual freedom of children. But their rhetoric has only one leg when it goes to see what *sexual freedom* for children really means! It means *free partner choice*. It means that a child can also choose *an adult as a partner* for play, including sex play.

When faced with that argument, all those researchers that from their books sound so well-bred, well-educated and well-groomed *block off*. They cease to argue and suddenly become dogmatic and declare that in such a case we could not speak any further of child sexuality but about *pedophilia*, and that the latter was invariably rape, violence and abuse. They assert this without having anything in hand for the backup of their unscientific rants. *That is how far our science goes, all science goes*.

We cannot access knowledge that we are not ready to get because we are *emotionally not mature for it*. These men, while they may have many letters behind their names and while they may be accredited at famous universities, are anxious children who, when it goes to open the forbidden door, shy away and declare that there are no forbidden doors because we lived in a *democratic* society. A society so democratic obviously that it incarcerates people for love and has genocided in its 250 years of existence more people than all other cultures around the world in the last 5000 years![208]

Published by Sirius-C Media Galaxy LLC, 2010

These apparently so liberal scientists would thus forbid their child to have an adult sex partner, while they would allow their child to have sex with a peer. What a high form of respect indeed! A slave is forbidden to make love with a noble while he is graciously given the right, by his master, to love and fuck his brothers-in-fate. *What, then, is that modern childhood else than slavery?*

And in what those liberal parents really differ from our patriarchal house tyrants of old? What Joseph Campbell calls a *solar worldview*, I call a worldview where *stupidity has become the order of the day*. To deny our shadow is suicidal, and it's exactly what a solar worldview is all about. It denies shadow.

A picture without shadows is a one-dimensional drawing of life, a shallow affair. It's sketching a life that is not worth to be lived, because all is on one level, without ups and downs, without excitements, the shallow boredom of a moralist who *goes to Church* at fixed hours or who bows to the ground to lick the feet of his cosmic monstermind.

In another publication I have demonstrated that all our major civilizations or *dominator cultures* since Babylonian times almost completely disregarded *eight fundamental patterns of living* that however pre-patriarchal cultures, and still today most tribal cultures respect.[209] This denial, I showed, makes for:

▸   our high destructivity;

▸   our judgmental and persecutory attitudes;

▸   our schizophrenic split between mind and body;

> ▸ our belief in the superiority of the mind;

> ▸ our belief in the inferiority of the body;

> ▸ our dysfunctional approach to living and loving;

> ▸ our hypocrite in-group and out-group morality;

> ▸ our disregard for the *yin* principle;

> ▸ our preference for death science and vivisection;

> ▸ our disregard for life science, integration and synthesis.

The terms *lunar* and *solar* world view that Campbell uses are derived from astrology. The bull enters the books of world mythology in both versions as the *mythic bull* and the *lunar bull*. And in both versions, according to Erich Neumann, the bull on which the goddess stands is the symbol of masculinity.[210] The bull is also said to have succumbed to the archetypal mother in *matriarchal incest*, which is a *phantasmagoric incest*, symbolizing the closeness and symbiotic aura of the mother-son relationship.

This phantasmagoric incest that in ancient matriarchal rituals was put on stage in dance and chant as a celebration of creation has quite little to do with actual incest, the prototypical father-daughter incest so ingrained in patriarchy. It is *not a sexual incest,* but symbolizes the need of the young male for a healthy symbiosis with his mother, if he is to develop his full psychosexual potential.

Published by Sirius-C Media Galaxy LLC, 2010

An important detail in those old representations of the Goddess and her Bull is that the bull is actually in a supportive role: the Goddess stands on him. He thus supports the Goddess. *The male supports the female.* That is the quintessential message of matriarchy.

That does not mean he's a servant of the female. In patriarchy these poles have not just be reversed. The female is not just supporting the male, but *serves* the male. That is the substantial difference.

In matriarchy, the son supports his mother, but he is not her servant and slave-partner. In patriarchy, the daughter is not just supporting her father, but she's supposed to be his sex servant. We have that incarnated both in the household-female and the love-female. The wife is supposed to serve her husband. The prostitute is supposed to serve her client. Both are in not just a supportive role, but hold actually slave roles.

That is why we can say that patriarchy has not just reversed matriarchy. It has distorted it to a caricature of life in which roles are no more naturally taken by people, but artificially forced upon people.[211]

Historical Turn

> If we look at the whole span of our cultural evolution
> from the perspective of cultural transformation theory,
> we see that the roots of our present global crises go back
> to the fundamental shift in our pre-history that brought
> enormous changes not only in social structure but also in
> technology. This was the shift in emphasis from tech-
> nologies that sustain and enhance life to the technologies
> symbolized by the Blade: technologies designed to de-
> stroy and dominate. This has been the technological em-
> phasis, rather than technology per se, that today threat-
> ens all life in our globe.
>
> – RIANE EISLER

To come back to our initial question, how was it possible
that science in dominator cultures *became aligned with the patriarchal or solar principle,* while the oldest, and most original science, the science of shamanism, was organized along the
lines of the lunar principle? This is really an intriguing question for it explains so many things, to mention only one here.

Shamanism's science is much more holistic, more right-
brain, because the lunar principle allows feminine values
with much more ease than the solar principle. This is why

Published by Sirius-C Media Galaxy LLC, 2010

shamanism is gaia-friendly and cares about *ecology* and harmony between man and nature. The solar principle, by contrast, is responsible for the fact that science in all dominator cultures, the modern industrial nations on top of the list, is left-brain, reductionist and nature-hostile, and has done more for destroying nature than for preserving it.

When we want to understand how the split came about between the original science and the modern sciences, we can make out certain events or a group of events that are typically considered to be the *turning point*, or historical turn. Riane Eisler Eisler usually associates it with the beginnings of patriarchy and the introduction of the school system. If we take Europe as an example, why and how schools came up? In the Middle Ages, when the Church tried to gain as much power over people as possible and indulged in human rights abuses of all kinds, monks and nuns opened the first schools. These schools were recruitment centers for the monasteries. From the beginning, boys and girls were separated in different schools. From the boys' classes, the monks the recruited, from the girls' schools the nuns.

When you read history books, the Church is cited as the great benefactor of mankind in implementing the school system. But the Church's intention was first of all an effort to sustain the power of its own worldly hierarchy and oppression system, and second, and most importantly, *direct perception of truth* was going to be wiped out from civilization from that point in history.

Before the existence of schools, children were raised by their parents and the other adults of the extended family.

They learned primarily by observation and *direct perception*. They *picked up* what they needed for their later career, from their early environment. It is interesting to remember, in this context, that early language learning takes place in exactly the same way. The young child picks up whole patterns from the language spoken around him or her.

Research in recent years provided evidence for the fact that this form of learning is much more holistic and adapted to the *passively organizing intelligence* of the human brain than any system that has so far been implemented in schools.

Human beings generally learn their mother tongue perfectly, whereas they cripple along learning a second or third language later in school or at college. Only relatively recent learning methods such as *Superlearning*® have taken serious the wisdom of nature present in every learning experience. Think tanks such as Edward de Bono have in addition shown us the relevance of the brain's functioning as a passively self-organizing system.[212]

From his experience as a corporate trainer, Edward de Bono found that our usual learning processes, such as curricula in schools, universities or, more specifically, in management training, are *awkwardly maladapted* to the way our brain organizes and stores information. De Bono, much in the same way as Dr. Georgi Lozanov, originator of the *Superlearning*® method, found that only in early childhood learning, and especially in the way young children learn their first language, we see nature's full intelligence at work.

It is a well-known fact that geniuses such as Albert Einstein, Pablo Picasso and many others among our creators

Published by Sirius-C Media Galaxy LLC, 2010

never finished school, dropped out or flew it. They knew that they knew better and followed their inner instinct rather than an artificial learning system which involves *a considerable waste of time and resources* and which violates human dignity in the most flagrant way. With one word, they followed their *soul* and thereby realized their soul reality without perhaps reflecting about it. They did not let society condition their inner mind to a point to crush their creative impulses. They were *marginal* in just the same way as a pedophile is marginal, or as an autistic child is marginal, and as we as creative souls are all marginal in front of the herd of school-fed morons.

It is not a matter of research or of statistics to draw out the human potential, and still less when we talk about the human soul. Every soul is marginal in the sense that it can't be measured on the lines of idiot science, the usual whitewash of complexity for the masses, the *perennial fascist cover-up of true human genius.*

Murder of the Goddess

> In every stoic was a stoic; but in Christendom where is the Christian?
>
> – JOSEPH CAMPBELL

Joseph Campbell's research on the *religious roots of culture* is not new, but for this reason it is not less a theme of the day.[213] For it is counteracting fascism, as it shows with such strong evidence not only how complex the human soul is, but also that this *has to be that way* if man wants to maintain psychic health, individually and collectively.

This astounding holistic information opens infinite insights into how to live peacefully, resourcefully and respectfully. The *historical shift toward stupidity* was a profound shift in human consciousness, opening a *deep schizoid split* that some explain esoterically with an alien manipulation of the human DNS. But even if we stay with the historical facts alone and see their symbolic and archetypal content, we must acknowledge that something went wrong at that point in human evolution.

Published by Sirius-C Media Galaxy LLC, 2010

Riane Eisler, in her best-selling study *The Chalice and the Blade (1995)* called it the *truncation of civilization*. It was the unwritten historical vow of many to deny their humanity and follow the course of atrocious violence that began with the slaughtering of peaceful and nature-abiding cultures by the new arrogant patriarchal hordes and their violent, jealous and blood-thirsty God Yahweh, and psychologically a turn from permissiveness to moralistic repression.

Wilhelm Reich called it the *irruption of compulsory sex-morality*[214], whereas Campbell qualifies it as 'the power impulse [being] the fundamental impulse in European history [that] got into our religious traditions.'[215]

As Reich and other psychoanalysts clearly showed, this power obsession, that lasts until today in our Western culture, was from the start a sort of cultural cancer as the result of the *denial of nature* and man's arrogant claim to *improve* creation and make it better, thereby destroying it. And that is exactly what I am saying earlier in this book, it is by repressing *primary power and breeding depression* that the thirst for power was taking immense dimensions in our culture, until this day. It was the denial of primary power that was at the origin of this cultural perversion. Joseph Campbell observed that the gravity of this historical shift was so deep and lasting that even the mythological and archetypical symbolism changed with it:

Joseph Campbell

The new age of the Sun God has dawned, and there is to follow an extremely interesting, mythologically confusing development known as solarization, whereby

> the entire symbolic system of the earlier age is to be
> reversed, with the moon and the lunar bull assigned to
> the mythic sphere of the female, and the lion, the solar
> principle, to the male.'[216]

It was the real beginning of the apocalypse, for all that came later and that we face today as *facts of life* are but results and consequences of the profound shift that took place at this time. It was the shift from matriarchy to patriarchy[217] or, in new terminology, the shift from a partnership culture to a dominator culture.

Joseph Campbell acknowledges the dominance of the patriarchal gods since then in our Western cultural paradigm, but he considers the goddess as *the counterplayer* in the collective unconscious and thus assigns her at least a shadow role.[218] This shadow role of the *Goddess* in Christianity is symbolized by the *Holy Virgin*, and it is sexually fantasized about as a secret wish to defile, debauch and rape virgins.[219]

In fact, how can a deity that originally stood for fertility become an eternal virgin? When we study Greek mythology, we see that the *original mother goddess* was Demeter, while the Church's virgin cult suggests that her daughter Persephone, a girl abducted by Hades, god of the underworld, became the new, castrated, Virgin Mother.[220] And Hades represents the psychologically and socially rejected sexual longing for *virgins*, for little girls, that became suppressed in our personal and collective *underworld*, the unconscious.

We can thus see that the virgin cult is a *direct consequence of patriarchy* and already well present in the Hellenistic and Roman cultures, and not an invention of the Christians.[221] The

Published by Sirius-C Media Galaxy LLC, 2010

female, to become acceptable in an entirely man-dominated world, had to be deprived of her own desire, castrated, relegated to the role of the 'obedient little girl'. Without desire herself, the girl-female became *undesirable* as an unconscious reflex of the superego's copulation prohibition. She was no more desired as a child to be lovingly procreated by her parents, but her birth *was largely considered an accident;* in many cases a man who procreated only girls or too many girls was considered weak in Antiquity and even through the Renaissance.

Hence, the undesirable girl-female became a sexual taboo; for the unconscious, in fact, there is no difference between the desire of a couple to have a child and the sexual desire for copulation with a child. In French, this is more obvious than in English. *Je désire un enfant* both means 'I wish to have a child (to be born)' and 'I desire a child (sexually)'. This linguistic particularity in French language reflects the fact that for the unconscious there is no difference between the desire to have a child born as a companion for a couple, on one hand, and the desire to have a child as a companion for bed, on the other. Both is *desire pro creation*, both is sexual desire in its larger cultural sense.

What I am saying is that with the psychological castration of the female and the moralistic prohibition of her being desirable sexually, human sexuality became forever damaged, and perverted into a voyeuristic cult that hypertrophies the visual and neglects the tactile: the psychological roots of pornography are to be found in the taboo to touch a female child sexually. (While she still could be looked at sexually).

The more sexuality with female children became tabooed, the more the sexual female child became a *haunting sex obsession* for males, and led to the criminal definition of rape, originally a property offense.[222]

Rape originally meant theft, and this can be well shown in French, where the word for theft is *vol* and for rape *viol*. In Antiquity, to possess a female child sexually meant in most cases to abduct her, a fact that is well established in Greek mythology in the story of Hades abducting *Persephone* for enjoying and possessing her sexually. In ancient patriarchy, the rape of a little girl was an offense against her father, a kind of *property damage* that could be repaired by paying an *indemnity* to the father, but not a crime against the person of the child.[223]

From the Church's modified goddess doctrine, its virgin cult becomes easily understandable. While the god-mother is a very old idea and existed long before Christianity, this god-mother, for the Christians, had to be a virgin, and even a Holy Virgin.[224]

Behold, the doctrine of *Immaculate Conception* was only valid for the conception of the *Son of God*. All other children were born in sin, from ordinary, non-virgin mothers. I can't think of a greater *perversion and distortion of nature* because this mental construct means in fact *that nature is wrong and faulty* and that the very denial of nature is right and holy. And yet, if the Church had really been consequent in their view, they would have needed to sanctify man-girl love as the ultimate modification or *culturalization* of nature. The pointe of this reasoning is that the Church's dogma, seen from this perspective, makes sense culturally. In addition, it would have helped men to *integrate their tender love for the small female*, a love every sensitive and cultured man knows to appreciate, instead of disintegrating this tender love and rendering it a shadow experience only to be experienced through the rape, abduction and sometimes even murder of little girls. The Church could have codified love of men for little women in its nonviolent and respectful dimension in a way that is socially useful, and even biologically logical. This would have had a peace-inducing effect on human sexuality and would have prevented widespread sexual crime against little girls. As in every *yang*, there is a small amount of *yin*, visually represented by the *yin-yang* symbol, in the large strong male there is a small weak female that the male can unconsciously project upon little girls. And if the Church had considered that it is important socially to integrate desires so as to avoid violence, she could have declared girllove a socially viable and acceptable sexual and cultural phenomenon.

If the Church had integrated the small female as a desirable love object, the consequences would have been positive. We would not face such terrible amounts of female children raped, abducted and killed every month in many countries that follow the patriarchal dominator paradigm.

And we would not have had such a raise of homosexuality in our culture because homosexuality is the result of an unconscious blinding out of the desire for small females.[225]

Thus, the Church, by the same token, would not have had to invest so much energy for fighting homosexuality during several centuries in its existence.

The large fallback into paganism today is the result of this *denial of responsibility* of the Church to integrate human sexual desire in all its forms. And by doing so the Church missed the sense of the Grail.[226]

The same can be projected for boylove, represented in the Church as the love of the priest for the altar boy, and the beautiful Sapphic love of adult females for little girls. In older documentations of the Church's annals, the love of the priest for his altar boys was never questioned.

The priest had to renounce females as a matter of religious dogma, but that never implied that priests had to live like eunuchs. Metaphorically, we may say that the undesirable little goddess was not less undesirable for the priest, but practically replaced by a desirable, and available, little god represented by the virgin altar boy. Only with the much more encompassing sex repression in modern times, and the fascist business of 'child protection' could it happen that *all* adult-child sexual relations became a matter of social disapproval, even within the Church. Many tragedies were the result. Under the original paradigm, the Church could well have promoted peaceful and respectful man-boy love relations as a social necessity for integrating homosexual pedophilia. The priest as an idealized figure of virtuous conduct could have served as example of a responsible and resourceful boylover.

Boylove is not fundamentally different from Girllove in that a small boy has predominantly *more feminine than masculine* characteristics. But the hot melting sensations experienced by a sensitive man toward a tender, smart and loving little boy was to be rejected by patriarchy as these natural feelings, as a

Published by Sirius-C Media Galaxy LLC, 2010

matter of *religious correctness*, had to be bulk-repressed together with all what even slightly reminded female power.

The Babylonic *Epic of Creation* amply demonstrates this fact:

Joseph Campbell

[It is] a forthright patriarchal document, where the female principle is devaluated, together with its point of view, and, as always happens when a power of nature and the psyche is excluded from its place, it has turned into its negative, as a demoness, dangerous and fierce. And we are going to find, throughout the following history of the orthodox patriarchal systems of the West, that the power of this goddess-mother of the world, whom we have here seen defamed, abused, insulted, and overthrown by her sons, is to remain as an ever-present threat to their castle of reason, which is founded upon a soil that they consider to be dead but is actually alive, breathing, and threatening to shift.[227]

On the other hand, Campbell reports in *Oriental Mythology* that in most non-Western cultures the very opposite paradigm was being in place, isolating Christian life-denial as something unique and atrociously perverse in human evolution:

Joseph Campbell

The dreamlike spell of this contemplative, metaphysically oriented tradition, where light and darkness dance together in a world-creating cosmic shadow play, carries into modern times an image that is of incalculable age. In its primitive form it is widely known among the jungle villages of the broad equatorial zone that extends from Africa eastward, through India, Southeast Asia, and Oceania, to Brazil, where the basic myth is of a dreamlike age of the beginning,

> when there was neither death nor birth, which, how-
> ever, terminated when a murder was committed.[228]

The synthesis is to be found in what the Taoists called *The Tao* and that Campbell calls '*the perfume, the flowering and fulfillment of human life, not a supernatural virtue imposed upon it.*'[229] And like the Taoists, Campbell says that '*heaven and hell are within us, and all the gods are within us.*'[230] Campbell makes his point succinctly by telling us to overcome the schizoid split so deeply rooted in the patriarchal shift that occurred five thousands years ago and reminds the myth of the Grail as a syncretic doctrine of love allowed to grow beyond all borders:

Joseph Campbell

The Grail becomes symbolic of an authentic life that is lived in terms of its own volition, in terms of its own impulse system, that carries itself between the pairs of opposites of good and evil, light and dark.[231]

And he emphasizes that love '*is not expressing itself in terms of the socially approved manners of life because it has nothing to do with the social order.*'[232] Even more clearly, Thomas Moore, in his book *Care of the Soul (1994)* states that '*[m]oralism is one of the most effective shields against the soul, protecting us from its intricacy*'.[233] Thomas Moore pursues:

Thomas Moore

The soul's complex means of self-expression is an aspect of its depth and subtlety. When we feel something soulfully, it is sometimes difficult to express that feeling clearly. At a loss of words, we turn to stories and images. Nicholas of Cusa concluded that we often have no alternative but to live with *enigmatic images*.

Published by Sirius-C Media Galaxy LLC, 2010

Since soul is more concerned with relatedness than intellectual understanding, the knowledge that comes from soul's intimacy with experience is more difficult to articulate than the kind of analysis that can be done at a distance. Soul is always in process, having, as Heraclitus says, its own principle of movement; so it is difficult to pin down with definition or a fixed meaning. When spirituality loses contact with soul and these values, it can become rigid, simplistic, moralistic, and authoritarian - qualities that betray a loss of soul.[234]

Reich stated this fact in similar terms in his book *Children of the Future (1950)*:

Wilhelm Reich

Moralism only increases the pressure of crime and guilt, and never gets at or can get at the roots of the problem.[235]

Finally, I wish to address that endlessly boring and senseless control paradigm of life-denying society that has brought about the split between so-called erós-inspired and agapé-inspired love.[236] Reich appears to anticipate Riane Eisler's research for almost a century:

Wilhelm Reich

The splitting of sexuality into debased sensuality and transfigured love, which generates entire systems of philosophy on the problem of *sexuality* and *eroticism* is nothing more than an expression of the dominant position of the man and, in addition, a consequence of the efforts of distinguished hypocrites to set themselves apart from the masses by adopting a special morality.[237]

The moralists, however, only have eyes for what occasionally, in their opinions, confirms their theory. They do not see and do not even want to see that their doctrines do not apply to the mass of young people, and they duck responsibility for what will happen in the future if people follow their teachings.[238]

Reich explains something very important for our quest to reunite with nature's wisdom and overcome our socially programmed and culturally sanctified alienation, our split existence; it is the fact that when the emotional nature of humans is not bent and has not been thwarted early and life, we are naturally sane, both emotionally and sexually.

To say it crudely, and with Reich, men and women who are sane and natural won't abduct, rape and murder children because of lust for child sex; hence, moralism's reasoning about the 'impossible human' is essentially a *perception error*, and so are our sex laws and the whole body of behavior rules that more or less implicitly assume that when people are unobserved, and let free, they will indulge in perverse acts of all kinds and jeopardize the friendly togetherness of the community.

Wilhelm Reich

Sexual responsibility is automatically present in a healthy, satisfying sexual life.[239]

It is this dependence on parental care and authority which the Church immediately enters the fray to defend, equipped with all the machinery of stultification and platitudes about an avenging God, his eternal will, and his wise foresight in its attempt to translocate marriage and family to divine regions far removed from the real world.[240]

Published by Sirius-C Media Galaxy LLC, 2010

Reich's position is clearly for a *free emotional and sexual life of children as a conditio sine qua non* for overcoming the life-denying patriarchal plague:

Wilhelm Reich

The means which such parental homes use to bring their children to heel consist essentially of sexually intimidating and crippling them and making them afraid of their sexual desires, thoughts, and deeds.[241]

When the Tao was lost, Lao-tzu wrote in the *Tao Te Ching*, the schizoid spirit of dualism began to build images of *ideal substance*. Instead of recognizing substance as eternal change, expressed in the *yin-yang alternation* of ever-changing evolution, the schizoid thinkers began to split the world into what they called *the miserable state of the world*, on one hand, and the *ideal paradise-like state of heavenly existence*, on the other. Despising the origin of their very existence, that is mother earth or *Gaia*, they began to despise and fear the essence of earth, the sparkling spirit of abundant creation, naming it *serpent* or *devil*. Having condemned the source of their bliss, the new inhabitants of a split world were making for the ground of their profound unhappiness, paranoia and the ultimate destruction of their basic life continuum.

Instead of striving for harmony and accepting all-that-is, they transformed in their madness peaceful togetherness into innumerable wars that they proudly proclaim as *war-of-the-sexes, war-for-survival, war-against-evil, war-against-perversion, war-against-drugs, war-against-pedophilia*, and so forth. The result is a world full of strife, war, destruction and a perverse rat-race for material gain and dominance. This new, and even larger, international dominator culture is currently spreading all over the world and our modern global consumer culture is its latest and most appalling offspring.

Taoism, by contrast, teaches that the seeker of truth does not will to consume or dominate the object of his love. The lover of the original state of existence who studies the Tao, the spontaneous principle of creation, is a *lover of small children* as they represent the original inhabitants of the non-split world. This was recognized in olden times all over the world, but has been forgotten and is today carefully veiled behind the lies of violent moralism. This is however not a new phenomenon, as the same, only regionally limited, was occurring when China became a feudal state. Lao-tzu then retired into the mountains and wrote in the *Tao Te Ching:*

Lao-tzu

The more laws and restrictions there are,
The poorer people become.
The more rules and regulations,
The more thieves and robbers.

The wise does not discriminate between the sexes, recognizing that *yin*, the female principle, was first in creation and is more encompassing than *yang*, the male principle.

That is why the wise man who is inspired by the Tao will celebrate and worship little girls as protector-goddesses of his universe. He will not nourish an exclusive preference for one of the sexes because he knows that in relating only to *yin*, his *yang* force will overflow and damage his inner *yin*, and that by

Published by Sirius-C Media Galaxy LLC, 2010

relating only to *yang*, his *yin* force will overflow and damage his inner *yang*. As a result of this fact, and because of his self-knowledge, the sage does *not reject* anything; he is not influenced by the split-paradigm that says 'There are adults, and there are children', as if talking about two different races. Knowing that all things are nourished by the Tao of spontaneous creation and change, the wistful lover recognizes the values of care, love and parenthood in other adults; he does not attack the family order nor the order of the state. The wistful lover does not reject the world, nor does he need to make for an ideal or paradise-like state of happiness. He is happy by accepting all-that-is, and the world as it is, by not trying to do creation better than the Tao.

We have created total confusion in our relationships, and put the *love principle* upside-down, demonized what is naturally beautiful and enriching and put up *false values* that render us shallow and mean, and full of suspicion and fear.

To justify what we see is producing still more confusion and destruction, and in order to veil our millenary stupidity, we blame nature and human nature. While it is so obvious that it's the perversion of nature and human nature that has created the mess and brings about the destruction, but not nature herself, we go on affirming that nature was wrong and not to be trusted and our so-called *scientific* mind could *correct the errors supposed to be inherent in nature*.

The Murder Culture

No murder can happen without being preceded by a murder inside of us way before we set out to kill in the first place. The very desire to kill comes about through the schizoid split created by killing something within ourselves.

Our past millennia of collective murder and genocide were preceded by the killing of one of our internal opposites and thus upsetting the *natural balance of yin and yang* within us: by condemning and tightly regulating *sexual pleasure or certain forms of it*, by achieving to interfere with and repress the natural emotional flow in the lives of our children, we have distorted the natural order, and turned upside-down the subtle energy flow, not only in the human being but, as all is connected, also in the stratosphere of the earth, the planetary energies within our galaxy and the intergalactic energy balance within the whole of the cosmos.

And we had no right doing so because the bodies of our children are not *our* bodies, as we do not own our bodies. The human body as the whole of life cannot be owned. Lao-tzu said the human body is 'the eternal adaptability of heaven.'

Other philosophers said that the universe as our mother earth lends us a body that we have to give back when we go back to the subtle realms of existence. In fact, even the dullest of the dull must admit that we cannot take our bodies into the afterlife. Minutes after our spirit has left our body, the body begins to decay and in a few days it is but a peace of rotten flesh that is virtually eaten up by a multitude of birds,

Published by Sirius-C Media Galaxy LLC, 2010

insects, worms, beetles and other animals and plants that mother earth sends out to embrace back in her substance what she has so generously granted us as a vehicle for our spiritual advancement.

Life is created by pleasure and natural death equally is pleasure as it opens an illuminated path into a subtle *vibrational* existence. Killing natural attraction was the foremost tool of dominator culture to get hold of humans and to manipulate and control them into the literal essence of their flesh and their bones. By the same token and with the same goal, dominator culture repressed the truth about the cyclic nature of birth-and-death, and invented the myth of a linear one-time life that supposedly ends in death as ultimate shock and destruction. The three dominator religions have coincided in suppressing the teaching of *natural reincarnation* that is an essential element of perennial philosophy.

There is a *new culture* now raising especially in highly civilized societies that refuses to stay with analyzing and blaming the terrible state of affairs we are in, and instead practices a new way of living. While these movements are very diverse, what they have in common is that they attempt to become germs or living cells of what could be realized on a larger scale within a new human, and truly humane, society and culture.

While communities were existing already in the 1960s, they were overruled by a new wave of fascism from about the 1980s, but the basic idea is familiar with all those who practice one or the other alternative lifestyle.

Young people today who subscribe to what could be called a *love culture* seem to be inspired by a deep quest for innocence. They tend to accept and understand the spiritual significance of matriarchy and respect what they call the *Gaia* principle, a deep veneration of *Mother Earth*. They are often involved in professions that either involve art, drama, dance and music, or the professions that deal with natural healing, body work, healthy diet and integral living, or else they are unconventional psychiatrists or psychoanalysts, astrologers or numerologists as well as those who engage in one or the other spiritual path such as Yoga or Zen. But there are also people from other professions who individually join these circles, temporarily or permanently.

This brings me to explain more in detail what I exactly mean when I am talking about the spiritual significance of matriarchy. In *Occidental Mythology*, Joseph Campbell observes:

Joseph Campbell

In the older mother myths and rites the light and darker aspects of the mixed thing that is life had been honored equally and together, whereas in the later, male-oriented, patriarchal myths, all that is good and noble was attributed to the new, heroic master gods, leaving to the native nature powers the character only of darkness - to which, also, a negative moral judgment now was added. For, as a great body of evidence shows, the social as well as mythic orders of the two contrasting ways of life were opposed. Where the goddess had been venerated as the giver and supporter of life as well as consumer of the dead, women as her representatives had been accorded a paramount position in society as well as in cult. Such an order of female-dominated social and cultic custom is termed, in a broad and general way, the order of Mother Right. And opposed to such, without quarter, is the

Published by Sirius-C Media Galaxy LLC, 2010

order of the Patriarchy, with an ardor of righteous
eloquence and a fury of fire and sword.[242]

In simple words, whenever we face a life paradigm that
does away with the changeability of life and thereby reduces
the concept of living to a monistic, monolithic principle, we
are facing not human saneness, but insanity at its peak, and
the result, invariably, is violence. All the eloquence of Biblical
preachers cannot betray the truth seeker's intuition of what is
naturally sane, and the more missionaries preach and exhort,
the more violent, the more dangerous, the more genocidal
they are. Human colonial history has given abundant factual
proof for that sad psychological reality.

Ours is undoubtedly a *murder culture* because those who
founded it were themselves based upon murder, the rape and
extinction of their surrounding out-groups which was, at that
time of much more limited population compared to today,
already a mass-murder to be qualified as genocide. I do not know on
which mountaintop today's conservatives gather to acknowledge that their worldview if founded on *goodness* or were inducing goodness in people? What *goodness*, the hell, comes
from a worldview in which only the in-group enjoys respect
and where everybody else, including their children, is subjected to torture, rape, murder and genocide? It is here where
the spiritual significance of matriarchy comes in as a regulatory principle. Joseph Campbell affirms:

Joseph Campbell

I am taking pains in this work to place considerable
stress upon the world age and symbolic order of the

goddess; for the findings both of anthropology and of archeology now attest not only to a contrast between the mythic and social systems of the goddess and the later gods, but also to the fact that in our own European culture that of the gods overlies and occludes that of the goddess – which is nevertheless effective as a counterplayer, so to say, in the unconscious of the civilization as a whole.[243]

What has the female become under patriarchy? A *virgin* to be defiled, raped, abducted and killed, on the subconscious level, and a *daughter of good breed*, the obedient slave-girl, at the apparent or outside level, the princess to be married off for material riches paid to the father. An *investment* at best, when older, a household item. When old, a nuisance.

Patriarchy instituted *correction homes* for the young, prisons for the free thinkers and retirement homes for the elder. All those who fall outside the in-group, which is the 20-40 majority, have to be taught that sex is a shame, and has to be repressed. They are deprived of it as a matter of social duty, just as prisoners are. That is the respect patriarchy has in front of the child and old age. The rest is lip service and sentimentalism. *The reality of life speaks in facts, not in cathedral speeches.* An old female, once the person of highest regard and social status in matriarchy, under patriarchy has become a double form of plague, the plague to be a female and the plague to be old and *useless* within the greed machinery of patriarchal making.

Published by Sirius-C Media Galaxy LLC, 2010

The Spiritual Laws of Matriarchy

The *spiritual laws of matriarchy* are the counterplayer in the collective subconscious that Campbell intuits, not the imaginative embodiment of the female in patriarchal minds because it will be debased until the, probably catastrophic end of patriarchy! But patriarchy cannot alter cosmic laws and this, and not human wit, is what I am talking about here.

Matriarchy is based upon a whole range of laws regulating the relationship between the human realm and the animal and plant realms. Patriarchy, by contrast, is based on the violation of these spiritual laws. It cannot last because it dethrones nature and, by doing so, debases creation itself. It is blasphemic, in last resort. And patriarchy's monotheism was born on the blood-soaked linen of raped and massacred nations and populations that have been *sacrificed* for Yahweh's *cool walk in the garden*.

Thomas Moore has spent more than a decade in monastic seclusion as a Catholic monk and he finally quit religious life with all its restrictions, only to discover that life within the busy world of modern international society can guard and purport the same soul values and the same sensitive and lucidly intelligent approach to life he once discovered and implemented by spending long years in monasteries. In his best-selling book *Care of the Soul (1994)*, Moore writes:

Thomas Moore

Moralism is one of the most effective shields against the soul, protecting us from its intricacy. (…) I would go even further. As we get to know the soul and fearlessly consider its oddities and the many different ways it shows itself among individuals, we may develop a taste for the perverse. We may come to appreciate its quirks and deviances. Indeed, we may eventually come to realize that individuality is born in the eccentricities and unexpected shadow tendencies of the soul, more so than in normality and conformity.[244]

Caring for our soul, being connected to all-that-is, implies to pay attention to detail – *all detail* in life. Moralism, in last resort, is a form of shallowness, an ingrained laziness to deal with all the stuff that makes our daily life, including our oddities and difficult-to-admit perversities. Moralism is the banner of patriarchy and for good reason it never had a stand in matriarchal cultures. Whoever is really soulful, and *spiritual*, pays attention to all-that-is and does not make up a phraseology of ought-to's that fills his mind in order to put his soul at rest, so that it does not become too virulent and inquisitive.

The highest spiritual law is *total attention*. The moralist does not deal with detail. He haughtily rushes over all detail in life, declaring that 'little daily matters' did not count for a 'spiritual' man, a man of word, of honor, of principles, a man whose life was based upon *values*. In reality, there are no *little daily matters* as all matters, as all is important for the one who pays attention to detail. Love is detail. For the truly spiritual person, there is no shame connected to talking about his perverse sides, fantasies, longings or deeds. He knows that the energy can flow in one direction and also in the other direction, that energy can retrograde and pent-up and that this

Published by Sirius-C Media Galaxy LLC, 2010

brings about perverse reactions, desires and needs. But to recognize this means to be freed of the obsession to follow-up to these delusional needs.

To admit perversity means to deal with it, while moralism entangles one who arrogantly wipes off the idea of admitting perverse desires, making him a slave of his repressed perversions. That is why non-judgmental, permissive cultures, which are those that are more matriarchal in their base setup really can deal with human perversity, and constructively so, while human history shows with all evidence that patriarchal moralism brings about emotional stuckness that puts on stage a *clean* reality, while behind the stage all the devils are playing hide-and-seek. Moore explains:

Thomas Moore

Care of the soul is interested in the not-so-normal, the way that soul makes itself felt most clearly in the unusual expressions of a life, even and maybe especially in the problematic ones. (...) Sometimes deviation from the usual is a special revelation of truth. In alchemy this was referred to as the *opus contra naturam*, an effect contrary to nature. We might see the same kind of artful unnatural expression within our own lives. When normality explodes or breaks out into craziness or shadow, we might look closely, before running for cover and before attempting to restore familiar order, at the potential meaningfulness of the event. If we are going to be curious about the soul, we may need to explore its deviations, its perverse tendency to contradict expectations. And as a corollary, we might be suspicious of normality. A facade or normality can hide a wealth of deviance, and besides, it is fairly easy to recognize soullessness in the standardizing of experience.[245]

The spiritual laws of matriarchy are of course no written laws. They are no worldly statutes or regulations. They are truths valid on a cosmic level, and on a soul level. But they are observable in the lives of those who live in close relation with nature, for example, tribal peoples who maintain a living spiritual contact with all natural forces through their *shamans*, and their long-standing traditions of dialogue with nature's wistful energies.

When I talk about the *spiritual laws of matriarchy*, I do not mean general spiritual laws such as the law of attraction, the law of prosperity, the law of harmony, and others. I am rather talking about *patterns of living*, directly observable in the lifestyle of wistful peoples and that are no secret knowledge, but to be verified by any serious researcher on shamanism. For there is no occult hermetic tradition to be studied, as these patterns are directly applied, by tribal peoples, in their daily life and relationships.

After several years of research on shamanism, I came to summarize these spiritual patterns of living.

The *Eight Dynamic Patterns of Living* are the result of an observation of tribal peoples and tribal lifestyle, to be applied for a *partnership oriented* and systemically[246] as well as emotionally intelligent[247],lifestyle and society. I namely found that all tribal populations who apply these *eight patterns* in their lives and societies swing in accord with the movement of life, are healthy, happy, peaceful and productive. I found the *Eight Dynamic Patterns of Living* to be valid as a nature-loving lifestyle that concords with universal laws and that is dynamically

Published by Sirius-C Media Galaxy LLC, 2010

pro-life, favoring mental, emosexual and economic health, happiness and peaceful togetherness.

Bull and Serpent

The spiritual significance of matriarchy is not just a matter of mythology, of energies, of symbols. Its meaning goes beyond the mythic bond of humans with nature and all the forces that have their imprint upon us and the whole of the universe. The matriarchal laws have a direct impact upon our soul. I dare to say that if the soul itself obeys to certain laws, then to these matriarchal laws or patterns that I was talking about in the previous chapter.

Our soul is at odds with the *normalcy* concept that is at the basis of patriarchal laws and their underlying morality code. As Moore expresses it:

Thomas Moore

> Care of the soul sees another reality altogether. It appreciates the mystery of human suffering and does not offer the illusion of a problem-free life. It sees every fall into ignorance and confusion as an opportunity to discover that the beast residing at the center of the labyrinth is also an angel. The uniqueness of a person is made up of the insane and the twisted as much as it is of the rational and the normal. To approach this paradoxical point of tension where adjustment and abnormality meet is to move closer to the realization of our mystery-filled, star-born nature.[248]

The soul really follows the *self-regulation* pattern of living; it cannot be forced to adopt other than its own rules, and its intelligence is not rational, but the *emotional intelligence of the heart*. The soul's major longing is balance, harmony, whole-

Published by Sirius-C Media Galaxy LLC, 2010

ness, and its major effort is the one of healing fragmentation. Carl Jung writes in *Religious and Psychological Problem of Alchemy*:

Carl Gustav Jung

But the right way to wholeness is made up, unfortunately, of fateful detours and wrong turnings. It is a *longissima via*, not straight but snakelike, a path that unites the opposites, in the manner of the guiding caduceus, and whose labyrinthine twists and turns are not lacking in terrors. It is on this longissima via that we meet with those experiences which are said to be *inaccessible*.[249]

The soul and its superior knowledge about life and happiness is indeed inaccessible if we *think* life, instead of living life, applying only our left brain hemisphere and considering only logical thought as being relevant for understanding life and living processes. It can be said that this kind of lifestyle, that today is widely adopted, is lacking shadow, rendering life as a one-dimensional drawing, a *solar* worldview in the words of Joseph Campbell, where shadows are *lighted away* by the sun rays of the purely rational mind. Thomas Moore explains:

Thomas Moore

A neurotic narcissism won't allow the time needed to stop, reflect, and see the many emotions, memories, wishes, fantasies, desires, and fears that make up the materials of the soul. As a result, the narcissistic person becomes fixed on a single idea of who he is, and other possibilities are automatically rejected.[250]

For the narcissist, all in life is about statistics. Love is expressed in percentages and probabilities. But what is the daily life taste of it? Never known, never seen. The narcissist talks about principles, rules and facts: he suggests love could be measured, quantified and scientifically *demonstrated*. Nay, such a thing cannot be, otherwise it would not be love, but the shallow soup that today is yelled from all megaphones of international consumer stupidity. *Love me forever!* The soul knows that love is not a concept and cannot endure according to the mind and will. It has its own life span, and it knows its own death.

The narcissist flies in the air and cherishes lofty Apollonian ideas. But where is their Dionysus? The truth is encoded in that part of them, and carefully hidden from their public appearance. Care of the Soul, for most of us, means care for the Dionysian principle in us, the *Sad King*, as I called it.

Narcissism is an inevitable by-product of patriarchy, and its etiology as *wrong relating*. Wrong relating to self, and as a result, wrong relating to others. It is built on the preclusion of the shadows of the soul – thereby ignoring its own shadow. Narcissists, therefore, are tragic figures. They are tragic in the sense that they run into the abyss without the slightest idea of what they are doing. Because they are not grounded and have their feet in the air, like the *Fool* in the Tarot. They are lunatics, because they have not integrated their own *Luna*, their Moon energy. They are the eternal Peter Pan's of sunshine movies, and present themselves to the public smiling, broadly smiling, most of the time, but in haphazard moments you see their true face – while they themselves ignore it.

Published by Sirius-C Media Galaxy LLC, 2010

Thomas Moore observes that narcissism, in our times, is a problem that by far surpasses the individual and has become a societal concern:

Thomas Moore

America has a great longing to be the New World of opportunity and a moral beacon for the world. It longs to fulfill these narcissistic images of itself. At the same time it is painful to realize the distance between the reality and that image. America's narcissism is strong. It is paraded before the world. If we were to put the nation on the couch, we might discover that narcissism is its most obvious symptom. And yet that narcissism holds the promise that this all-important myth can find its way into life. In other words, America's narcissism is its refined puer spirit of genuine new vision. The trick is to find a way to that water of transformation where hard self-absorption turns into loving dialogue with the world.[251]

Narcissism is so destructive because it eternally believes in shortcuts, quick fixes and once-for-all solutions. The soul however, and evolution in general, proceeds in a spiraled movement, which is something like a circled forward movement. Thomas Moore describes it in alchemical terms:

Thomas Moore

All work on the soul takes the form of a circle, a rotatio. (…) I keep in mind the alchemical circulatio. The life of the soul, as the structure of dreams reveals, is a continual going over and over the material of life.[252]

The spiraled movement is more holistic than the linear movement because it carries our base all along from here to there. That means we do not leave our origins, but remain firmly rooted in where we came from. These roots are essential for providing us with *living energy*. It's a serpentine movement, the movement of a snake. However, our quick efficiency, our stress on immediacy, our lack of time, our focus on straight solutions prevent us from integrating, rooting, personal evolution in the soul. As a result, our progresses are merely peripheral and remain at the surface of the personality.

All this, while it sounds commonplace, is the inevitable result of patriarchal morality because it circumvents the soul. To explain this on a mythological level, let me introduce another symbol, as important or even more important in world mythology than the bull: it's the *serpent*. Ralph Metzner, in the introduction to his reader *Ayahuasca: Human Consciousness and the Spirits of Nature (1999)*, observes:

Ralph Metzner

Not only among Amazonian shamans, but throughout the world, in Asia, the Mediterranean, Australia, serpent images are used to represent the basic life force and regarded as a source of knowledge – the wisdom of the serpent. The serpent image is seen often as a link between heaven and earth, and in this regard the snake is often found in association with other images of ascent.[253]

Joseph Campbell reports two crucial turning points for the cosmic serpent in world mythology. The first occurs in the context of the Iron Age Hebrews of the first millennium B.C where the mythology became inverted, so as to represent the opposite to its origin, the second is to be found in the creation myth where the serpent who had been revered in the Levant for at least seven thousand years before the composition of the *Book of Genesis*, plays the part of the villain.

Yahweh, who replaces it in the role of the creator, ends up defeating *the serpent of the cosmic sea, Leviathan*. For Campbell, the second turning point occurs in Greek mythology where Zeus was initially represented as a serpent, but then, when the myths changed, Zeus became a *serpent-killer*.

From that time, Zeus was depicted to secure the reign of the patriarchal gods of Mount Olympus by defeating *Typhon*, the enormous serpent-monster who is the child of the earth goddess Gaia and the incarnation of the forces of nature.[254]

It is in accordance with this fundamental change in mythology that, as I mentioned earlier, the significance of the bull equally *changed from a matriarchal to a patriarchal meaning*, and the Hispanic tradition of *bull fighting* clearly reflects the perverted patriarchal tradition rather than, as some pretend, representing a matriarchal base structure in Hispanic machismo. While the female principle in the Babylonian epic of creation has been devalued, we can still find it associated with the serpent, the boa, the *Great Mother*, in the natural philosophy of most tribal peoples, as reflected by shamanism.

The *Ayahuasca* reader by Ralph Metzner, already quoted, contains a range of reports contributed by people from all walks of life who have taken the traditional Ayahuasca brew in order to encounter the plant teachers or *spirits* of the plant.

Two of them report visions of the cosmic serpent. Raoul Adamson entitled his experience *Initiation into Ancient Lineage of Visionary Healers*[255] and writes:

Raoul Adamson

I become aware of a morphic resonance between serpent and intestines: the form of the snake is more or less a long intestinal tract, with a head and a tail end; and conversely, our gut is serpentine, with its twists and turns and its peristaltic movement. So the serpent, winding its way through my intestinal tract was *teaching* my intestines how to be more powerful and effective – certainly a gut-level experience![256]

Ganesha, in her *Vision of Sekhmet*[257], reports:

Ganesha

As I read about Sekhmet and assimilated my experience with her, the understanding that formed in my consciousness was that Sekhmet is a Great Mother Goddess, one that spans all time. With the sun disk at her head and the snake around it, she symbolizes the serpent power of the root chakra having risen to the crown. Thus, she encompasses both heaven and Earth, and demonstrates the way to unite the heaven and Earth of our own nature, Spirit and Form, through the awakening of the kundalini power in the muladhara chakra and its arising to the sahasrara chakra.[258]

Raimundo D., in his *The Great Serpentine Dance of Life*[259], writes:

Raimundo D.

The plumed serpent is masculine, involves outer impression and show of power; the unplumbed serpent is feminine, involving inner expression and statement of strength. (...) I experienced my entire body being reprogrammed and rearranged, even reconstituted at the deep cellular level. This resulted in an incredible feeling of openness, solidity, wholeness and openness.[260]

Published by Sirius-C Media Galaxy LLC, 2010

We can thus summarize that the association of the serpent as a matriarchal symbol for the Mother Goddess is not only a recurring theme of world mythology, but can be experienced, with the use of entheogens, as a spiritual vision that impacts directly on the soul and super-consciousness.

As it can be argued that what is seen in psychedelic visions is but the content of the collective unconscious of humanity, there is truth in Campbell's statement that the Goddess still today acts as a *counterplayer* to patriarchy, on the level of the unconscious, and this independently of personal beliefs or intellectual understanding of shamanism or nature religions.

CHAPTER SEVEN

A Scientific-Shamanic Approach to Religion

> The fact that modern physics, the manifestation of an extreme specialization of the rational mind, is now making contact with mysticism, the essence of religion and manifestation of an extreme specialization of the intuitive mind, shows very beautifully the unity and complementary nature of the rational and intuitive modes of consciousness; of the yang and the yin.
>
> – FRITJOF CAPRA

Introduction

> Eternal, the Self,
> Source of Contemplation
> Pure and Joyful,
> Free of ideas and convictions,
> Free of theories and ideologies,
> Free of concepts and categories,
> The unique Self is in itself.

American selfhelp literature is certainly useful, and there are people who were completely transformed through one or the other of those methods, provided they applied it seriously, consistently over a certain period of time, and without giving up. The demand for such kind of guides is considerable. The books written by Dale Carnegie, Joseph Murphy, Anthony Robbins, Stephen Covey or Deepak Chopra were translated in more than twenty languages and continue to appear on bestseller lists around the world.

In the present chapter, I will present a somewhat different approach. It is markedly different from most of these publications, if not by its modest tone, but anyway because the approach I propose for *religio* is completely intuitive, yet scientific. It is scientific because it proceeds empirically and without any preliminary assumptions.

I have not in mind to help anybody achieve brilliant results, as those results may not last; what I am going to show is that there is a way to our inner resources that is accessible to each and everybody. To get there, no casebook is needed, no classes need to be attended, and no library needs to be researched. All that is to know is contained, in form of holographic patterns, within the unique Self. *It is a shamanic quest.*

If then the unique Self is the guide, as for example Ramana Maharshi affirmed it, I don't know why anybody should spend their money on expensive workshops for healing and self-transformation? Most of those books only repeat verities that you possess, literally, within yourself.

After all, a guru can only speak about his or her own experience, but yours will be different, for sure. You can never do what the guru did, simply because your action is a different one, because you are a different person. Thus, how can you think a guru can be of help to you, given that the guru can only speak for his own spiritual pathway, not yours? Why should you not ask your inner guide, the unique self? This is what I am going to write about in this chapter of the book that is dedicated to all those who run around in the world searching for what they will never find – because they never lost it, and all their outside search will not help them to see the treasure they bear inside!

Now, how to access this inner realm, how to communicate with it? Is it at all possible to communicate it, or is religion bound to remain ultimately an intimate experience that cannot be shared?

I do not give an answer to this question here, and for good reason, as our little self-inquiry will result in an overall insight that *intuition bears its own veracity* that need not be corroborated by rational mental concept making. On the contrary, we are going to see that when we talk about religion, that is, contact with our inner source, such concept making is

Published by Sirius-C Media Galaxy LLC, 2010

a fallacy as it leads to a highly distorted communication. As the terms and phrases used to describe religious experience are always bound to be personal, their sharing leads to an impasse, *when such sharing is verbal.* It may be shared through silence, sitting together in silence, yes, as in such a case the sharing is on a deeper, telepathic, ethereal or metaphorical level, on a level of shared beingness, on a level of one may call 'total communication'.

But what our big dominator religions have always done is to encourage verbal sharing of religious experience, through scriptures, holy books, and *preaching*, that is, the verbal transmission of such content. And here is where the maze opens up because thought comes in and with it, mental concepts, which are in turn conditioned by previous experience and by beliefs of all kinds. I have questioned all belief, and making of belief in this text, as belief is another obstacle to the religious experience, as it distorts direct perception and may even stand in the way of it.

Eventually, as many religious distinguish between belief and faith, we may ask if faith is the trigger of the religious experience? But faith is but an inner attitude, it's a certain openness to the miraculous, and in that sense it may be conducive. However, in most cases, when people say they have Christian faith, or Muslim faith or Buddhist faith, their faith is not just mere openness to experiencing the unknown, but conditioned by their particular religious belief system.

In such a case, faith is not conducive to experiencing the unknown. Krishnamurti said it very lucidly in telling us that we always find what we are searching for. The Christian will encounter the Virgin Mary, the Muslim will experience the Prophet, the Buddhist will see the Buddha when they make such transcendental experiences. Hence, they do not meet the unknown; they encounter projections of their own mind when they are in that state of trance where we experience the content of our subconscious mind. This is not what the religious experience is about. And that is why I conclude that faith is not conclusive to that purpose either.

What really leads us there cannot be expressed verbally which is why I cannot answer the questions I am asking. Answers anyway are temporary, while the questions, the basic and great questions of life, remain. So we need not bother about answers, and don't need to search for them. When we leave the questions open, subtle answers will come through all kinds of circumstances, *visions, intuitions, hunches, encounters, dreams, spontaneous insights or sudden realization of transcendental truth.* But as long as we search for answers, this whole process is blocked and cannot unfold.

It is intuition, pure intuition, that leads us there, by *not* leading us there. We cannot approach the divine, we cannot step out of blindness and into total enlightenment in a second. But the divine can do that, and approach us that way when we remain supple, open, and innocent in the right sense of the word. Hence, true *religio*, religion is more of a state of innocence than anything else; it is pretty much the contrary of what religions do, it's more on the passive side of life, more on the contemplative side of life, and requires *very little or no ritual,* other than inner silence and a certain poised attitude that is basically nonviolent and non-demanding.

Published by Sirius-C Media Galaxy LLC, 2010

The Unique Self

Self-development is not a modern idea. Truly it is age-old, and was practiced by the sages of ancient times; it was an integral part of the Egyptian and later the Greek initiation rites, and it was taught by the Celtic seers and many other peoples. This kind of work is really at the basis of shamanism, which is one of many pathways to human perfection. It is also to be found in our holy books around the world, as well as in Sufi literature and the books written by masters and gurus from the Far East, and last not least, the pamphlets of yogis and alchemists.

When there is an excessive thirst for novelty, as it's the sign of the times in our modern society, there is no consistent line of systematized wisdom about life and living, but trends, fashions, and interest groups that bombard and influence our national and international media. What happens here is that simply an old vocabulary is being re-baptized and millenary teachings are sold under new names and fashionable slogans that fit the addiction of the day.

What we need instead is to define the role of man in the cosmos on both the individual and the collective, transpersonal levels. What are the sources of knowledge that lead us there? The danger in such a situation is that we try to respond to our present problems with old recipes that we find on the flea market, or the new age bazaar, instead of finding answers from deep inside us.

This is what is called the 'vulgarization of knowledge', which leads to not an evolution of culture, but rather to the decline of culture.

When one is unable to see the common roots behind the multiple problems, and behind the thousand and one manifestations of those problems, one is quickly caught by the number one ghost of our times, *boomeritis*. The temple merchants get those, mostly from our younger generations, who have lost their souls and navigate on the waves of one or the other 'spiritual' addiction, and they make huge profits with this kind of eye-wiping.

I do not talk about serious and respected authors in this context, who, before they propose their novel ideas, take explicit reference to appropriate perennial teachings or native wisdom. Unfortunately, in the big confusing new age bazaar, these authors are rather the exception that confirms the rule. It seems to me that books that deny the origin of their ideas are sold much better, while their appalling arrogance in allegedly producing 'striking novelty' is all-too-typical for our modern media culture.

To read about concepts, old or new, is nourishing our left brain, but it does not serve the synthetic thirst of our right brain hemisphere. Hence, for developing *holistic thinking capability*, which is the way the unique self 'reasons', we need to establish a *harmonious synching of our two brain hemispheres*. This is best done by listening to music and by relaxation, to enter a *receptive* state of mind.

In this particular state of mind that is known under the term of *alpha state*, we become creative and can begin to ex-

Published by Sirius-C Media Galaxy LLC, 2010

press our inner talents. A guide then is readily at our disposition, the unique self. It directs not only our spiritual evolution but also our creative process. It doesn't obey to any guru for the unique self is itself the guru. Any guru you find crossing your path, in whatever dimension this happens, is an incarnation of the unique self.

Before I come to the essence of my intuitive journey toward true *religio*, I would like to give you some practical advice how you can strengthen the bond with your unique self.

First Advice

Observe

Observe what is in your way, here and now, to reach a complete state of inner peace, a state of total happiness, of divine bliss! What is it you are lacking out on to get there? Take a sheet of paper and list all the points that come to mind. This is a *negative list*. In this list, you note all the obstacles to your happiness, be it financial hurdles, but it personal issues, be it a historical or global situation you find is oppressing you. If you believe that because you are Jewish, you can't get what you deserve in this world, write on the paper: 'It is because I am Jewish that I cannot be rich, recognized and happy.' If you are a young girl and you can't find a lover, write: 'It is because I am too timid [stupid, fat …] that I do not find a man who desires me.'

Why should you do this? When you write down your negative thoughts, the negativity that is attached to those thoughts slowly vanishes off and disappears. Why?

It's because you focus your attention on what doesn't work in your life. What is the result? You are going to dissolve what doesn't work in your life. And why that?

Because all dissolves under the light of consciousness. Is that uncanny? Is that disturbing as an insight? Are you interested to know why this is so? The answer is that consciousness is fluid, and problems are static. This is an insight from bioenergetics. Speak of the devil, and the devil comes to you.

Speak constantly of the devil, and you will build a shield against the devil. Or, to express it in Buddhist terms: if you press too hard when you have a fish in your hand, the fish is going to escape your grasp. When you affirm that you are stupid, one moment later, you are going to be less stupid!

And wait a moment, look at this. You are going to write down what you always thought and once of a sudden, this affirmation appears to be ridiculous to you. It is namely because it was not always that way, because you were not always negative. You will then see that some of those affirmations, once you have fixed them on paper and reread them, will appear to be so stupid that you will begin to question them. It is exactly in this moment of doubt that the truth can begin to operate in your psyche, and that you can begin to hear the voice of the unique self.

Once you have done this list, do another list, an even more important one. Do a *positive list!* This list will contain all your desires, material or immaterial, all you wish to have or to be. This intellectual work on preparing and programming your future, it's well with your left brain that you do it. Then another task is waiting for you, this time of a totally different nature.

Second Advice

Focus

All religion results from connecting with the unique self. Are you aligned with it? How to get there? In watching inside of you. Introspection is an age-old technique for getting in touch with your source, and that's why this kind of work is called religion. For religion means *religio*, derived from the Latin verb *relinquere*, linking back. The term correctly connotes the meaning.

However, and unfortunately so, the practice of so-called 'religious organizations' has perverted the original meaning of the term and has reversed it to its contrary. Natural religion namely proceeds from inside-out, through acquiring self-knowledge, while religious organizations work at the periphery of our being, and thus by using *indoctrination*.

432 | Chapter Seven

The Secret and the Real

An inner approach to life, an exam of what it means to be religious, an introspection that is meant to lead to the discovery of the eternal mysteries, must it not be extremely rigorous?

You are going to ask me if this approach is mystical, or rather scientific? And my answer is that our soul is an infamous garden where there are the deepest abysses of our beingness but also the greatest resemblance with the divine, our inner god, our unique self. There is no difference between mystical and scientific reality. All inquisitiveness is gradual, all knowledge comes in junks and quanta; the distinction between mysticism and science stems from pure ignorance.

The more you know about the mysteries of life, the more you move from the *mythical-personal* to the *verifiable-transpersonal,* and as a result, the more you can share with others.

If you see only the outside shell, if you remain at the periphery of your being, you will lose yourself in endless projections, in endless mental concepts, in endless rationalizations and superstitions.

It is by questioning that you gain knowledge, not by answering your questions. I am not giving answers here. I ask questions. If the question stays without an answer, it germs, it develops, it transforms itself and becomes creative.

That's why it is better you stay with your questions than giving temporary answers to them. For you must see that all answers are temporary. Only the questions remain.

Published by Sirius-C Media Galaxy LLC, 2010

To begin with, how can we know what another person expresses? Do you know your own language? Do you know to correctly use the words you put on things? Do you know how the way you see the world is accurately transcribed by the language you are using? I am not talking about musicians here, nor painters, but people who express themselves and their philosophical views primarily through verbal language.

Have you observed how you create a personal vocabulary over time, and how you change it, and how you recreate it from time to time? Have you observed to what point the language you are using is the result of your mental, social, cultural, emotional, psychic and sexual conditioning?

Everybody has their language, right? But this language, is it not the final outcome of a long apprenticeship? Everyone thus has his or her personal language, using certain notions, words, phrases, which depend on their particular personal, cultural or religious conditioning in the sense of a cult, a ritual, an organized something. For giving an example of this essential problem in all human relations, I will talk about a man that people use to call a saint. We go there and we listen to him. This man speaks of *reality*. He made certain spiritual experiences, people say. And he talks, almost constantly, of reality. And we don't know what to think of that term and wonder *which reality* he conveys?

What do we know about how this person perceives the 'reality' he is talking about? Is it at all something that can be communicated? How can you know the meaning of that word he uses, the word 'reality', as you are different, and thus live in a different reality? Can you ever know exactly the meaning of that term he uses, given that his verbal description of that reality is but a transcription of his sensorial or extrasensorial perception of it, and given also that you did not share in his experience? But even if you had shared that experience, you would certainly have perceived things in a different way, because you are *other*, because you are the *summa summaris* of your own conditioning.

Everyone of us, it seems to me, is conditioned in a different way, even if we have gone through very similar life experiences. What essentially is different is not the stimulus but how we *react* to the stimulus, which means how we react to the conditioning influences. This reaction is always genuinely personal for it depends on the state of our body and our soul.

So after you see this, how can you imagine any possible communication between people about religious experience and thus about reality?

Published by Sirius-C Media Galaxy LLC, 2010

Body and Soul

We are distinct because our bodies and souls are distinct. There is desire. We find certain bodies beautiful, attractive, seductive, wonderful and sweet. We love those bodies, and our own body, through our desire. We want to unify our body with theirs, through loving copulation. And it is desire that makes us grow when we are still in childhood. Desire gives us satisfaction, joy, ecstasy, and when we understand it intelligently, we develop our intrinsic beauty and grace. Desire is at the origin of all life; it creates, procreates, constructs, transforms. Desire animates us and ensures the pulsation of the vital energy, so essential for our physical and psychic health and our longevity.

It is this same desire that lets us choose certain groups of people, or a specific person emotionally, as a matter of predilection, and as a result we feel that *this desire sexualizes somehow.* We feel that love is both, the emotional and the sexual attraction, and some higher form of attraction that is neither emotional nor sexual, nor even related to pleasure. We feel that love is all of that and that we cannot reduce it to any part of that whole.

We may ask then, if there are different desires, emotional, sexual desires, or only one desire? In other words, the essential question is if we can just desire a body? When we desire to make love with a certain person, does that mean we desire only their body? This seems to be the prevailing view. But how is that possible, given that there is no body that is not animated by an incarnated soul? Only those who desire cadavers, which is a very rare perversion, desire only a body.

More generally put, we see now that we cannot *not* desire the whole of that person, that we cannot desire 'just their body'; it is logically not possible to be attracted only by the body of that person, and not by their soul and their whole being. For the soul expresses itself through the body, and forms the body, giving it its beauty, its vibrancy, its erotic flair, its charm, its aura.

A body, as a body only, can it possibly desire anything? If you follow the dualistic principle that postulates a distinction between body and soul, a Platonic and later Christian concept, body and soul can exist at different levels, so to speak.

But a body without soul is dead! A body without soul is a cadaver for death is defined exactly by the fact that the soul leaves the body in a fluid subtle shell, that goes elsewhere, to another field of vibration. *The body as such can thus not desire anything.* It's always the soul or, as psychologists say, the psyche that desires. Desire is thus located in the psyche, and it's only from there that it gets to be felt in the body and becomes visible through physiological reactions. And love, we have seen, is equally related to the soul.

Sexual desire and love, how can they be separate, as the dualistic concept assumes? How, when there is union with another through loving embrace, can there be separation at the same time? Does the body desire another body because it is not content with itself?

There is an ancient myth, recounted in Plato's *Banquet*, which explicates that desire comes from the fact that in olden

Published by Sirius-C Media Galaxy LLC, 2010

times, we have been androgynous and bisexual, more complete thus, as we had both sexes within us. Then we somehow lost the other half, which is why we run after it. But desire, is it always desire for the other sex? *And homosexual desire, is it not desire?*

Psychoanalysis reveals that we are all bisexual on the psychic level. The Cabala, Tantra, Taoism and other ancient religions affirm this since times immemorial. In the Tarot, the arcana of the *Monde* (World) represents the end of the voyage of the Fool, the androgynous being, the superior, integrated form of human accomplishment. The archetype of complementarity, the completeness of being, they can only be conceived hermaphroditically. By the same token, the terms of *Naljorpa* in Tibetan language and *Yogi* in Sanskrit mean 'the person who has unified the male and females principles'.

This archetype is not at the root of desire, it is rather at the root of the sublimation of desire, in the form of self-contentment. It is at the same time the beginning, the childhood, and the end of life, old age. Desire thus cannot be said to be per se projected upon the other sex. However, we can say this, desire is always projected upon a person, and it always, as we saw, embraces both the body and the soul of that person. It doesn't really matter in this context how we explain the origin of desire, if we recur to that old myth, or otherwise. What about staying with the question? It is obvious that without desire, there is no life. Desire thus belongs to life. This simple observation, can it not suffice as an answer?

Desire and Morality

Morality, be it of religious or ideological origin, has arrogated itself the right to regulate desire, to regulate the choice of love objects, and the choice how desire may be satisfied, or not be satisfied. It went as far as creating a concept called *chastity* which boils down to a metaphorical castration of desire, if not a condemnation of it, or a sacrifice of it for some dubious notion of 'purity'.

We may rightly ask where compulsive morality or moralism takes its legitimacy from? Is it based upon natural right? But how could nature grant a right that basically betrays and defeats nature itself, and its own survival mechanism? Moralities have made laws, first ecclesiastic laws, later state laws that regulate desire according to certain criteria such as sex or age of the partner, and the way the sexual embrace is carried through.

Morality gives answers. Life asks questions. One of these questions is how love and desire have come to be separated, as they are, in fact, in our days? What has morality done to love? Has it not profoundly sullied love, has it not dissected love into loves – parental love, self-love, sexual love, child love, charitable love, and so on and so forth? What remains of that body called love after it has been vivisected that way? Has it become a mental concept, or a commercial concept? Does it bear a label now 'Love such-and-such'? Has it become respectable? Or has it become a hypocrite lie that has not deserved the word *love?*

Published by Sirius-C Media Galaxy LLC, 2010

Were the images desire is clouded in not always chosen among our primary archetypes? In all ancient religions as in Tantra, Taoism or the Cabala, sexuality, the embrace of a noble man with his young wife have been considered as an essential element of the Divine in man, the vital flow, the mystery of the deep unity of all life. Can we go from there, now at the beginning of the 21st century and recognize the goodness of *all desire*, in all its forms? Would we thereby not embrace and honor the wholeness of life, and its vast space of freedom? Would we not honor humanity within that creation, given that human desire is *not an animal-like automatism or instinct*, but is living flow, emotional flow, which swings, like a lotos flower, in the wind of destiny? Would we not be able to draw more happiness from our existence, and reach the peak of *wellbeing*, and at the same time the peak of transcendence?

Why has this wisdom been buried down so deep in the course of history, in the course of what we call civilization? Why does moralism destroy life, pleasure, contentment, happiness? Why has it destroyed the primal continuum of childhood, to replace it with endless taboos, moral terror and shallowness?

Why have we come to crucify the person for expanding the collective, and collective guilt, by subduing the individual under pseudo-religious doctrines that only bring about violence? Why, under the pretext of progress, have we created murderous ideologies, by they capitalist, militarist, socialist, fascist, communist or other? Why crucifying natural life functions, why creating guild and shame for doing what is natural and feels good? Why regulating, condemning, vivisecting desire, love? Where does that *hubristic arrogance* originate from that pretends to know what kind of desire is 'natural' and what kind of desire is 'unnatural'? How did this life hate come about, this very denial of happiness, of pleasure?

And why have effectiveness, duty and utilitarian concepts become more important that love in our society? Why have we allowed violent moralism to kill love, desire, life?

Why do our highly industrialized postmodern cultures stink like cadavers? Why is their emotional flow locked? Why is there so much ice in human relations, and so much misery, so much suspicion and persecution? The emotional plague, as Reich said, it is that, perhaps. It's the agony of a culture that is dying to its old skin, and all the *false values*, all the *pseudo values* it was based upon and that have brought about nothing of real value.

The signs of global sociocultural transformation are visible to everyone today. But our regard should rather go inside and contemplate our inevitable inner cataclysms. Where we can act, where we have the power to act, inside of us, there is really a lot to do!

Published by Sirius-C Media Galaxy LLC, 2010

Approaching the Divine?

Can we approach the divine? Can you approach what is unknown, by forging a mental concept about it, an intellectual rationalization of it, or by calling it Gee Oh Dee or otherwise? Can you get there by belief?

What is belief? Is it not a mental concept as well? Is it not a construct of thought just as ideals are, as ideologies are, as religious or scientific theories are? Faith, is that the answer and the way to go? Can go go toward the divine at all? Or do you have to wait patiently until the divine approaches you?

The divine in yourself, can you cultivate it. Can you become a religious person, free of all concepts, free of all conditioning, free of all belief?

Can you become a saint – and sain? Or are we chosen people in the sense that we are chosen by the divine, and cannot choose ourselves? Are we predestined to be vessels or vassals?

What is a religious life? Is it to follow a certain belief system, a sect, a religious or spiritual dogma? *Is to to believe anything at all?* Is it reasoning? Or is it to sense things on an intuitive level? Is it spiritual or esoteric knowledge to get there, or is it done by rigorous daily practice, yoga, asceticism, abnegation of self, chastisement of ego? Is it to follow the teaching of Christ, the Prophet or Buddha?

Is it discipline for carrying out certain rituals, or a particular cult? Is it sacrifice, mental torture or the appeasement of thought, the quest for inner silence? Is it to follow Eastern wisdom or a particular guru who rejects one thing by indoctrinating another? Is it all of this, or nothing of this?

Is it not rather a personal path that, in all freedom, leads to an abyss? Is it not living in abundant joy and love for all beings, a love so strong that it seems to tear apart our heart? Is it not to remain open and supple, to accept our oscillation, to accept change and constant transformation as a pattern of life? Is it not to live with all our senses, and sense all, not just ourselves, but all other beings, all other living things, and to embrace them sensually and lovingly?

It is not to see and hear what goes on around us, and in the world, instead of closing our eyes for 'not being hurt' by human suffering? Is it not to get out of our walls of silence, out of our prejudice, out of our mental concepts, and theories – and start to really communicate with others? Is it not to give up all self-serving security and to give oneself in the hand of destiny? Is it not to observe, to question, to be critical, while being humble, avoiding to become arrogant, cynical or cold? Is it not to develop true empathy for self and others, true patience with self and others, true latitude with self and others?

Is it not embracing all-that-is, both joy, and suffering, desire and satisfaction, and the whole dualism of life, while being observant of the unity behind that apparent dualism? Is it not to accept life as it is, instead of striving for what should be?

Is it not to remain childlike and joyful, creative and spontaneous? Is it not to observe all attachments without trying to brutally tear them off, but to understand their significance? Is

Published by Sirius-C Media Galaxy LLC, 2010

it not to recognize that attachment is related to a certain form of consciousness, and that we can grow beyond it?

Is it not, first of all, to love and not flee it?

CHAPTER EIGHT

The Integrative Function of Shamanism and Channeling

> The religious idea of God cannot do full duty for the metaphysical infinity.
>
> – ALAN WATTS

Introduction

In the last chapter, I shall give an account of *intuitive insights* that came about after years of meditation and assiduous study of both *spiritual and channeled literature* from sources around the world, including native cultures. These insights, as paradoxical as it appears, show that the ancient shamanic understanding of the universe, and reality, was basically scientific, not prelogical, nor mythical.

It was based upon observation of nature, while this observation was much more holistic than our today's 'scientific' worldview, in that the observer was not just a logical freak, but a full human who was absorbing reality on all levels, the rational, the emotional and the spiritual level.

I also contend that the science of shamanism was *methodological* in just the sense modern science is, in that observed phenomena were meticulously catalogued and systematized.

This is hardly known because the shamanic tradition is, as most ancient traditions, an oral tradition. It was possible only because at that time, scholars, be they shamans or alchemists within our own tradition, had to have huge memories; it was virtually in most cases only on memory that they could rely upon when, for example, concocting a healing tincture composed of a *mix of collateral plants* and where the composition had to be precise.

Generally speaking, the overall impression I gained from these insights is that our human consciousness today is rather limited, and not worth the great human that was originally set in this world, as a unique and self-regulating creation. What we have today is the 'small human', an almost grotesque reduction of its original plan, a dwarf being spiritually, mentally, emotionally and even sexually. A dwarf being not rendered a dwarf but *having himself restricted to a dwarf universe*, by putting all kinds of limitations to its original, unbounded and very versatile and flexible structure.

Nature created us totally free and we have done all we could to do away with this freedom. Nature created us totally intelligent and self-reliant and we have done all we could to do away with raising children in truth and intelligence, molding them to our hundreds of concepts that are not worth the ink spent on writing them, perpetuating them, in our books of wisdom, bibles and other holy (and dusty) books.

Nature has been honest with us; we are not born clothed, and we are not holy enough to not being urged for toilet visits once in a while where we face and smell the not-so-nice nature of our bodily universe. We have done everything to pay nature her honesty back, by being largely and irresponsibly *dishonest*, hypocrite and abject over millennia, and in a manner that I can only call criminal. And the most dishonest creatures are paid, in our society, for being 'professionally dishonest', our politicians and members of parliaments, and those scientists that are top of the list when it goes to justify the past or present abuses, and that are generously funded by the military.

Nature has been humble with us, to cloth all its supreme wisdom and power in a human skin, and to embody in the human being the greatest of intelligence among all creatures

Published by Sirius-C Media Galaxy LLC, 2010

on the globe. We have paid nature this humility back by having become the most *hubristic, fragmented, violent and devastating creature on the globe* and probably within our planetary system, the creature that has accumulated so many weapons of mass destruction that it can unleash the overkill of the planet *more than four thousand times*, and that has brought about, through its acclaimed and highly intelligent science, the ecological ruin of the entire planet.

Nature has been taking care of us, to preserve our species, and we have paid nature this care back by ruthlessly eradicating thousands of species, and, worse, to reduce our own species drastically through large-scale murder, persecution and genocide, and this independent of time and space, in virtually all epochs, throughout human written history.

Instead of repeating or paraphrasing the insights given in this chapter, I encourage the reader to just re-read it over and over again, for something is likely to happen on the consciousness level when doing this. It is much less likely to happen by paraphrasing the originally received insights, as they were coming from not my intellect and not even my being, but from a source I qualify as 'intuition' and that others would qualify as 'channeled' or 'spiritual'. The word is not the thing.

What is important to retain after having read this last chapter, is that you and me are equal in that we can at any time walk a path of self-criticism and freedom, unaffected by the cultural and social hypnosis, and look over the fence.

I have tried to do this with the present book, and I went even a step farther, showing that tribal cultures, native peoples around the world, and their wisdom, can be a real guideline for the development of our own culture – which namely is no culture – as shamanic science can for our own science.

Of course, to be so humble and recognize our need for change and for new learning needs courage, and persistence, and while only very few humans do that on a consistent basis, that doesn't mean it's impossible. It means that nature, despite all the perversion the human being has brought about, holds the backdoor open for those who do not agree with the lies and fairy tales that are served in our lukewarm media soup with every day to come, and who begin to think for themselves.

Our culture has never fostered an integrative worldview, while this was since times immemorial the natural, native point of view and regard upon life. Our own cultural tradition, by contrast, was one of dissection, separation, in an attempt to intellectually grasp the world, and put it in our mental drawers – where it died. We have arrived at an impasse and it is time for a change of our basic life paradigm, and develop an *integrative vision of all living* that is based, physiologically speaking, upon a higher coordination of our two brain hemispheres. And on the level of religion, law and world order, of course, the change is one from punishing life to embracing life, and one from preferring the 'possible human' over the 'impossible human'. We are all naturally neither good nor bad, as these terms are bringing confusion and strife to creation which is perfect in its imperfection, and imperfect in its perfection.

Published by Sirius-C Media Galaxy LLC, 2010

It simply *is*, existentially so, and moral values have no place when it's about comprehending creation. Instead of judging creation, we could then get ready to eventually start with *understanding creation*. And from there it's but a step to understanding ourselves and our role, as shamans or simple humans, with a truly scientific mind, which is a mind that is neither purely rational, or mechanical, nor merely artistic or poetic, but *visionary*. It's the visionary mindset that best defines the shamanic experience, and the visionary mindset is one that is both deductive-logical and integrative-emotional.

Or, put simply, and in the words of Baron d'Holbach, 'knowledge becomes comprehension when it swings together with emotion'.

On Consciousness

§1

Reality is wake reality plus dream reality. Dreaming is not only, as it is assumed by mainstream psychologists, a form of psychological digestion and integration of conscious events and experiences[261], but also a direct collection of knowledge and thus, a special form of perception.[262]

In dream, our perception shifts and runs on altered frequencies allowing us to perceive reality not in a logical and sequential manner, but in a *synchronistic way* that allows time to stretch and dissolve. In dreams we have access to past and future events without the limits set by the rational and time-bound mind.

In dream, we perceive reality in a way more akin to the right brain hemisphere; as a result, the rationality of dreams is one of *associative logic*. In addition, dreams are a test forum for ideas, plans and projects. In dreams, we can create unlimited virtual realities in which we can act at will, take roles and act out our ideas on a virtual stage in order to see how they will affect others and the world.

Nowhere in nature the Darwinian theory of the surplus power of the stronger is to be found. It was through this fundamental error that dominator philosophies, until today, have been backed up and justified. In nature the main organizing principles are harmony, not strife, order, not chaos, balance, not the psychotic power abuse that humanity developed as a secondary drive structure because of repression and denial.

Published by Sirius-C Media Galaxy LLC, 2010

§2

Freud said: 'Wherever I went, a poet was there before me.' Many poets are more scientific than scientists because they see more of what is essential in living. Scientists, in most cases, are defined *more by what they blind out* from the abundance of life's appearances than by what they understand about them.

§3

No outside circumstances have an impact on the state of the world; it is our own beliefs that create our reality and consequently it's by changing our beliefs that we are going to change this somewhat ungainly reality that we collectively created through a mindset of limitation, prejudice, the belief in scarcity, and spiritual ignorance.

§4

Every change in life comes about first on a psychic or etheric level and only thereafter on the level of physical or material density.

§5

All our emotions have a direct bioenergetic impact on other humans, animals and plants, the environment and even the weather. There is a direct link between emotions and cosmic bioenergetic events through the natural streaming of the cosmic energy field that penetrates all and animates all.

§6

Our life experiences are not 'falling from heaven' but are the direct consequence of our beliefs and convictions. In this sense, all in life is *synchronistic* ass part of an all-connected web of higher logic. Nothing in our universe is single and unconnected.

§7

Consciousness consists both of an *outer and an inner reality.* The inner reality is as important as the outer, and its main functioning mode is intuition. Intuition is the highest form of intelligence. When we follow our intuition instead of acting according to our beliefs or, worse, because of outside signals and commands, we always act right.[263]

§8

Psychology and religion have to meet because they are complementary halves of one and the same science: the science of *total awareness.* It will not end war when we hate war and it will not end racism and discrimination when we hate racism and discrimination. We have to begin to switch our inner polarity from negative to positive. Then our vital energies will be used much more economically and we will dispose of a higher bioenergetic potential. With this *higher energy potential* we will naturally strive to connect more with the positive than with the negative forces of the universe.

We cannot *filter out of our consciousness* undesired events, situations, perceptions, insights or generally any knowledge

Published by Sirius-C Media Galaxy LLC, 2010

that will burden us. But what we can do is to see the *relativity of all information* and counterbalance burdening knowledge with positive, uplifting information. The world always offers both sides of the spectrum, and we are limiting our powers when we focus only on the negative, or the positive side of life or of people.

§9

All beings are born innocently and there is no original sin. Empowerment is a consciousness-enhancing process that acknowledges the natural absence of any guilt or shame; all existing guilt and shame are a result of *negative childhood conditioning* and came about through hypnotic spells and abuse.

§10

The belief in illness and faultiness of nature or the human body is the result of *guilt*. A guilt-free worldview looks upon life and studies health, and not death, and sickness.

§11

Health and longevity depend on the capacity to render beliefs conscious and to feed the mind with positive and integrative beliefs. The body is as conscious and as intelligent as the mind and works in synch with a mind that is feedbacking well with the body.

Nobody dies without having taken the decision to pass over to another dimension and move on with his or her spiritual development in the other, more subtle, dimensions. By the same token, all those who die by accidents or catastrophes get to know in advance about their anticipated destiny, and can change it if they wish to. Their involvement in the event was a choice they have done on a subconscious level.

§12

We live many lives simultaneously. Reincarnation is a misleading belief as it suggests that we lived lives sequentially. In reality, we live several lives at the same time, but on different consciousness levels and dimensions.

§13

Most diseases are the result of repressing our natural emotions, especially our hot and aggressive ones. When aggressiveness cannot be expressed in constructive ways, it acts like poison inside of us and, in the long run, blocks our *emotional flow* which in last instance means to destroy inner organs.

§14

Our dreams have a major task in self-healing. Our psyche and our soma are both cleansed through healing dreams, many of which we are not conscious of because we forget them in the moment or before we wake up.

§15

The identification with either the body or the mind is unhealthy and brings about imbalances. Identification with

Published by Sirius-C Media Galaxy LLC, 2010

the body leads to the wrong belief that life ends with death, while identification with the mind locks us out from our body and therefore in the long run destroys the body, which leads to premature death. Only the healthy integration of mind and body enhances vitality and generally leads to longevity.

§16

Music balances the bioenergetic polarities in the mind-body; in addition, it stimulates and enhances our natural self-healing capacities. However, effective self-healing is blocked to a large extent when the person cherishes beliefs that belittle or deny nature's inherent self-healing potential.

§17

Consciousness and science are linked to each other in the process of *perception*. Perception proceeds through *sensation*. Sensation is something like a window through which consciousness accesses nature. Through the gates of perception we realize the depth of knowledge that with quantitative science alone is not accessible. A holistic and *whole-brain process of perception* is needed for accurately conceiving and comprehending reality in a scientific manner.

On Love

§1

Love is part of living like breathing and drinking; love is self-expression. This means that it is inherent in love that it wants to express itself. It is difficult to live love naturally in a pornographic society.

§2

A society that is pornographic is a society that represses emotions and allows to express them only in a very limited range of so-called *sexual behavior*; however, emotions are universal and by far not limited to sexual expression.

§3

Sexual love within the family has often karmic reasons. Parents and children are often *complementary* in that they incarnate highly different or even opposing characteristics. For similar reasons, a couple may get a psychotic child because they are too stiff while the child incarnates the spontaneity that is lacking in their lives.

§4

All sexuality is an 'emosexuality' in the sense that emotions cannot be excluded from sexual attraction, and also in the sense that sexual attraction follows emotional attraction, and *not vice versa*. Western society has throughout history suf-

Published by Sirius-C Media Galaxy LLC, 2010

fered from a schizoid split of love into *erós* and *agapé*, while this split does not exist in nature.

§5

Most people follow sexual roles that are *cliché models* and not the natural expression of their soul desire; many have not enough self-knowledge to know what they really want. That is why the majority of people are unable to really live their love.

Naturally, our love and sex desires go together and are not to be split on different partners, one for love, and one for sex. Many men suffer from a schizoid split of love and sex projected upon different partners. This schizoid split of love and sex has brought about and brings about war, destruction, civil war and genocide on a worldwide scale. It is largely the result of moralism.

§6

Sexuality is by no means fixated and rather of a moving, changing nature. It brings about strife, unhappiness and neurosis when people are exclusively heterosexual, homosexual or pedophile. All exclusivity in love and sex is a matter of rigid conditioning, of rigidity and neurosis, and not of intelligence and emotional, erotic and sexual sanity.

For the future of humanity, it is absolutely essential that our bisexual nature as well as our *pedoemotions* be acknowledged and socially recognized and validated.

§7

Children are from birth emotionally and sexually *conscious and awake*, but their sexuality has hardly anything to do with what most psychologists project on it. To condition children into early gender separation means to destroy their innate bisexuality.

§8

Both lesbian and homosexual tendencies are natural with children, as pedophile tendencies are with adolescents, and are not to be interfered with. However, because sexual matters are generally little understood in our society, when children display lesbian, homosexual or pedophile attraction, such desires more often than not are feared and repressed.

One of the results is that natural heterosexuality is a rare exception within Western culture because it is regularly destroyed by fearful and ignorant parents and educators during the psychosexual growth period of the child.

That is how psychosexual imbalances such a homosexuality or pedophilia, and similar patterns are created as bioenergetic imprints early in childhood. This is because natural psychosexual growth processes are interfered with by ignorance instead of letting nature its course. Homosexuality and pedophilia as *exclusive love options* are unnatural, but the paradox is that they arise with higher frequency when early homosexual or pedophile experiences have been repressed and sexual feelings were thus sullied by guilt and shame. All early

Published by Sirius-C Media Galaxy LLC, 2010

sexuality that is lived in freedom develops into genuine and lasting heterosexuality.

§9

Most children are conditioned upon toys and pets in order to get away from *feeling* and *being with their own body*; this alienation and conditioning upon body replacements is the condition for international consumer culture to function, and shows *how destructive this society model is in truth*. An overwhelming part of children's natural love potential is thus wasted and vested into material possessions.

§10

Elders have the right to live their love life and sexuality without limits until old age. Sexuality is not limited to a period between birth and menopause but extends into old age without any limits but those set by people's own minds, habits and beliefs. The Bible and other spiritual books abound of examples of couples having had late childbirth, up until in their eighties. Natural fertility depends but on the uninhibited flow of the bioenergy. The body even after menopause is able to regenerate completely so that something like a second puberty can naturally come about with people whose bioplasmatic streaming is intact and strong. When both partners are in this lucky condition, childbirth even in high age is no miracle and brings about happiness for parents and child.

§11

All those who have achieved self-knowledge and are in vivid dialogue with their inner selves have discovered their bisexual nature and have integrated their other sexual persona. This work is the starting point of true spirituality. Sexual paraphilias are not sexual orientations *sui generis* but *fake identities* that offer those with lacking identity a point of reference for *social or group acceptance,* even though this very acceptance may be actually reject. On the other hand, what society projects upon paraphilias is inaccurate, because it consists of *repressed perverse thoughts and longings* and therefore lacks insight into the nature of these emotional and sexual attractions.

§12

With observing some basic rules of conduct, every human life can be lived without destructive somatic disease in the form of cancer, aids, heart disease, rheumatism, arthritis, leukemia and others of the same kind, and without mental or emotional diseases such as neurosis, psychosis, schizophrenia, epilepsy or Alzheimer.

§13

Psychoanalysis and most psychotherapeutic treatments are not natural and effective methods for healing emotional entanglement and early trauma, but conform with the cultural credo to reeducate, if not to brainwash people according to certain cultural beliefs. Effective trauma healing can only be done by *healing the luminous body* and by honest shamans and clairvoyants. A developmental course is open for modern medicine to integrate these practices and wistful traditions in the future.

§14

A new society will have to take responsibility for socially coding sexual paraphilias. The social code is the rule of life for all. Uncoded behavior creates individual and social chaos and brings about conflictual and harmful behavior patterns as well as black markets. Coding behavior requires social acceptance.

On Power

§1

For constructive living and a growing network of fruitful and mutually beneficial relationships, we need to develop our natural soul power, which is an acute awareness of our spiritual human potential and all its possible expressions. Soul power typically manifests as a deep and lasting trust in the goodness, intelligence and integratedness of cosmic life and love, as well as one's own unique potential and creativity.[264]

Soul power is the ability to live *one's unique form of love*, whatever our loving energies are attracted to and wish to be close with. Soul power is developed mainly through the integration of our dream knowledge and wisdom.

This source of knowledge is unlimited and depends only on our skill to read the language of dreams so that it is intelligible for us. Dreams also are a most genuine source of creativity.

§2

Self-affirmation as a source of strength and goes along with soul power. Self-affirmation starts from the premise that we all have a unique potential and mission and do not need to accept others that are disagreeable for us or impeding us from realizing our mission.

Published by Sirius-C Media Galaxy LLC, 2010

§3

Self-acceptance is a spiritual, psychic and biological necessity and the primary condition for success in life and love. We have a right to say *no*. To accept all in life without wavering can indicate a *lack of self-acceptance*. Nobody is supposed to accept worldviews, beliefs and ideologies that deny the value of the individual.

§4

Traditional patriarchal education is based upon *smashing the genuine soul power of the child*, raising disempowered emotional cripples that, once the emosexual castration called education is completed, are proudly called *citizens*. Only by a revolution in education, which is a *psychological revolution*, can this sad state of affairs gradually be changed and consciousness be raised about the need and the necessity of an education that respects the child as a person in her own right, as a spiritual entity and unique soul that incarnates with a definite mission as a learning process for his or her lifetime.

On Science

§1

Science as unbiased observation of nature and comprehension of the cosmic laws and functionalities has been misunderstood and distorted throughout patriarchy in all dominator cultures of the world.

Still in Antiquity, when science was in the hands of philosophers who had a meta-cognitive view upon nature, science gradually developed into a quantitative concept, out for construing measuring devices and defining reality more and more along the lines of what could be measured with these devices, declaring nonsense or banning from perception what could not be measured.

Unfortunately, the unresolvable rest regarded as trash and blinded out from reality was the greater half of nature.

As a result, the true science and its effective and functional counterpart, natural healing, had to go underground in Western civilization and survive under the threat of widespread official denial, slander and persecution.

This science was first called *philosophy* during Antiquity, then hermetic science or alchemy, and then magnetism, spiritism, and finally perennial science or new age science.

§2

True science is first of all energy science, the intelligent understanding and use of the life force or cosmic life energy that is called *ch'i* in China, *ki* in Japan, *mana* or *aka* with the

Published by Sirius-C Media Galaxy LLC, 2010

Kahunas from Hawaii and with most aboriginal peoples, *prana* in India, *ka* in old Egypt and that in Western alternative science was called *vis vitalis* by Paracelsus, *spirit energy* by Swedenborg, *animal magnetism* by Mesmer, *Odic force* by Reichenbach, *élan vital* by Coué and with most hypnotists, *universion* by Lakhovsky and *orgone* by Reich.

A future society's science agenda should put the *research on the human energy field*, the intelligent study of the use and functionality of the bioenergy at the first place of its science agenda. Modern science has done a first step in the right direction with the abandonment of the rigid observer standpoint and the acknowledgment of *probabilities* and *paradoxes* in quantum physics.

The next step can only be the acknowledgment of the existence of the ether, respectively the vacuum, or plenum as an energetic continuum, and the all-pervading cosmic energy field as the one single creator principle in the universe.[265]

On Health

§1

For effective healing and medical care, we need to look at what is *health*, and not at what is sickness, at what is *wholeness* and not what is fragmentation, at what is *sane* and not what is insane, at what is *life* and not what is death.

§2

Western science, since Aristotle, has looked at death to find out about life. Medical science was and is a *death science*. It has accumulated knowledge about decaying processes by vivisecting cadavers. It knows about the skeleton, tissues and blood vessels, but it has no idea what it is that distinguishes a living body from a dead body. To get to know that, it would have had to look at the life force, the *bioplasmatic energy* contained in living organisms, and which vanishes from the dead and decaying body shortly after the moment of death.

§3

Traditional Western medical science has asked the wrong questions and therefore got the wrong answers. It is merely palliative and cures symptoms *without having the slightest idea about the true causes of illness*, be it somatic, be it psychic illness.

It knows nothing about the energetic equilibrium of a healthy organism, and as a result ignores that all illness simply is a state of affairs where the natural equilibrium between

Published by Sirius-C Media Galaxy LLC, 2010

positive and negative energetic polarities, called *yang* and *yin*, is out of balance. It is as if somebody, in order to see the sun, builds a telescope, points it at the moon, looks through it and affirms: 'The sun is a dry planet, without life, without heat, without energy, a simple mass of stone; it appears heavy, stupid and useless'.

This is how Western medical science looks at the human body and what it understands about it. In reality, the human body is a luminous sun-like energy egg that irradiates a powerful bioplasmatic energy and is synergistically connected to, and resonating with, all other beings that vibrate at about the same frequency range, including animals, plants, and inanimate matter.

§4

The complete human body contains *twenty-four disc-shaped energy nods called chakras,* seven of which are around the upper half of the human body and five of which are higher up, floating in the ether, and bioenergetically resonating with the human organism. These twelve basic chakras are mirrored by another twelve *aura chakras* that are contained in the luminous body and which have similar biotransformational functions.

All sickness is to be seen, long before it manifests in the physical body, in the aura, the etheric body or bodies that build something like an egg-shaped luminous shell around the physical body. Illness typically manifests as a certain irregularity in the color spectrum of the aura; this was scientifically identified and corroborated as early as in the 1930s by a photographic technique invented by Dr. Walter J. Kirlian, a British physician of Russian origin. The technique is known as *Kirlian Photography*. In addition, Tibetan medicine is specialized in measuring the pulse with such accuracy that it equals aura reading; illnesses can be predicted by this highly sophisticated traditional method several years in advance of their first somatization.

§5

Already today, with the knowledge we possess about energy science, especially acupuncture, magnetic and cell resonance healing, aura treatment, homeopathy and orgone refreshment as well as macrobiotic diet, psychosomatic healing techniques and psychosynthesis, we can *establish a body of futurist medicine* that is effective, affordable, human, organic and sustainable.

§6

The only reason why this has not yet been done so far is the worldwide medical and pharmaceutical establishment with their strong mercantile power monopoly and their resistance against any progress in holistic medical science.

§7

Each and every human being can be healthy from birth to death, *without any use of chemical pharmaceuticals,* without any visits to a medical doctor or hospital, and without any opera-

Published by Sirius-C Media Galaxy LLC, 2010

tions. This is true for the regular case; for accidents, especially road, train and airplane accidents highly effective urgency aid is of course needed, and in so far modern Western medicine has proven high effectiveness, indeed.

§8

Childbirth is a sexual, orgasmic and first of all pleasurable experience for mother, father and child and should be enjoyed at home, in the familiar setting, as a celebration of life and as a welcome party for the new soul that incarnates in the physical shell. For this to happen, the old profession of the *sage femme* could be re-instituted. Childbirth has to be taken completely out of hospitals because it has nothing to do with sickness, and giving birth to a child is not the task of a medical doctor but of a specialized helper who by preference is a woman.

Why this profession disappeared is clearly the result of the Church's witch hunts over several centuries and the cruel persecution, torture and murder of thousands and thousands of natural obstetricians all over Europe.

What is true for childbirth is true for contraception and abortion as well. As it is the responsibility of the mother to carry her baby to birth or to decide to stop pregnancy, or else to avoid pregnancy altogether, a large body of methods existed over time in different cultures that was in the hands of natural obstetricians. Effective natural means for contraception were available for affordable prices that ensured parents or a single mother got their children only in case they really wanted them, and thus were able to take care of them. This knowledge has been scattered for the most part if it is not completely forgotten.

§9

The *condom* is one of the most decrepit, ineffective and body-hostile devices man ever came up with to regulate sexual intercourse, as this device almost nullifies sexual pleasure for the man as it desensitizes the gland, and in addition forces largely built men who have tiny built partners to use an unusual amount of force and pressure to realize orgasm – which puts unnecessary strain and in many cases also vaginal soreness and genital abrasions on the female partner, not to talk about the displeasing smell and handling of condoms that must put off any sensitive human from sex altogether.

It is likely that behind the invention of the condom a sex-denying Puritan *Weltanschauung* has built up enough imbecility about natural functions to make the thought forms that eventually resulted in the invention of that murky, ineffective and unaesthetic solution for 'safer' intercourse, while the truth is that *real safety* comes from healthy living, not from plastics and idiotic devices that are but money-making machines for the multinationals that market them worldwide.

§10

As long as international medical care is in the hands of physicians and pharmacy multinationals, human health, on a worldwide scale, will surely deteriorate with every year to

Published by Sirius-C Media Galaxy LLC, 2010

come, while the costs for this gigantic gambling machine of white-coated nonsense will astronomically increase.

On Emotions

§1

Present society confuses natural aggressiveness with violence and brings about violence through repressing healthy aggressiveness. Rage and anger are *positive emotions* because they help us to become conscious of unhealthy fusionary attachments and be grounded through our first chakra energy. In aggressiveness contained is a high communication potential that cannot be used when aggressiveness is totally repressed. Aggressiveness is not violence in that it is outgoing and creative, and as such an *expression of our soul power*, while violence is coming about through *suicidal and disempowering feelings* and is an expression of our weakness.

§2

Our emotions are naturally complete and intelligent, and they are kaleidoscopically linked to each other. All emotions have a place in life and should be lived consciously and without controlling or repressing any of them. When one emotion is blocked, the natural cycle of emotions which can be visualized as a kaleidoscope, is broken with the result that the natural bioenergetic balance of our organism becomes impaired.

Published by Sirius-C Media Galaxy LLC, 2010

§3

Knowledge about our *primary power* or *soul power* frees us of all our fears and also from rage and hate feelings. Natural soul power *cannot be gained when we deny our natural goodness,* and the creativeness of our subconscious mind. Even small children have a natural soul power potential that is regularly destroyed in patriarchal education, bringing about fearful and perversely obedient children that do not question the status quo and go along with every kind of abuse without defending themselves.

§4

Sex laws should be abandoned and sexual experience be completely decriminalized and taken out of the hands of the state. Personal intimacy has no place on police agendas and in newspaper columns; what this leads to is that love and intimacy are screwed up, defiled and degraded into sexual acts without meaning and without soul. So save human love from vulgarity and perversion, and for fighting sexual crime, there is only one way: *freedom.* Hence, a body of independent consultants that function as *public welfare agencies* should step in and provide emotional and sexual consultancy in accordance with special statutes and regulations that establish their social function and usefulness within a responsible medical and psychic health care system.[266]

§5

Children have to be raised in a peaceful, natural and comprehensive environment, where their whole person, including their bodies and emotions are truly understood and accepted. For this to happen, any form of educational violence called corporal or physical punishment has to be definitely, by law, prohibited and abandoned.[267] No adult has the right to exert violence upon a child, and education is no exception from this rule.

To justify educational violence and at the same time *wonder why violence is one of humanity's major problems* testifies of the shortsightedness and thwarted emotional intelligence of most humans raised under patriarchy. It is for this reason that changes in this important area are so slow to come and that so many, even educated people, continue beating their children as masters are beating their slaves.

Children who are regularly beaten cannot integrate their bodies and emotions into a non-fragmented and peaceful mindset because this very process is impaired through the presence of strong fear, guilt and shame in the child's psyche.

This means that children raised in a violent manner are crippled and handicapped for life!

Educational violence therefore should be prohibited by law for both parents and persons who act in loco parentis, such as educators, tutors and other caretakers. Ideally physical and sexual child abuse in the form of violence acted out upon a child should be treated in one and the same legal bill,

Published by Sirius-C Media Galaxy LLC, 2010

and not, as it is now the case, in diverse and difficult-to-find separate legislations.[268]

On Peace

§1

World peace can be attained only by changing our basic paradigms about emotions and sexuality for they condition our views upon violence and how we handle this major problem in its manifold manifestations as domestic violence, street violence, gang violence, in-group versus out-group violence, racial and ideological violence, official governmental, military or police violence and structural violence.

§2

All research done over the last decades on the roots of violence points to violence being a direct result of repressing our natural *pleasure function* and the establishment, within patriarchy, of rigid and compulsive behavior rules.

Pleasure, joy and sexual excitement are *essential for bringing about novelty* in individual and collective life patterns and the bioenergetic exchange between organisms. This is independent of sex and age and it is valid for people from all walks of life and belonging to all races and cultures.

§3

When natural body pleasure is thwarted and man thus deprived of tactile stimulation, a *secondary drive structure builds up* that is violent and destructive. This secondary drive structure that is fed by stuck and pent-up bioenergy rigidifies over

Published by Sirius-C Media Galaxy LLC, 2010

time into a character armor that impedes man from feeling and thus inhibits important functions in the human character, namely empathy and compassion. This is how humanity could become as violent as it is today, ruthlessly building in-group empires that shut out millions of people and dozens of nations and that expand on the price of genocide and public insanity.

§4

Man is not by nature violent and destructive. Those who believe that build their worldview upon myths and a manipulative history science that denies the fact that before patriarchy came up about five thousands years ago, most peoples of the world *lived in peace and fruitful exchanges*, while they maintained partnership economies instead of dominator economies.

These cultures, one of them being the Kahuna natives from Hawaii, do not know strict moralistic behavior rules; they simply comply with the golden rule to not do to another what one wishes not to suffer oneself. Anthropological and cross-cultural research has shown that a residue of these cultures luckily survived until this day, one of them being the Trobriand culture in Papua New Guinea, another being the Muria tribe in India. And there are certainly others that I did not have the chance to study.

These cultures have in common that they do not interfere with the emotions and sexual behavior of their children and thus practice what today is called *permissive education*.

They also tend to raise their children not as a result of economic constraints but with great respect for natural biological growth cycles. Thus, for example, puberty is naturally the period where the child takes a leave from his or her parents, marries and engages in a professional career, as this was still the case in the Middle-Ages in Europe but afterwards was completely eroded by the extended educational cycles as part of mechanistic and alienating industrial culture.

§5

Today's postmodern international consumer culture or *Oedipal Culture,* with its worldwide exportation of traditional Western values, is an utterly perverse upside-down movement that replaces life by a form of *collective psychosis* and rampant violence and that is based on denial and ignorance: denial of the most part of our natural and important pleasure function and ignorance about the most fundamental scientific and spiritual patterns of living.

For the peace agenda of a future society the following essential points have inter alia to be considered:

▸ Emosexual freedom for children and adolescents;

▸ Emosexual freedom for the elder;

▸ Respect and recognition of our natural bisexuality;

▸ Recognition of the cosmic energy field and parallel universes;

Published by Sirius-C Media Galaxy LLC, 2010

- ▸ A unified field theory for the cosmic energy field;

- ▸ Expert advice for contraception and abortion;

- ▸ Respect of every state's territorial and political integrity;

- ▸ Worldwide training of cross-cultural communication;

- ▸ Worldwide free-of-charge student exchanges;

- ▸ Scientific study of life-after-death and cosmic life cycles.

BIBLIOGRAPHY

General Bibliography

A

Abrams, Jeremiah (Ed.)

Reclaiming the Inner Child
New York: Tarcher/Putnam, 1990

Die Befreiung des Inneren Kindes
Die Wiederentdeckung unserer ursprünglichen kreativen Persönlichkeit
und ihre zentrale Bedeutung für unser Erwachsenwerden
München: Scherz Verlag, 1993

Adrienne, Carol

The Numerology Kit
New American Library, 1988

Agni Yoga Society

COEUR : Signes de l'Agni Yoga
Toulon: Sté Edipub, 1985
Publication originale date de 1932

Albrecht, Karl

The Only Thing That Matters
New York: Harper & Row, 1993

Alston, John P. / Tucker, Francis

The Myth of Sexual Permissiveness
The Journal of Sex Research, 9/1 (1973)

Appleton, Matthew

A Free Range Childhood
Self-Regulation at Summerhill School
Foundation for Educational Renewal, 2000

Summerhill
Kindern ihre Kindheit zurückgeben
Demokratie und Selbstregulierung in der Erziehung
Hohengehren: Schneider Verlag, 2003

Arcas, Gérald, Dr

Guérir le corps par l'hypnose et l'auto-hypnose
Paris: Sand, 1997

Ariès, Philippe

L'enfant et la famille sous l'Ancien Régime
Paris, Seuil, 1975

Centuries of Childhood
New York: Vintage Books, 1962

Geschichte der Kindheit
Frankfurt/M: DTV, 1998

Arntz, William & Chasse, Betsy

What the Bleep Do We Know
20th Century Fox, 2005 (DVD)

Down The Rabbit Hole Quantum Edition
20th Century Fox, 2006 (3 DVD Set)

Bleep
An der Schnittstelle von Spiritualität und Wissenschaft
Verblüffende Erkenntnisse und Anstösse zum Weiterdenken
Berlin: Vak Verlag, 2007

Arroyo, Stephen

Astrology, Karma & Transformation
The Inner Dimensions of the Birth Chart
Sebastopol, CA: CRSC Publications, 1978

Published by Sirius-C Media Galaxy LLC, 2010

Astrologie, Karma und Transformation
Die Chancen schwieriger Aspekte
Frankfurt/M: Heyne Verlag, 1998

Relationships and Life Cycles
Astrological Patterns of Personal Experience
Sebastopol, CA: CRCS Publications, 1993

Handbuch der Horoskop-Deutung
Berlin: Rowohlt, 1999

Atlee, Tom

The Tao of Democracy
Using Co-Intelligence to Create a World That Works for All
North Charleston, SC: Imprint Books / WorldWorks Press, 2003

B

Bachelard, Gaston

The Poetics of Reverie
Translated by Daniel Russell
Boston: Beacon Press, 1971

Poetik des Raumes
Frankfurt/M: Fischer Verlag, 2001

Bachofen, Johann Jakob

Gesammelte Werke, Band II
Das Mutterrecht
Basel: Benno Schwabe & Co., 1948
Erstveröffentlichung im Jahre 1861

Baggins, David Sadofsky

Drug Hate and the Corruption of American Justice
Santa Barbara: Praeger, 1998

Bagley, Christopher

Child Abusers
Research and Treatment
New York: Universal Publishers, 2003

Balter, Michael

The Goddess and the Bull
Catalhoyuk, An Archaeological Journey
to the Dawn of Civilization
New York: Free Press, 2006

Bandler, Richard

Get the Life You Want
The Secrets to Quick and Lasting Life Change
With Neuro-Linguistic Programming
Deerfield Beach, Fl: HCI, 2008

Barbaree, Howard E. & Marshall, William L. (Eds.)

The Juvenile Sex Offender
Second Edition
New York: Guilford Press, 2008

Barnes, A. James, Dworkin, Terry and Richards Eric L.

Law for Business, 9th Edition
New York: McGraw-Hill, 2006

Barnes, J. (Ed.)

The Complete Works of Aristotle, Vol. 1
Princeton: Princeton University Press, 1971

Barron, Frank X., Montuori, et al. (Eds.)

Creators on Creating
Awakening and Cultivating the Imaginative Mind
(New Consciousness Reader)
New York: P. Tarcher/Putnam, 1997

Bateson, Gregory

Steps to an Ecology of Mind
Chicago: University of Chicago Press, 2000
Originally published in 1972

Bender Lauretta & Blau, Abram

The Reaction of Children to Sexual Relations with Adults
American J. Orthopsychiatry 7 (1937), 500-518

Benkler, Yochai

The Wealth of Networks
How Social Production Transforms Markets and Freedom
New Haven, CT: Yale University Press, 2007

Bennion, Francis

Statutory Interpretation
London: Butterworths, 1984

Bernard, Frits

Paedophilia
A Factual Report
Amsterdam: Enclave, 1985

Pädophilie ohne Grenzen
Theorie, Forschung, Praxis
Frankfurt/M: Foerster Verlag, 1997

Kinderschänder?
Pädophilie, von der Liebe mit Kindern
3. Auflage
Frankfurt/M: Foerster Verlag, 1982

Bertalanffy, Ludwig von

General Systems Theory
Foundations, Development, Applications
New York: George Brazilier Publishing, 1976

Besant, Annie

An Autobiography
New Delhi: Penguin Books, 2005
Originally published in 1893

Karma
4e édition
Paris: Adyar, 1923

Bettelheim, Bruno

A Good Enough Parent
New York: A. Knopf, 1987

The Uses of Enchantment
New York: Vintage Books, 1989

Kinder brauchen Märchen
Frankfurt/M: DTV, 2002

Beutler/Bieber/Pipkorn/Streil

Die Europäische Gemeinschaft
Rechtsordnung und Politik
2. Auflage
Baden-Baden: Nomos, 1982

Block, Peter

Stewardship
Choosing Service Over Self-Interest
San Francisco: Berrett-Koehler, 1996

Blofeld, J.

The Book of Changes
A New Translation of the Ancient Chinese I Ching
New York: E.P. Dutton, 1965

Published by Sirius-C Media Galaxy LLC, 2010

Blum, Ralph H. & Laughan, Susan

The Healing Runes
Tools for the Recovery of Body, Mind, Heart & Soul
New York: St. Martin's Press, 1995

Boadalla, David

Wilhelm Reich, Leben und Werk
Frankfurt/M: Fischer, 1980

Bodin, Jean

On Sovereignty (1576)
Six Books of the Commonwealth
Edited by Professor Julian Franklin
New York: Seven Treasures Publications, 2009

Böhm, Wilfried

Maria Montessori
2. Auflage
Bad Heilbrunn: Julius Klinkhardt, 1991

Bohm, David

Wholeness and the Implicate Order
London: Routledge, 2002

Die implizite Ordnung
Grundlagen eines dynamischen Holismus
München: Goldmann Wilhelm, 1989

Thought as a System
London: Routledge, 1994

Quantum Theory
London: Dover Publications, 1989

La plénitude de l'univers
Paris: Rocher, 1992

La conscience de l'univers
Paris: Rocher, 1992

Boldt, Laurence G.
Zen and the Art of Making a Living
A Practical Guide to Creative Career Design
New York: Penguin Arkana, 1993

How to Find the Work You Love
New York: Penguin Arkana, 1996

Zen Soup
Tasty Morsels of Zen Wisdom From Great Minds East & West
New York: Penguin Arkana, 1997

The Tao of Abundance
Eight Ancient Principles For Abundant Living
New York: Penguin Arkana, 1999

Das Tao der Fülle
Vom Reichtum, der uns glücklich macht
Mittelberg: Joy Verlag, 2001

Bordeaux-Szekely, Edmond
Teaching of the Essenes from Enoch to the Dead
Sea Scrolls
Beekman Publishing, 1992

Gospel of the Essenes
The Unknown Books of the Essenes
& Lost Scrolls of the Essene Brotherhood
Beekman Publishing, 1988

Gospel of Peace of Jesus Christ
Beekman Publishing, 1994

Gospel of Peace, 2d Vol.
I B S International Publishers

Published by Sirius-C Media Galaxy LLC, 2010

Das Friedensevangelium der Essener
Saarbrücken: Neue Erde/Lentz, 2002

Évangile essénien de la paix
La vie biogénique
Genève: Éditions Soleil, 1978

Die unbekannten Schriften der Essener
Saarbrücken: Neue Erde/Lentz, 2002

Branden, Nathaniel
How to Raise Your Self-Esteem
New York: Bantam, 1987

Die 6 Säulen des Selbstwertgefühls
Erfolgreich und zufrieden durch ein starkes Selbst
München: Piper Verlag, 2009

Brant & Tisza
The Sexually Misused Child
American J. Orthopsychiatry, 47(1)(1977)

Brassai
Conversations with Picasso
Chicago: University of Chicago Publications, 1999

Brennan, Barbara Ann
Hands of Healing
A Guide to Healing Through the Human Energy Field
New York: Bantam, 1988

Brongersma, Edward
Aggression against Pedophiles
7 International Journal of Law & Psychiatry 82 (1984)

Loving Boys
Amsterdam, New York: GAP, 1987

Das verfemte Geschlecht
Berlin: Lichtenberg Verlag, 1970

Bruce, Alexandra

Beyond the Bleep
The Definite Unauthorized Guide to 'What the Bleep Do we Know!?'
New York: Disinformation, 2005

Bullough & Bullough (Eds.)

Human Sexuality
An Encyclopedia
New York: Garland Publishing, 1994

Sin, Sickness and Sanity
A History of Sexual Attitudes
New York: New American Library, 1977

Burgess, Ann Wolbert

Child Pornography and Sex Rings
New York: Lexington Books, 1984

Burwick, Frederick

The Damnation of Newton
Goethe's Color Theory and Romantic Perception
New York: Walter de Gruyter, 1986

Butler-Bowden, Tom

50 Success Classics
Winning Wisdom for Work & Life From 50 Landmark Books
London: Nicholas Brealey Publishing, 2004

Published by Sirius-C Media Galaxy LLC, 2010

50 Klassiker des Erfolgs
Die wichtigsten Werke von Kenneth Blanchard, Warren Buffet,
Andrew Carnegie, Stephen R. Covey, Spencer Johnson,
Benjamin Franklin, Napoleon Hill, Nelson Mandela, Anthony Robbins,
Brian Tracy, Sun Tsu, Jack Welch und vielen anderen
Frankfurt/M: MVG Verlag, 2005

50 Lebenshilfe Klassiker
Frankfurt/M: MVG Verlag, 2004

50 Klassiker der Psychologie
Die wichtigsten Werke von Alfred Adler, Sigmund Freud,
Daniel Goleman, Karen Horney, William James, C.G. Jung, Jean Piaget,
Viktor Frankl, Howard Gardner, Alfred Kinsey, Abraham Maslow, Iwan
Pawlow, Stanley Milgram, Martin Seligman und vielen anderen
Frankfurt/M: MVG Verlag, 2004

50 Klassiker der Spiritualität
Die wichtigsten Werke von Augustinus, Khalil Gibran, Mahatma Ghandi,
Dag Hammarskjölkd, Hermann Hesse, C. G. Jung, Eckhart Tolle,
J. Krishnamurti, Thich Nhat Hanh, Mutter Teresa, Dan Millman
und vielen anderen
Frankfurt/M: MVG Verlag, 2006

Buxton, Richard
The Complete World of Greek Mythology
London: Thames & Hudson, 2007

C

Cain, Chelsea & Moon Unit Zappa
Wild Child
New York: Seal Press (Feminist Publishing), 1999

Calderone & Ramey
Talking With Your Child About Sex
New York: Random House, 1982

Campbell, Herbert James

The Pleasure Areas
London: Eyre Methuen Ltd., 1973

Der Irrtum mit der Seele
München: Scherz Verlag, 1973

Les principes du plaisir
Paris: Stock, 1974

Campbell, Jacqueline C.

Assessing Dangerousness
Violence by Sexual Offenders, Batterers and Child
Abusers
New York: Sage Publications, 2004

Campbell, Joseph

The Hero With A Thousand Faces
Princeton: Princeton University Press, 1973
(Bollingen Series XVII)
London: Orion Books, 1999

Der Heros in Tausend Gestalten
München: Insel Verlag, 2009

Occidental Mythology
Princeton: Princeton University Press, 1973
(Bollingen Series XVII)
New York: Penguin Arkana, 1991

The Masks of God
Oriental Mythology
New York: Penguin Arkana, 1992
Originally published 1962

Mythologie des Ostens
Die Masken Gottes Bd. 2
Basel: Sphinx Verlag, 1996

Published by Sirius-C Media Galaxy LLC, 2010

The Power of Myth
With Bill Moyers
ed. by Sue Flowers
New York: Anchor Books, 1988

Die Kraft der Mythen
Düsseldorf: Patmos Verlag, 2007

Cantelon, Philip L. (Ed.)

The American Atom
A Documentary History of Nuclear Policies from the
Discovery of Fission to the Present
With Richard G. Hewlett (Ed.) and Robert C. Williams (Ed.)
Philadelphia, PA: University of Pennsylvania Press, 1992

Capacchione, Lucia

The Power of Your Other Hand
North Hollywood, CA: Newcastle Publishing, 1988

Capra, Bernt Amadeus

Mindwalk
A Film for Passionate Thinkers
Based Upon Fritjof Capra's *The Turning Point*
New York: Triton Pictures, 1990

Capra, Fritjof

The Turning Point
Science, Society And The Rising Culture
New York: Simon & Schuster, 1987
Original Author Copyright, 1982

Wendezeit
Bausteine für ein neues Weltbild
München: Droemer Knaur, 2004

Le temps du changement
Science, société et nouvelle culture
Paris: Rocher, 1994

The Tao of Physics
An Exploration of the Parallels Between Modern
Physics and Eastern Mysticism
New York: Shambhala Publications, 2000
(New Edition) Originally published in 1975

Das Tao der Physik
Die Konvergenz von westlicher Wissenschaft und östlicher Philosophie
Neue und erweiterte Auflage
München: O.W. Barth bei Scherz, 2000
Ursprünglich erschienen 1975 bei Droemersche Verlagsanstalt
in Hamburg

Le tao de la physique
Paris: Sand & Tchou, 1994

The Web of Life
A New Scientific Understanding of Living Systems
New York: Doubleday, 1997
Author Copyright 1996

Lebensnetz
Ein neues Verständnis der lebendigen Welt
München: Scherz Verlag, 1999

The Hidden Connections
Integrating The Biological, Cognitive And Social
Dimensions Of Life Into A Science Of Sustainability
New York: Doubleday, 2002

Verborgene Zusammenhänge
München: Scherz, 2002

Steering Business Toward Sustainability
New York: United Nations University Press, 1995

Uncommon Wisdom
Conversations with Remarkable People
New York: Bantam, 1989

Published by Sirius-C Media Galaxy LLC, 2010

The Science of Leonardo
Inside the Mind of the Great Genius of the Renaissance
New York: Anchor Books, 2008
New York: Bantam Doubleday, 2007 (First Publishing)

Complete List of Publications
http://www.fritjofcapra.net/publishers.html

Cassou, Michelle & Cubley, Steward

Life, Paint and Passion
Reclaiming the Magic of Spontaneous Expression
New York: P. Tarcher/Putnam, 1996

Castaneda, Carlos

The Teachings of Don Juan
A Yaqui Way of Knowledge
Washington: Square Press, 1985

Journey to Ixtlan
Washington: Square Press: 1991

Tales of Power
Washington: Square Press, 1991

The Second Ring of Power
Washington: Square Press, 1991

Castel, Robert

L'ordre psychiatrique, l'âge d'or de l'aliénisme
Paris: Éditions de Minuit, 1977

Cayce, Edgar

Modern Prophet
Four Complete Books
'Edgar Cayce On Prophecy'
'Edgar Cayce On Religion and Psychic Experience'
'Edgar Cayce On Mysteries of the Mind'

'Edgar Cayce On Reincarnation'
By Mary Ellen Carter
Ed. by Hugh Lynn Cayce
New York: Random House, 1968

Chaplin, Charles

My Autobiography
New York: Plume, 1992
Originally published in 196

Chevalier, Jean & Gheerbrant, Alain

A Dictionary of Symbols
Translated from the French by John Buchanan-Brown
New York: Penguin, 1996

Cho, Susanne

Kindheit und Sexualität im Wandel der Kulturgeschichte
Eine Studie zur Bedeutung der kindlichen Sexualität unter besonderer
Berücksichtigung des 17. und 20. Jahrhunderts
Zürich, 1983 (Doctoral thesis)

Chopra, Deepak

Creating Affluence
The A-to-Z Steps to a Richer Life
New York: Amber-Allen Publishing (2003)

Life After Death
The Book of Answers
London: Rider, 2006

Leben nach dem Tod
Das letzte Geheimnis unserer Existenz
Berlin: Allegria Verlag, 2008

Synchrodestiny
Discover the Power of Meaningful Coincidence to Manifest Abundance
Audio Book / CD
Niles, IL: Nightingale-Conant, 2006

Published by Sirius-C Media Galaxy LLC, 2010

The Seven Spiritual Laws of Success
A Practical Guide to the Fulfillment of Your Dreams
Audio Book / CD
New York: Amber-Allen Publishing (2002)

Die Sieben Geistigen Gesetze des Erfolgs
Berlin: Ullstein Verlag, 2004

The Spontaneous Fulfillment of Desire
Harnessing the Infinite Power of Coincidence
New York: Random House Audio, 2003

Cicero, Marcus Tullius

Selected Works
New York: Penguin, 1960 (Penguin Classics)

Clarke, Ronald

Einstein: The Life and Times
New York: Avon Books, 1970

Clarke-Steward, S., Friedman, S. & Koch, J.

Child Development, A Topical Approach
London: John Wiley, 1986

Cleary, Thomas

The Taoist I Ching
Translated by Thomas Cleary
Boston & London: Shambhala, 1986

Constantine, Larry L.

Children & Sex
New Findings, New Perspectives
Larry L. Constantine & Floyd M. Martinson (Eds.)
Boston: Little, Brown & Company, 1981

Treasures of the Island
Children in Alternative Lifestyles
Beverly Hills: Sage Publications, 1976

Where are the Kids?
in: Libby & Whitehurst (ed.)
Marriage and Alternatives
Glenview: Scott Foresman, 1977

Open Family
A Lifestyle for Kids and other People
26 FAMILY COORDINATOR 113-130 (1977)

Cook, M. & Howells, K. (Eds.)
Adult Sexual Interest in Children
Academic Press, London, 1980

Coudenhove-Kalergi, Richard N.
Paneuropa
Wien-Leipzig: Paneuropa Verlag, 1926

Covey, Stephen R.
The 7 Habits of Highly Effective People
Powerful Lessons in Personal Change
New York: Free Press, 2004
15th Anniversary Edition
First Published in 1989

Die 7 Wege zur Effektivität
Prinzipien für persönlichen und beruflichen Erfolg
Offenbach: Gabal Verlag, 2009

The 8th Habit
From Effectiveness to Greatness
London: Simon & Schuster, 2004

Der 8. Weg
Von der Effektivität zur wahren Grösse
Offenbach: Gabal Verlag, 2006

Published by Sirius-C Media Galaxy LLC, 2010

Covitz, Joel

Emotional Child Abuse
The Family Curse
Boston: Sigo Press, 1986

Cox, Geraldine

The Home is Where the Heart is
Sydney: Macmillan, 2000

Craze, Richard

Feng Shui
Feng Shui Book & Card Pack
London: Thorsons, 1997

Cross, Sir Rupert

Cross on Evidence
5th ed.
London: Butterworths, 1979

Introduction to Criminal Law
10th Edition
London: Butterworths, 1984

Currier, Richard L.

Juvenile Sexuality in Global Perspective
in : Children & Sex, New Findings, New Perspectives
Larry L. Constantine & Floyd M. Martinson (Eds.)
Boston: Little, Brown & Company, 1981

D

Daco, Pierre

Les triomphes de la psychanalyse de Pierre Daco
Bruxelles: Éditions Gérard & Co., 1965 (Marabout)

Dalai Lama

Ethics for the New Millennium
New York: Penguin Putnam, 1999

David-Neel, Alexandra

Magic and Mystery in Tibet
New York: Dover Publications, 1971

The Secret Oral Teachings in Tibetan Buddhist Sects
New York: Secrets of Light Publishers, 1981

Initiations and Initiates in Tibet
New York: Dover Publications, 1993

Immortality and Reincarnation
Wisdom from the Forbidden Journey
New York: Inner Tradition, 1997

Davidson, Gustav

A Dictionary of Angels
Including Fallen Angels
New York: Free Press, 1967

Davis, A. J.

Sexual Assaults in the Philadelphia Prison System and Sheriff's Van
Trans-Action 6, 2, 8-16 (1968)

Dean & Bruyn-Kops

The Crime and the Consequences of Rape
New York: Thomas, 1982

De Bono, Edward

The Use of Lateral Thinking
New York: Penguin, 1967

The Mechanism of Mind
New York: Penguin, 1969

Published by Sirius-C Media Galaxy LLC, 2010

Sur/Petition
London: HarperCollins, 1993

Tactics
London: HarperCollins, 1993
First published in 1985

Taktiken und Strategien erfolgreicher Menschen
Frankfurt/M: MVG Verlag, 1995

Serious Creativity
Using the Power of Lateral Thinking to Create New Ideas
London: HarperCollins, 1996

Delacour, Jean-Baptiste

Glimpses of the Beyond
New York: Bantam Dell, 1975

Deleuze, Gilles, Guattari, Felix

L'Anti-Oedipe
Capitalisme et Schizophrénie
Nouvelle Édition Augmentée
Paris: Éditions de Minuit, 1973

DeMause, Lloyd

The History of Childhood
New York, 1974

Foundations of Psychohistory
New York: Creative Roots, 1982

DeMeo, James

Heretic's Notebook
Emotions, Protocells, Ether-Drift and Cosmic Life Energy
with New Research Supporting Wilhelm Reich
Pulse of the Planet, #5 (2002)
Ashland, Oregon: Orgone Biophysical Research Laboratories, Inc., 2002

Nach Reich, Neue Forschungen zur Orgonomie
Sexualökonomie / Die Entdeckung der Orgonenergie
Herausgegeben zusammen mit Professor Bernd Senf, Berlin
Frankfurt/M: Zweitausendeins Verlag, 1997

Saharasia
The 4000 BCE Origins of Child Abuse, Sex-Repression,
Warfare and Social Violence in the Deserts of the Old World
Ashland, Oregon: Orgone Biophysical Research Laboratories, Inc., 1998

Deshimaru, Taisen

Zen et vie quotidienne
Paris: Albin Michel, 1985

Diamond, Stephen A., May, Rollo

Anger, Madness, and the Daimonic
The Psychological Genesis of Violence, Evil and Creativity
New York: State University of New York Press, 1999

DiCarlo, Russell E. (Ed.)

Towards A New World View
Conversations at the Leading Edge
Erie, PA: Epic Publishing, 1996

Dicta et Françoise

Tarot de Marseille
Paris: Mercure de France, 1980

Dolto, Françoise

La Cause des Enfants
Paris: Laffont, 1985

Mein Leben auf der Seite der Kinder
Ein Plädoyer für eine kindgerechte Welt
Hamburg: Lübbe Verlagsgruppe, 1993

Published by Sirius-C Media Galaxy LLC, 2010

Psychanalyse et Pédiatrie
Paris: Seuil, 1971

Psychoanalyse und Kinderheilkunde
Frankfurt/M: Suhrkamp, 1997

Séminaire de Psychanalyse d'Enfants, 1
Paris: Seuil, 1982

Séminaire de Psychanalyse d'Enfants, 2
Paris: Seuil, 1985

Séminaire de Psychanalyse d'Enfants, 3
Paris: Seuil, 1988

Praxis der Kinderanalyse. Ein Seminar.
Hamburg: Klett-Cotta, 1985

Alles ist Sprache
Kindern mit Worten helfen
Berlin: Quadriga, 1996

Über das Begehren
Die Anfänge der menschlichen Kommunikation
2. Auflage
Hamburg: Klett-Cotta, 1996

Kinder stark machen
Die ersten Lebensjahre
Berlin: Beltz Verlag, 2000

L'évangile au risque de la psychanalyse
Paris: Seuil, 1980

Dover, K.J.

Greek Homosexuality
New York: Fine Communications, 1997

504 | The Science of Shamanism

Strafgesetzbuch und Nebengesetze
42. Aufl.
München: Beck, 1985

Dürckheim, Karlfried Graf

Hara: The Vital Center of Man
Rochester: Inner Traditions, 2004

Hara
Die Erdmitte des Menschen
Neuausgabe
München: O.W. Barth bei Scherz, 2005

Zen and Us
New York: Penguin Arkana 1991

The Call for the Master
New York: Penguin Books, 1993

Absolute Living
The Otherworldly in the World and the Path to Maturity
New York: Penguin Arkana, 1992

The Way of Transformation
Daily Life as a Spiritual Exercise
London: Allen & Unwin, 1988

Der Alltag als Übung
Vom Weg der Verwandlung
Bern: Huber, 2008

The Japanese Cult of Tranquility
London: Rider, 1960

Kultur der Stille
Frankfurt/M: Weltz Verlag, 1997

Published by Sirius-C Media Galaxy LLC, 2010

E

Eden, Donna & Feinstein, David

Energy Medicine
New York: Tarcher/Putnam, 1998

The Energy Medicine Kit
Simple Effective Techniques to Help You Boost Your Vitality
Boulder, Co.: Sounds True Editions, 2004

The Promise of Energy Psychology
With David Feinstein and Gary Craig
Revolutionary Tools for Dramatic Personal Change
New York: Jeremy P. Tarcher/Penguin, 2005

Edmunds, Francis

An Introduction to Anthroposophy
Rudolf Steiner's Worldview
London: Rudolf Steiner Press, 2005

Edwardes, A.

The Jewel of the Lotus
New York, 1959

Einstein, Albert

The World As I See It
New York: Citadel Press, 1993

Mein Weltbild
Berlin: Ullstein, 2005

Out of My Later Years
New York: Outlet, 1993

Ideas and Opinions
New York: Bonanza Books, 1988

Einstein sagt
Zitate, Einfälle, Gedanken
München: Piper, 2007

Albert Einstein Notebook
London: Dover Publications, 1989

Eisler, Riane

The Chalice and the Blade
Our history, Our future
San Francisco: Harper & Row, 1995

Kelch und Schwert, Unsere Geschichte, unsere Zukunft
Weibliches und männliches Prinzip in der Geschichte
Freiburg: Arbor Verlag, 2005

Sacred Pleasure: Sex, Myth and the Politics of the Body
New Paths to Power and Love
San Francisco: Harper & Row, 1996

The Partnership Way
New Tools for Living and Learning
With David Loye
Brandon, VT: Holistic Education Press, 1998

The Real Wealth of Nations
Creating a Caring Economics
San Francisco: Berrett-Koehler Publishers, 2008

Eliade, Mircea

Shamanism
Archaic Techniques of Ecstasy
New York: Pantheon Books, 1964

Ellis, Havelock

Sexual Inversion
Republished
New York: University Press of the Pacific, 2001
Originally published in 1897

Published by Sirius-C Media Galaxy LLC, 2010

Analysis of the Sexual Impulse
Love and Pain
The Sexual Impulse in Women
Republished
New York: University Press of the Pacific, 2001
Originally published in 1903

The Dance of Life
New York: Greenwood Press Reprint Edition, 1973
Originally published in 1923

Elwin, V.

The Muria and their Ghotul
Bombay: Oxford University Press, 1947

Emerson, Ralph Waldo

The Essays of Ralph Waldo Emerson
Cambridge, Mass.: Harvard University Press, 1987

Emoto, Masaru

The Hidden Messages in Water
New York: Atria Books, 2004

Die Botschaft des Wassers
Burgrain: Koha Verlag, 2008

The Secret Life of Water
New York: Atria Books, 2005

Die Heilkraft des Wassers
Burgrain: Koha Verlag, 2004

Encyclopédies d'Aujourd'hui

Encyclopédie de la Franc-Maçonnerie
Paris: Librairie Générale Française, 2000
(La Pochothèque)

Erickson, Milton H.

My Voice Will Go With You
The Teaching Tales of Milton H. Erickson
by Sidney Rosen (Ed.)
New York: Norton & Co., 1991

Complete Works 1.0, CD-ROM
New York: Milton H. Erickson Foundation, 2001

Erikson, Erik H.

Childhood and Society
New York: Norton, 1993
First published in 1950

Erman/Ranke

Ägypten und Ägyptisches Leben im Altertum
Hildesheim: Gerstenberg, 1981

Evans-Wentz, Walter Yeeling

The Fairy Faith in Celtic Countries
London: Frowde, 1911
Republished by Dover Publications
(Minneola, New York), 2002

F

Farson, Richard

Birthrights
A Bill of Rights for Children
Macmillan, New York, 1974

Feinberg, Joel

Harmless Wrongdoing
The Moral Limits of the Criminal Law, Vol. 4
New York: Oxford University Press, 1990

Published by Sirius-C Media Galaxy LLC, 2010

Fensterhalm, Herbert

Don't Say Yes When You Want to Say No
With Jean Bear
New York: Dell, 1980

Fericla, Josep M.

Al trasluz de la Ayahuasca
Antropología cognitiva, oniromancia y consciencias alternativas
Barcelona: La Liebre de Marzo, 2002

Finkelhor, David

Sexually Victimized Children
New York: Free Press, 1981

Finkelstein, Haim N. (Ed.)

The Collected Writings of Salvador Dali
Cambridge: Cambridge University Press, 1998

Flack, Audrey

Art & Soul
Notes on Creating
New York: E P Dutton, Reissue Edition, 1991

Forte, Robert (Ed.)

Entheogens and the Future of Religion
Council on Spiritual Practices, 2nd ed., 2000

Fortune, Mary M.

Sexual Violence
New York: Pilgrim Press, 1994

Foster/Freed

A Bill of Rights for Children
6 FAMILY LAW QUARTERLY 343 (1972)

Foucault, Michel

The History of Sexuality, Vol. I : The Will to Knowledge
London: Penguin, 1998
First published in 1976

The History of Sexuality, Vol. II : The Use of Pleasure
London: Penguin, 1998
First published in 1984

The History of Sexuality, Vol. III : The Care of Self
London: Penguin, 1998
First published in 1984

Fourcade, Jean-Michel

Analyse transactionnelle et bioénergie
Paris: Delarge, 1981

Foxwood, Orion

The Faery Teachings
Arcata, CA: R.J. Steward Books, 2007

Franz Anton Mesmer

Franz Anton Mesmer und die Geschichte des Mesmerismus
Beiträge zum internationalen wissenschaftlichen Symposium
anlässlich des 250. Geburtstages von Mesmer
Stuttgart, 1985

Freud, Anna

War and Children
London: 1943

Published by Sirius-C Media Galaxy LLC, 2010

Freud, Sigmund

Three Essays on the Theory of Sexuality
in: The Standard Edition of the Complete Psychological
Works of Sigmund Freud
London: Hogarth Press, 1953-54
Vol. 7, pp. 130 ff
(first published in 1905)

Drei Abhandlungen zur Sexualtheorie
Frankfurt/M: Fischer, 1991

The Interpretation of Dreams
New York: Avon, Reissue Edition, 1980
and in: The Standard Edition of the Complete Psychological
Works of Sigmund Freud , (24 Volumes) ed. by James Strachey
New York: W. W. Norton & Company, 1976

Die Traumdeutung
Frankfurt/M: Fischer, 2005

Totem and Taboo
New York: Routledge, 1999
Originally published in 1913

Totem und Tabu
Einige Übereinstimmungen im Seelenleben der Wilden und
der Neurotiker
Frankfurt/M: Fischer Verlag, 1972

Freund, Kurt

Assessment of Pedophilia
in: Cook, M. and Howells, K. (eds.)
Adult Sexual Interest in Children
Academic Press, London, 1980

Frisch, Max

Biedermann und die Brandstifter
München: Suhrkamp, 1996
Erstmals 1955 als Hörspiel veröffentlicht

Fromm, Erich

The Anatomy of Human Destructiveness
New York: Owl Book, 1992
Originally published in 1973

Anatomie der menschlichen Destruktivität
Berlin: Rowohlt, 1977

Escape from Freedom
New York: Owl Books, 1994
Originally published in 1941

Die Furcht vor der Freiheit
München: DTV Verlag, 1993

To Have or To Be
New York: Continuum International Publishing, 1996
Originally published in 1976

Haben oder Sein
Die seelischen Grundlagen einer neuen Gesellschaft
München: DTV Verlag, 2005

The Art of Loving
New York: HarperPerennial, 2000
Originally published in 1956

Die Kunst des Liebens
Berlin: Ullstein, 2005

G

Gates, Bill

The Road Ahead
New York, Penguin, 1996
(Revised Edition)

Published by Sirius-C Media Galaxy LLC, 2010

Gawain, Shakti

Creative Visualization
Use the Power of Your Imagination to Create What You Want
Novato, CA: New World Library, 1995

Creative Visualization Meditations (Reader)
Novato, CA: New World Library, 1997

Geldard, Richard

Remembering Heraclitus
New York: Lindisfarne Books, 2000

Gerber, Richard

A Practical Guide to Vibrational Medicine
Energy Healing and Spiritual Transformation
New York: Harper & Collins, 2001

Geller, Uri

The Mindpower Kit
Includes Book, Audiotape, Quartz Crystal And Meditation Circle
New York: Penguin, 1996

Gesell, Izzy

Playing Along
37 Group Learning Activities Borrowed from Improvisational Theater
Whole Person Associates, 1997

Ghiselin, Brewster (Ed.)

The Creative Process
Reflections on Invention in the Arts and Sciences
Berkeley: University of California Press, 1985
First published in 1952

Gibson, Ian

The Shameful Life of Salvador Dali
New York: Norton, 1998

Gil, David G.

Societal Violence and Violence in Families
in: David G. Gil, Child Abuse and Violence
New York: Ams Press, 1928

Gimbutas, Marija

The Language of the Goddess
London: Thames & Hudson, 2001

Glucksmann, André, Wolton, Thierry

Silence On Tue
Paris: Grasset, 1986

Goethe, Johann Wolfgang von

The Theory of Colors
New York: MIT Press, 1970
First published in 1810

Goethes Farbenlehre
Leipzig: Seemann-Henschel Verlag, 1998

Goldenstein, Joyce

Einstein: Physicist and Genius
(Great Minds of Science)
New York: Enslow Publishers, 1995

Goldman, Jonathan & Goldman, Andi

Tantra of Sound
Frequencies of Healing
Charlottesville: Hampton Roads, 2005

Tantra des Klanges
Mehr Liebe und Intimität in der Partnerschaft
Mit CD
Hanau: Amra Verlag, 2009

Healing Sounds

Published by Sirius-C Media Galaxy LLC, 2010

The Power of Harmonies
Rochester: Healing Arts Press, 2002

Klangheilung
Die Schöpferkraft des Obertongesangs
Hanau: Amra Verlag, 2008

Healing Sounds
Principles of Sound Healing
DVD, 90 min.
Sacred Mysteries, 2004

Goldstein, Jeffrey H.
Aggression and Crimes of Violence
New York, 1975

Goleman, Daniel
Emotional Intelligence
New York, Bantam Books, 1995

EQ. Emotionale Intelligenz
München: DTV Verlag, 1997

Goodwin, Matthew O.
The Complete Numerology Guide
New York: Red Wheel/Weiser, 1988

Gordon, Rosemary
Pedophilia: Normal and Abnormal
in: Kraemer, The Forbidden Love
London, 1976

Gordon Wasson, R.
The Road to Eleusis
Unveiling the Secret of the Mysteries
With Albert Hofmann, Huston Smith, Carl Ruck and Peter Webster
Berkeley, CA: North Atlantic Books, 2008

Goswami, Amit

The Self-Aware Universe
How Consciousness Creates the Material World
New York: Tarcher/Putnam, 1995

Das Bewusste Universum
Wie Bewusstsein die materielle Welt erschafft
Stuttgart: Lüchow Verlag, 2007

Gottlieb, Adam

Peyote and Other Psychoactive Cacti
Ronin Publishing, 2nd edition, 1997

Grant

Grant's Method of Anatomy
10th ed., by John V. Basmajian
Baltimore, London: Williams & Wilkins, 1980

Greene, Liz

Astrology of Fate
York Beach, ME: Red Wheel/Weiser, 1986

Saturn
A New Look at an Old Devil
York Beach, ME: Red Wheel/Weiser, 1976

The Astrological Neptune and the Quest for Redemption
Boston: Red Wheel Weiser, 1996

The Mythic Journey
With Juliet Sharman-Burke
The Meaning of Myth as a Guide for Life
New York: Simon & Schuster (Fireside), 2000

Die Mythische Reise
Die Bedeutung der Mythen als ein Führer durch das Leben
München: Atmosphären Verlag, 2004

Published by Sirius-C Media Galaxy LLC, 2010

The Mythic Tarot
With Juliet Sharman-Burke
New York: Simon & Schuster (Fireside), 2001
Originally published in 1986

Le Tarot Mythique
Une nouvelle approche du Tarot
Paris: Solar, 1988

The Luminaries
The Psychology of the Sun and Moon in the Horoscope
With Howard Sasportas
York Beach, ME: Red Wheel/Weiser, 1992

Sonne und Mond
Die Bedeutung der grossen Lichter in der Mythologie und im Horoskop
Saarbrücken: Neue Erde/Lentz, 2000

Greer, John Michael

Earth Divination, Earth Magic
A Practical Guide to Geomancy
New York: Llewellyn Publications, 1999

Grinspoon, Lester

Marihuana
The Forbidden Medicine
With James B. Bakalar
New Haven, CT: Yale University Press, 1997
First published in 1971

Groeben/Boeckh/Thiesing/Ehlermann

Kommentar zum EWG-Vertrag
Band 2, Dritte Auflage
Baden-Baden: Nomos, 1983

Grof, Stanislav

Ancient Wisdom and Modern Science
New York: State University of New York Press, 1984

Beyond the Brain
Birth, Death and Transcendence in Psychotherapy
New York: State University of New York, 1985

LSD: Doorway to the Numinous
The Groundbreaking Psychedelic Research into Realms of the
Human Unconscious
Rochester: Park Street Press, 2009

Psychologie transpersonnelle
Paris: Rocher, 1984

Realms of the Human Unconscious
Observations from LSD Research
New York: E.P. Dutton, 1976

The Cosmic Game
Explorations of the Frontiers of Human Consciousness
New York: State University of New York Press, 1998

The Holotropic Mind
The Three Levels of Human Consciousness
With Hal Zina Bennett
New York: HarperCollins, 1993

When the Impossible Happens
Adventures in Non-Ordinary Reality
Louisville, CO: Sounds True, 2005

Wir wissen mehr als unser Gehirn
Die Grenzen des Bewusstseins überschreiten
Freiburg: Herder, 2007

Groth, A. Nicholas
Men Who Rape
The Psychology of the Offender
New York: Perseus Publishing, 1980

Published by Sirius-C Media Galaxy LLC, 2010

Grout, Pam

Art & Soul
New York: Andrews McMeel Publishing, 2000

Gunn, John

Violence
New York/Washington, 1973

Gurdjieff, George Ivanovich

The Herald of Coming Good
London: Samuel Weiser, 1933

H

Hall, Manly P.

The Pineal Gland
The Eye of God
Article extracted from the book: Man the Grand Symbol of the Mysteries
Kessinger Publishing Reprint

The Secret Teachings of All Ages
Reader's Edition
New York: Tarcher/Penguin, 2003
Originally published in 1928

Hameroff, Newberg, Woolf, Bierman et al.

Consciousness
20 Scientists Interviewed
Director: Gregory Alsbury
5 DVD Box Set, 540 min.
New York: Alsbury Films, 2003

Hargous, Sabine

Les appeleurs d'âmes
L'univers chamanique des Indiens des Andes
Paris: Albin Michel, 1985

Harner, Michael

Ways of the Shaman
New York: Bantam, 1982
Originally published in 1980

Der Weg des Schamanen
Das praktische Grundlagenbuch zum Schamanismus
Genf: Ariston, 2007

Chamane
Les secrets d'un sorcier indien d'Amérique du Nord
Paris: Albin Michel, 1982

Hasegawa, Tsuyoshi

Racing the Enemy
Stalin, Truman, and the Surrender of Japan
Cambridge, MA: Belknap Press of Harvard University Press, 2006

Henkin/Pugh/Schachter/Smit

International Law
Cases and Materials
St. Paul (West): American Casebook Series, 1980

Herman, Dean M.

A Statutory Proposal to Prohibit the Infliction of Violence upon Children
19 FAMILY LAW QUARTERLY, 1986, 1-52

Hermes Trismegistos

Corpus Hermeticum
New York: Edaf, 2001

Héroard, J.

Journal de Jean Héroard sur l'Enfance et la Jeunesse de Louis XIII
Paris: Soulié/Barthélemy, 1868

Published by Sirius-C Media Galaxy LLC, 2010

Herrigel, Eugen

Zen in the Art of Archery
New York: Vintage Books, 1999
Originally published in 1971

Hicks, Esther and Jerry

The Amazing Power of Deliberate Intent
Living the Art of Allowing
Carlsbad, CA: Hay House, 2006

Hobbes, Thomas

Leviathan (1651)
New York: Longman Library, 2006

Hofmann, Albert

LSD, My Problem Child
Reflections on Sacred Drugs, Mysticism and Science
Santa Cruz, CA: Multidisciplinary Association for Psychedelic Studies, 2009
Originally published in 1980

LSD, Mein Sorgenkind
Die Entdeckung der 'Wunderdroge'
München: DTV Verlag, 1999

Holmes, Ernst

The Science of Mind
A Philosophy, A Faith, A Way of Life
New York: Jeremy P. Tarcher/Putnam, 1998
First Published in 1938

Holstiege, Hildegard

Montessori Pädagogik und soziale Humanität
Freiburg: Herder, 1994

Hood, J. X.

Scientific Curiosities of Love, Sex and Marriage
A Survey of Sex Relations, Beliefs and Customs of Mankind in
Different Countries and Ages
New York, 1951

Houston, Jean

The Possible Human
A Course in Enhancing Your Physical, Mental, and Creative Abilities
New York: Jeremy P. Tarcher/Putnam, 1982

Howells, Kevin

Adult Sexual Interest in Children
Considerations Relevant to Theories of Aetiology in:
Cook, M. and Howells, K. (eds.): Adult Sexual Interest in Children
Academic Press, London, 1980

Huang, Alfred

The Complete I Ching
The Definite Translation from Taoist Master Alfred Huang
Rochester, NY: Inner Traditions, 1998

Hunt, Valerie

Infinite Mind
Science of the Human Vibrations of Consciousness
Malibu, CA: Malibu Publishing, 2000

Huxley, Aldous

The Doors of Perception and Heaven and Hell
London: HarperCollins (Flamingo), 1994
(originally published in 1954)

The Perennial Philosophy
San Francisco: Harper & Row, 1970

I

Innocenti Declaration

Declaration on the Protection, Promotion and Support of Breastfeeding
http://www.innocenti15.net/inno.htm

J

Jackson, Nigel

The Rune Mysteries
With Silver RavenWolf
St. Paul, Minn.: Llewellyn Publications, 2000

Jackson, Stevi

Childhood and Sexuality
New York: Blackwell, 1982

Jaffe, Hans L.C.

Picasso
New York: Abradale Press, 1996

James, William

Writings 1902-1910
The Varieties of Religious Experience / Pragmatism / A Pluralistic
Universe / The Meaning of Truth / Some Problems of Philosophy / Essays
New York: Library of America, 1988

Jampolsky, Gerald

Aimer c'est se libérer de la peur
Genève: Éditions Soleil, 1986

Janov, Arthur

Primal Man
The New Consciousness
New York: Crowell, 1975

Das Neue Bewusstsein
Frankfurt/M: Fischer Verlag, 1988
Urausgabe 1975

Johnson, Paul

A History of the Jews
New York: Harper & Row, 1987

Johnston & Deisher

Contemporary Communal Child Rearing: A First Analysis
52 PEDIATRICS 319 (1973)

Jones, W.H.S., Litt, D.

Pliny Natural History
Cambridge, Mass.: Harvard University Press, 1980

Jung, Carl Gustav

Archetypen
München: DTV Verlag, 2001

Archetypes of the Collective Unconscious
in: The Basic Writings of C.G. Jung
New York: The Modern Library, 1959, 358-407

Collected Works
New York, 1959

Dialectique du moi et de l'inconscient
Paris, Gallimard, 1991

Published by Sirius-C Media Galaxy LLC, 2010

On the Nature of the Psyche
in: The Basic Writings of C.G. Jung
New York: The Modern Library, 1959, 47-133

On the Psychogenesis of Schizophrenia
in: The Basic Writings of C.G. Jung
New York: The Modern Library, 1959, 474-494

Psychological Types
Collected Writings, Vol. 6
Princeton: Princeton University Press, 1971

Psychologie und Religion
München: DTV Verlag, 2001

Psychology and Religion
in: The Basic Writings of C.G. Jung
New York: The Modern Library, 1959, 582-655

Religious and Psychological Problems of Alchemy
in: The Basic Writings of C.G. Jung
New York: The Modern Library, 1959, 537-581

Symbol und Libido
Freiburg: Walter Verlag, 1987

Synchronizität, Akausalität und Okkultismus
Frankfurt/M: DTV, 2001

The Basic Writings of C.G. Jung
New York: The Modern Library, 1959

The Development of Personality
Collected Writings, Vol. 17
Princeton: Princeton University Press, 1954

The Meaning and Significance of Dreams
Boston: Sigo Press, 1991

The Myth of the Divine Child
in: Essays on A Science of Mythology
Princeton, N.J.: Princeton University Press Bollingen
Series XXII, 1969. (With Karl Kerenyi)

Traum und Traumdeutung
München: DTV Verlag, 2001

Two Essays on Analytical Psychology
Collected Writings, Vol. 7
Princeton: Princeton University Press, 1972
First published by Routledge & Kegan Paul, Ltd., 1953

Zur Psychologie westlicher und östlicher Religion
Fünfte Auflage
Olten: Walter Verlag, 1988

K

Kahn, Charles (Ed.)
The Art and Thought of Heraclitus
Cambridge: Cambridge University Press, 2008

Kaiser, Edmond
La Marche aux Enfants
Lausanne: P.-M. Favre, 1979

Kalweit, Holger
Shamans, Healers and Medicine Men
Boston and London: Shambhala, 1992
Originally published with Kösel Verlag, Munich, in 1987

Kant, Immanuel
Kant's Werke
Band VIII, Abhandlungen nach 1781 (Neudruck)
Berlin und Leipzig: Walter de Gruyter, 1923

Published by Sirius-C Media Galaxy LLC, 2010

Kapleau, Roshi Philip

Three Pillars of Zen
Boston: Beacon Press, 1967

Karagulla, Shafica

The Chakras
Correlations between Medical Science and Clairvoyant Observation
With Dora van Gelder Kunz
Wheaton: Quest Books, 1989

Die Chakras und die feinstofflichen Körper des Menschen
Mit Dora van Gelder-Kunz
Grafing: Aquamarin Verlag, 1994

Karremann, Manfred

Es geschieht am helllichten Tag
Die Verborgene Welt der Pädophilen
und wie wir unsere Kinder vor Missbrauch Schützen
Köln: Dumont, 2007

Kerner Justinus

F.A. Mesmer aus Schwaben
Frankfurt/M, 1856

Kiang, Kok Kok

The I Ching
An Illustrated Guide to the Chinese Art of Divination
Singapore: Asiapac, 1993

Kiesewetter, Carl

Franz Anton Mesmer's Leben und Lehre
Leipzig, 1893

Kingston, Karen

Creating Sacred Space With Feng Shui
New York: Broadway Books, 1997

Kinski, Klaus

Kinski Uncut: The Autobiography of Klaus Kinski
New York: Penguin, 1997

Klein, Melanie

Love, Guilt and Reparation, and Other Works 1921-1945
New York: Free Press, 1984
(Reissue Edition)

Envy and Gratitude and Other Works 1946-1963
New York: Free Press, 2002
(Reissue Edition)

Klimo, Jon

Channeling
Investigations on Receiving Information from Paranormal Sources
New York: North Atlantic Books, 1988

Koestler, Arthur

The Act of Creation
New York: Penguin Arkana, 1989.
Originally published in 1964

Kraemer

The Forbidden Love
London, 1976

Krafft-Ebing, Richard von

Psychopathia sexualis
New York: Bell Publishing, 1965
Originally published in 1886

Krause, Donald G.

The Art of War for Executives
London: Nicholas Brealey Publishing, 1995

Published by Sirius-C Media Galaxy LLC, 2010

Krishnamurti, J.

Freedom From The Known
San Francisco: Harper & Row, 1969

The First and Last Freedom
San Francisco: Harper & Row, 1975

Education and the Significance of Life
London: Victor Gollancz, 1978

Commentaries on Living
First Series
London: Victor Gollancz, 1985

Commentaries on Living
Second Series
London: Victor Gollancz, 1986
Krishnamurti's Journal
London: Victor Gollancz, 1987

Krishnamurti's Notebook
London: Victor Gollancz, 1986

Beyond Violence
London: Victor Gollancz, 1985

Beginnings of Learning
New York: Penguin, 1986

The Penguin Krishnamurti Reader
New York: Penguin, 1987

On God
San Francisco: Harper & Row, 1992

On Fear
San Francisco: Harper & Row, 1995

The Essential Krishnamurti
San Francisco: Harper & Row, 1996

The Ending of Time
With Dr. David Bohm
San Francisco: Harper & Row, 1985

Kwok, Man-Ho

The Feng Shui Kit
London: Piatkus, 1995

L

Labate, Beatriz Caluby

Ayahuasca Religions
A Comprehensive Bibliography and Critical Essays
Santa Cruz, CA: Maps, 2009

Laing, Ronald David

Divided Self
New York: Viking Press, 1991

R.D. Laing and the Paths of Anti-Psychiatry
ed., by Z. Kotowicz
London: Routledge, 1997

The Politics of Experience
New York: Pantheon, 1983

Sagesse, déraison et folie
Paris: Seuil, 1986

Lakhovsky, Georges

La Science et le Bonheur
Longévité et Immortalité par les Vibrations
Paris: Gauthier-Villars, 1930

Le Secret de la Vie
Paris: Gauthier-Villars, 1929

Published by Sirius-C Media Galaxy LLC, 2010

Secret of Life
New York: Kessinger Publishing, 2003

L'étiologie du Cancer
Paris: Gauthier-Villars, 1929

L'Universion
Paris: Gauthier-Villars, 1927

Lanouette, William

Genius in the Shadows
A Biography of Leo Szilard, the Man behind the Bomb
With Bela Silard
Chicago: University of Chicago Press, 1994

Laszlo, Ervin

Holos. Die Welt der neuen Wissenschaften
Petersberg: Via Nova Verlag, 2002

Science and the Akashic Field
An Integral Theory of Everything
Rochester: Inner Traditions, 2004

Macroshift
Die Herausforderung
Frankfurt/M: Insel Verlag, 2003

Quantum Shift to the Global Brain
How the New Scientific Reality Can Change Us and Our World
Rochester: Inner Traditions, 2008

Science and the Reenchantment of the Cosmos
The Rise of the Integral Vision of Reality
Rochester: Inner Traditions, 2006

The Akashic Experience
Science and the Cosmic Memory Field
Rochester: Inner Traditions, 2009

The Chaos Point
The World at the Crossroads
Newburyport, MA: Hampton Roads Publishing, 2006

Laud, Anne & Gilstrop, May

Violence in the Family
A Selected Bibliography on Child Abuse, Sexual Abuse of
Children & Domestic Violence
June 1985
University of Georgia Libraries
Bibliographical Series, No. 32

Lauterpacht, E., Q.C.

International Law Reports
Cambridge: Grotius Publishers

Lauterpacht, Hersch

International Law
Ed. By E. Lauterpacht, Q.C.
Vol. 3
London: Cambridge University Press, 1977

LaViolette, Paul A.

*Secrets of Antigravity Propulsion: Tesla, UFOs, and
Classified Aerospace Technology*
New York: Bear & Company, 2008

The U.S. Antigravity Squadron
In: Thomas Valone, Ed., *Electrogravitics Systems,
Reports on a New Propulsion Methodology*
Washington, D.C.: Integrity Research Institute, 1993, 78-96

Leadbeater, Charles Webster

Astral Plane
Its Scenery, Inhabitants and Phenomena
Kessinger Publishing Reprint Edition, 1997

Dreams
What they Are and How they are Caused
London: Theosophical Publishing Society, 1903
Kessinger Publishing Reprint Edition, 1998

The Inner Life
Chicago: The Rajput Press, 1911
Kessinger Publishing

Leary, Timothy

Our Brain is God
Berkeley, CA: Ronin Publishing, 2001
Author Copyright 1988

Über die Kriminalisierung des Natürlichen
Löhrbach: Werner Pieper Verlag, 1990

Leboyer, Frederick

Birth Without Violence
New York, 1975

Pour une Naissance sans Violence
Paris: Seuil, 1974

Geburt ohne Gewalt
München: Kösel 1981

Cette Lumière d'où vient l'Enfant
Paris: Seuil, 1978

Inner Beauty, Inner Light
New York: Newmarket Press, 1997

Weg des Lichts
München: Kösel, 1991

Loving Hands
The Traditional Art of Baby Massage
New York: Newmarket Press, 1977

Sanfte Hände
Die Kunst der indischen Baby-Massage
München: Kösel, 1979

The Art of Breathing
New York: Newmarket Press, 1991

Le Crapouillot

Les pédophiles
Nouvelle série, n°73, Janvier 1984
Vincent Acker, Le Vilain Manège du Coral, pp. 36-42

LeCron, Leslie M.

L'auto-hypnose
8e édition
Genève: Ariston, 1984

Leggett, Trevor P.

A First Zen Reader
Rutland: C.E. Tuttle, 1980
Originally published in 1972

Lenihan, Eddie

Meeting the Other Crowd
The Fairy Stories of Hidden Ireland
With Carolyn Eve Green
New York: Jeremy P. Tarcher/Penguin, 2004
Authors Copyright 2003

Published by Sirius-C Media Galaxy LLC, 2010

Leonard, George, Murphy, Michael

The Live We Are Given
A Long Term Program for Realizing the
Potential of Body, Mind, Heart and Soul
New York: Jeremy P. Tarcher/Putnam, 1984

Leopardi, Angelo (Hrsg.)

Der Pädosexuelle Komplex
Frankfurt/M: Foerster Verlag, 1988

Licht, Hans

Sexual Life in Ancient Greece
New York: AMS Press, 1995

Liedloff, Jean

Continuum Concept
In Search of Happiness Lost
New York: Perseus Books, 1986
First published in 1977

Auf der Suche nach dem verlorenen Glück
Gegen die Zerstörung der Glücksfähigkeit in der frühen Kindheit
München: C.H. Beck Verlag, 2006

Lip, Evelyn

The Design & Feng Shui of Logos, Trademarks and Signboards
Singapore: Prentice Hall, 1995

Lipgens, Walter

Europa-Föderationspläne der Widerstandsbewegungen 1940-1945
München, 1968

Lipton, Bruce

The Biology of Belief
Unleashing the Power of Consciousness, Matter and Miracles
Santa Rosa, CA: Mountain of Love/Elite Books, 2005

Intelligente Zellen
Wie Erfahrungen unsere Gene steuern
Burgrain: Koha Verlag, 2006

Liss, Jérôme

Débloquez vos émotions
Lausanne: Éditions Far, 1988

Locke, John

Some Thoughts Concerning Education
London, 1690
Reprinted in: The Works of John Locke, 1823
Vol. IX., pp. 6-205

Gedanken über Erziehung
Ditzingen: Reclam Verlag, 1986

Long, Max *Freedom*

The Secret Science at Work
The Huna Method as a Way of Life
Marina del Rey: De Vorss Publications, 1995
Originally published in 1953

Geheimes Wissen hinter Wundern
Die Entdeckung der HUNA-Lehre
Darmstadt: Schirner Verlag, 2006

Growing Into Light
A Personal Guide to Practicing the Huna Method,
Marina del Rey: De Vorss Publications, 1955

Lowen, Alexander

Angst vor dem Leben
Über den Ursprung seelischen Leides und den Weg zu
einem reicheren Dasein
München: Goldmann Wilhelm, 1989

Published by Sirius-C Media Galaxy LLC, 2010

Bioenergetics
New York: Coward, McGoegham 1975

Bioenergetik
Therapie der Seele durch Arbeit mit dem Körper
Berlin: Rowohlt, 2008

Depression and the Body
The Biological Basis of Faith and Reality
New York: Penguin, 1992

Fear of Life
New York: Bioenergetic Press, 2003

Honoring the Body
The Autobiography of Alexander Lowen
New York: Bioenergetic Press, 2004

Joy
The Surrender to the Body and to Life
New York: Penguin, 1995

Liebe und Orgasmus
Persönlichkeitserfahrung durch sexuelle Erfüllung
München: Goldmann Wilhelm, 1993

Love and Orgasm
New York: Macmillan, 1965

Love, Sex and Your Heart
New York: Bioenergetics Press, 2004

Narcissism: Denial of the True Self
New York: Macmillan, Collier Books, 1983

Narzissmus
Die Verleugnung des wahren Selbst
München: Goldmann Wilhelm, 1992

Pleasure: A Creative Approach to Life
New York: Bioenergetics Press, 2004
First published in 1970

The Language of the Body
Physical Dynamics of Character Structure
New York: Bioenergetics Press, 2006

Luna, Luis Eduardo & Amaringo, Pablo

Ayahuasca Visions
North Atlantic Books, 1999

Lusk, Julie T. (Editor)

30 Scripts for Relaxation Imagery & Inner Healing
Whole Person Associates, 1992

Lutyens, Mary

Krishnamurti: The Years of Fulfillment
New York: Avon Books, 1983

Krishnamurti: Die Biographie
München: Aquamarin Verlag, 1997

The Life and Death of Krishnamurti
Chennai: Krishnamurti Foundation India, 1990

Lutzbetak, Louis J.

Marriage and the Family in Caucasia
Vienna, 1951, first reprinting, 1966

M

Machiavelli, Niccolo

The Prince
New York: Soho Books, 2009
Written in 1513
First posthumous publishing 1531

Der Fürst
Frankfurt/M: Insel Verlag, 2009

Mack, Carol K. & Mack, Dinah

A Field Guide to Demons, Fairies, Fallen Angels, and Other Subversive Spirits
New York: Owl Books, 1998

Maharshi, Ramana

The Collected Works of Ramana Maharshi
New York: Sri Ramanasramam, 2002

The Essential Teachings of Ramana Maharshi
A Visual Journey
New York: Inner Directions Publishing, 2002
by Matthew Greenblad

Sei was du bist!
München: O.W. Barth, 2001

Nan Yar? Wer bin ich?
München: Kamphausen, 2002

Maisel, Eric

Fearless Creating
A Step-By-Step Guide to Starting and Completing
Work of Art
New York: Tarcher & Putnam, 1995

Malachi, Tau

Gnosis of the Cosmic Christ
A Gnostic Christian Kabbalah
St. Paul: Llewellyn Publications, 2005

Malinowski, Bronislaw

Crime und Custom in Savage Society
London: Kegan, 1926

Sex and Repression in Savage Society
London: Kegan, 1927

The Sexual Life of Savages in North West Melanesia
New York: Halycon House, 1929

Das Geschlechtsleben der Wilden in Nordwest-Melanesien
Liebe, Ehe und Familienleben bei den Eingeborenen
der Trobriand Inseln,
Britisch-Neuguinea
Eschborn: Klotz Verlag, 2005

Mallet, Carl-Heinz
Das Einhorn bin ich
Das Bild des Menschen im Märchen
Hamburg: Hoffmann & Campe Verlag, 1982

Untertan Kind
Nachforschungen über Erziehung
München: Max Hueber Verlag, 1987

Mann, Edward W.
Orgone, Reich & Eros
Wilhelm Reich's Theory of Life Energy
New York: Simon & Schuster (Touchstone), 1973

Mann, Sally
At Twelve
Portraits of Young Women
New York: Aperture, 1988

Immediate Family
New York: Phaidon Press, 1993

Published by Sirius-C Media Galaxy LLC, 2010

Marciniak, Barbara

Bringers of the Dawn
Teachings from the Pleiadians
New York: Bear & Co., 1992

Boten des Neuen Morgens
Lehren von den Pleiaden
Freiburg: Hermann Bauer Verlag, 1995

Martinson, Floyd M.

Sexual Knowledge
Values and Behavior Patterns
St. Peter: Minn.: Gustavus Adolphus College, 1966

Infant and Child Sexuality
St. Peter: Minn.: Gustavus Adolphus College, 1973

The Quality of Adolescent Experiences
St. Peter: Minn.: Gustavus Adolphus College, 1974

The Child and the Family
Calgary, Alberta: The University of Calgary, 1980

The Sex Education of Young Children
in: Lorna Brown (Ed.), *Sex Education in the Eighties*
New York, London: Plenum Press, 1981, pp. 51 ff.

The Sexual Life of Children
New York: Bergin & Garvey, 1994

Children and Sex, Part II: Childhood Sexuality
in: Bullough & Bullough, Human Sexuality (1994)
Pp. 111-116

Master Lam Kam Chuen

The Way of Energy
Mastering the Chinese Art of Internal
Strength with Chi Kung Exercise
New York: Simon & Schuster (Fireside), 1991

Master Liang, Shou-Yu & Wu, Wen-Ching

Tai Chi Chuan
24 & 48 Postures With Martial Applications
Roslindale, Mass.: YMAA Publication Center, 1996

Masters, R.E.L.

Forbidden Sexual Behavior and Morality
New York, 1962

McCarey, William A.

In Search of Healing
Whole-Body Healing Through the Mind-Body-Spirit Connection
New York: Berkley Publishing, 1996

McCormick

McCormick on Evidence
by Edward W. Cleary, 3d ed.
Lawyers Edition (Homebook Series)
St. Paul: West, 1984

McKenna, Terence

The Archaic Revival
San Francisco: Harper & Row, 1992

Food of The Gods
A Radical History of Plants, Drugs and Human Evolution
London: Rider, 1992

Die Speisen der Götter
Berlin: Synergia/Syntropia, 1996

The Invisible Landscape
Mind Hallucinogens and the I Ching
New York: HarperCollins, 1993
(With Dennis McKenna)

Published by Sirius-C Media Galaxy LLC, 2010

True Hallucinations
Being the Account of the Author's Extraordinary
Adventures in the Devil's Paradise
New York: Fine Communications, 1998

McLeod, Kembrew

Freedom of Expression
Resistance and Repression in the Age of Intellectual Property
Minneapolis, MN: University of Minnesota Press, 2007

McNiff, Shaun

Art as Medicine
Boston: Shambhala, 1992

Art as Therapy
Creating a Therapy of the Imagination
Boston/London: Shambhala, 1992

Trust the Process
An Artist's Guide to Letting Go
New York: Shambhala Publications, 1998

McTaggart, Lynne

The Field
The Quest for the Secret Force of the Universe
New York: Harper & Collins, 2002

Mead, Margaret

Sex and Temperament in Three Primitive Societies
New York, 1935

Meadows, Donella H.

Thinking in Systems
A Primer
White River, VT: Chelsea Green Publishing, 2008

Mehta, Rohit

J. Krishnamurti and the Nameless Experience
A Comprehensive Discussion of J. Krishnamurti's Approach to Life
Delhi: Motilal Banarsidass Publishers, 2002

Méric, de, Philippe

Le Yoga sans postures
Paris: Livre de Poche, 1967

Merle, Roger & Vitu, André

Traité de Croit Criminel
Droit Pénal Spécial
Vol. II, par André Vitu
Paris: Editions Cujas, 1982

Merleau-Ponty, Maurice

Phenomenology of Perception
London: Routledge, 1995
Originally published 1945

Phénoménologie de la perception
Paris: Gallimard, 1945

Metzner, Ralph (Ed.)

Ayahuasca, Human Consciousness and the Spirits of Nature
ed. by Ralph Metzner, Ph.D
New York: Thunder's Mouth Press, 1999

The Psychedelic Experience
A Manual Based on the Tibetan Book of the Dead
With Timothy Leary and Richard Alpert
New York: Citadel, 1995

Miller, Alice

Four Your Own Good
Hidden Cruelty in Child-Rearing and the Roots of Violence
New York: Farrar, Straus & Giroux, 1983

Published by Sirius-C Media Galaxy LLC, 2010

Am Anfang war Erziehung
München: Suhrkamp Verlag, 2008
Erstmals publiziert im Jahre 1986

Pictures of a Childhood
New York: Farrar, Straus & Giroux, 1986

The Drama of the Gifted Child
In Search for the True Self
translated by Ruth Ward
New York: Basic Books, 1996

Das Drama des Begabten Kindes
Und die Suche nach dem wahren Selbst
München: Suhrkamp Verlag, 1983

Der gemiedene Schlüssel
München: Suhrkamp, 2007

Das verbannte Wissen
Frankfurt/M: Suhrkamp, 1988

Thou Shalt Not Be Aware
Society's Betrayal of the Child
New York: Noonday, 1998

Du Sollst Nicht Merken
Variationen über das Paradies-Thema
Neuauflage
München: Suhrkamp, 2005

The Political Consequences of Child Abuse
in: The Journal of Psychohistory 26, 2 (Fall 1998)

Miller, Mary & Taube, Karl
*An Illustrated Dictionary of the Gods and Symbols of
Ancient Mexico and the Maya*
London: Thames & Hudson, 1993

Moll, Albert

The Sexual Life of the Child
New York: Macmillan, 1912
First published in German as
Das Sexualleben des Kindes, 1909

Monroe, Robert

Ultimate Journey
New York: Broadway Books, 1994

Monsaingeon, Bruno

Svjatoslav Richter
Notebooks and Conversations
Princeton: Princeton University Press, 2002

Richter
Écrits, conversations
Paris: Éditions Van de Velde, 1998

Richter The Enigma / L'Insoumis / Der Unbeugsame
NVC Arts 1998 (DVD)

Montagu, Ashley

Touching
The Human Significance of the Skin
New York: Harper & Row, 1978

Körperkontakt
8. Auflage
Stuttgart: Klett/Cotta, 1995

Monter, E. William

Witchcraft in France and Switzerland
Ithaca & London: Cornell University Press, 1976

Published by Sirius-C Media Galaxy LLC, 2010

Montessori, Maria

The Absorbent Mind
Reprint Edition
New York: Buccaneer Books, 1995
First published in 1973

Das Kreative Kind
Der absorbierende Geist
Freiburg: Herder, 2007

Moody, Raymond

The Light Beyond
New York: Mass Market Paperback (Bantam), 1989

Moore, Thomas

Care of the Soul
A Guide for Cultivating Depth and Sacredness in Everyday Life
New York: Harper & Collins, 1994

Die Seele Lieben
Tiefe und Spiritualität im täglichen Leben
München: Droemer Knaur, 1995

Moser, Charles Allen

DSM-IV-TR and the Paraphilias: an argument for removal
With Peggy J. Kleinplatz
Journal of Psychology and Human Sexuality 17 (3/4), 91-109 (2005)

Murdock, G.

Social Structure
New York: Macmillan, 1960

Murphy, Joseph

The Power of Your Subconscious Mind
West Nyack, N.Y.: Parker, 1981, N.Y.: Bantam, 1982
Originally published in 1962

Die Macht Ihres Unterbewusstseins
München: Hugendubel, 2000

La puissance de votre subconscient
Genève: Ramón Keller, 1967

The Miracle of Mind Dynamics
New York: Prentice Hall, 1964

Miracle Power for Infinite Riches
West Nyack, N.Y.: Parker, 1972

The Amazing Laws of Cosmic Mind Power
West Nyack, N.Y.: Parker, 1973

Secrets of the I Ching
West Nyack, N.Y.: Parker, 1970

Think Yourself Rich
Use the Power of Your Subconscious Mind to Find True Wealth
Revised by Ian D. McMahan, Ph.D.
Paramus, NJ: Reward Books, 2001

Das Erfolgsbuch
Wie sie alles im Leben erreichen können
Hamburg: Heyne Verlag, 2002

Wahrheiten die ihr Leben verändern
Dr. Joseph Murphys Vermächtnis
München: Hugendubel, 1996

Murphy, Michael

The Future of the Body
Explorations into the Further Evolution of Human Nature
New York: Jeremy P. Tarcher/Putnam, 1992

Der Quanten-Mensch
München: Ludwig Verlag, 1996

Published by Sirius-C Media Galaxy LLC, 2010

Myers, Tony Pearce

The Soul of Creativity
Insights into the Creative Process
Novato, CA: New World Library, 1999

Myss, Caroline

The Creation of Health
The Emotional, Psychological, and Spiritual Responses that Promote
Health and Healing
New York: Three Rivers Press, 1998

N

Naparstek, Belleruth

Your Sixth Sense
Unlocking the Power of Your Intuition
London: HarperCollins, 1998

Staying Well With Guided Imagery
New York: Warner Books, 1995

Narby, Jeremy

The Cosmic Serpent
DNA and the Origins of Knowledge
New York: J. P. Tarcher, 1999

Die Kosmische Schlange
Auf den Pfaden der Schamanen zu den Ursprüngen modernen Wissens
Stuttgart: Klett-Cotta, 2007

Nau, Erika

Self-Awareness Through Huna
Virginia Beach: Donning, 1981

Selbstbewusst durch Huna
Die magische Weisheit Hawaiis
2. Auflage
Basel: Sphinx Verlag, 1989

Neill, Alexander Sutherland

Neill! Neill! Orange-Peel!
New York: Hart Publishing Co., 1972

Neill! Neill! Birnenstiel!
Berlin: Rowohlt, 1973

Summerhill
A Radical Approach to Child Rearing
New York: Hart Publishing, Reprint 1984
Originally published 1960

Theorie und Praxis der Antiautoritären Erziehung
Das Beispiel Summerhill
Berlin: Rowohlt Verlag, 1969

Summerhill School
A New View of Childhood
New York: St. Martin's Press
Reprint 1995

Das Prinzip Summerhill
Berlin: Rowohlt, 1971

Neuhaus, Heinrich

The Art of Piano Playing
London: Barrie & Jenkins, 1973
Reprinted 1997, 2001, 2002, 2006
First published in 1958

Neumann, Erich

The Great Mother
Princeton: Princeton University Press, 1955
(Bollingen Series)

Published by Sirius-C Media Galaxy LLC, 2010

Die Grosse Mutter
Die weiblichen Gestaltungen des Unterbewussten
Düsseldorf: Patmos Verlag, 2003

Newton, Michael

Life Between Lives
Hypnotherapy for Spiritual Regression
Woodbury, Minn.: Llewellyn Publications, 2006

Ni, Hua-Ching

I Ching
The Book of Changes and the Unchanging Truth
2nd edition
Santa Barbara: Seven Star Communications, 1999

Esoteric Tao The Ching
The Shrine of the Eternal Breath of Tao
Santa Monica: College of Tao and Traditional
Chinese Healing, 1992

The Complete Works of Lao Tzu
Tao The Ching & Hua Hu Ching
Translation and Elucidation by Hua-Ching Ni
Santa Monica: Seven Star Communications, 1995

Nichols, Sallie

Jung and Tarot: An Archetypal Journey
New York: Red Wheel/Weiser, 1986

Die Psychologie des Tarot
Interlaken: Ansata Verlag, 1996

Nin, Anaïs

The Diary of Anaïs Nin (7 Volumes)
New York, 1966

Volume 1 (1931-1934)
New York: Harvest Books, 1969

Volume 2 (1934-1939)
New York: Harvest Books, 1970

O

O'Brian, Shirley

Child Pornography
2nd edition
New York: Kendall/Hunt, 1992

Odent, Michel

Birth Reborn
What Childbirth Should Be
London: Souvenir Press, 1994

The Scientification of Love
London: Free Association Books, 1999

Die Wurzeln der Liebe
Wie unsere wichtigsten Emotionen entstehen
Olten: Walter Verlag, 2001

Primal Health
Understanding the Critical Period Between Conception
and the First Birthday
London: Clairview Books, 2002
First Published in 1986 with Century Hutchinson in London

La Santé Primale
Paris: Payot, 1986

Die sanfte Geburt
Die Leboyer-Methode in der Praxis
Bergisch-Gladbach: Lübbe Verlag, 2001

Published by Sirius-C Media Galaxy LLC, 2010

The Functions of the Orgasms
The Highway to Transcendence
London: Pinter & Martin, 2009

Ollendorf-Reich, Ilse

Wilhelm Reich, A Personal Biography
New York, St. Martins Press, 1969

Wilhelm Reich
Vorwort von A.S. Neill
München, Kindler, 1975

Ong, Hean-Tatt

Amazing Scientific Basis of Feng-Shui
Kuala Lumpur: Eastern Dragon Press, 1997

Oppenheim, Lassa

International Law
4th Edition, by Sir Arnold D. McNair
New York, 1928

Ostrander, Sheila & Schroeder, Lynn

Superlearning 2000
New York: Delacorte Press, 1994

Superlearning
Die revolutionäre Lernmethode
München: Scherz Verlag, 1979

Supermemory
New York: Carroll & Graf, 1991

SuperMemory
Der Weg zum optimalen Gedächtnis
München: Goldmann, 1996

Ouspensky, Pyotr Demianovich

In Search of the Miraculous
New York: Mariner Books, 2001
First published in 1949

P

Pearce, John A. II and Robinson B. Jr.

Strategic Management
Formulation, Implementation and Control
Tenth Edition
New York: McGraw-Hill, 2007

Pearce Myers, Tony (Editor)

The Soul of Creativity
Insights into the Creative Process
Novato: New World Library, 1999

Pert, Candace B.

Molecules of Emotion
The Science Behind Mind-Body Medicine
New York: Scribner, 2003

Petrash, Jack

Understanding Waldorf Education
Teaching from the Inside Out
London: Floris Books, 2003

Plato

Complete Works
Ed. by John M. Cooper
New York: Hackett Publishing Company, 1997

Published by Sirius-C Media Galaxy LLC, 2010

Plummer, Kenneth

Pedophilia
Constructing a Sociological Baseline
in: in: Cook, M. and Howells, K. (Eds.):
Adult Sexual Interest in Children
Academic Press, London, 1980, pp. 220 ff.

Plutarch

Plutarch's Lives
The Dryden Translation
New York: Bantam Books, 2006

Ponder, Catherine

The Healing Secrets of the Ages
Marine del Rey: DeVorss, 1985

Porteous, Hedy S.

Sex and Identity
Your Child's Sexuality
Indianapolis: Bobbs-Merrill, 1972

Prescott, James W.

Affectional Bonding for the Prevention of Violent Behaviors
Neurobiological, Psychological and Religious/Spiritual Determinants
in: Hertzberg, L.J., Ostrum, G.F. and Field, J.R., (Eds.)
Violent Behavior
Vol. 1, Assessment & Intervention, Chapter Six
New York: PMA Publishing, 1990

Alienation of Affection
Psychology Today, December 1979

Body Pleasure and the Origins of Violence
Bulletin of the Atomic Scientists, 10-20 (1975)

Deprivation of Physical Affection as a Primary Process in the
Development of Physical Violence A Comparative
and Cross-Cultural Perspective,
in: David G. Gil, ed., Child Abuse and Violence
New York: Ams Press, 1979

Early somatosensory deprivation as an ontogenetic
process in the abnormal development of the brain and behavior,
in: Medical Primatology, ed. by I.E. Goldsmith and J. Moor-Jankowski,
New York: S. Karger, 1971

Genital Mutilation of Children
Failure of Humanity and Humanism
Unprinted Essay (2005)
http://www.violence.de/prescott/letters/
CIRC_CONGRESS_MONTAGUE_9.30.05.html

Genital Pain vs. Genital Pleasure
Why the One and not the Other
The Truth Seeker, July/August 1989, pp. 14-21
http://www.violence.de/prescott/truthseeker/genpl.html

How Culture Shapes the Developing Brain and the Future of Humanity
A Brief Summary of the research which links brain
abnormalities and violence to an absence of nurturing and bonding
very early in childhood,
in: Touch the Future: Optimum Learning Relationships
for Children & Adults
Spring 2002 (Ed. by Michael Mendizza)
Nevada City, CA, 2002

Invited Commentary: Central nervous system functioning in altered
sensory environments,
in: M.H. Appley and R. Trumbull (Eds.), *Psychological Stress*,
New York: Appleton-Century Crofts, 1967

Our Two Cultural Brains: Neurointegrative and Neurodissociative
http://www.violence.de/prescott/letters/Our_Two_Cultural_Brains.pdf

Published by Sirius-C Media Galaxy LLC, 2010

Phylogenetic and ontogenetic aspects of human affectional development,
in: Progress in Sexology, Proceedings of the 1976 International Congress of Sexology,
ed. by R. Gemme & C.C. Wheeler
New York: Plenum Press, 1977

Prevention or Therapy and the Politics of Trust
Inspiring a New Human Agenda
in: *Psychotherapy and Politics International*
Volume 3(3), pp. 194-211
London: John Wiley, 2005

Sex and the Brain
Midcontinent & Eastern Regions, June 13-16, 2002
Big Rapids, MI: Society for Cross-Cultural Research,
32nd Annual Meeting, 2005
http://www.violence.de/archive.shtml

Sixteen Principles for Personal, Family and Global Peace
The Truth Seeker, March/April 1989
http://www.violence.de/prescott/letters/Sixteen_Principles.pdf

Somatosensory affectional deprivation (SAD) theory of drug and alcohol use,
in: Theories on Drug Abuse: Selected Contemporary Perspectives,
ed. by Dan J. Lettieri, Mollie Sayers and Helen Wallenstien Pearson,
NIDA Research Monograph 30, March 1980
Rockville, MD: National Institute on Drug Abuse,
Department of Health and Human Services, 1980

The Origins of Human Love and Violence
Pre- and Perinatal Psychology Journal
Volume 10, Number 3:
Spring 1996, pp. 143-188The Origins of Love and Violence
Sensory Deprivation and the Developing Brain
Research and Prevention (DVD)
http://ttfuture.org/store/origins_orders

http://violence.de
http://ttfuture.org/violence
http://montagunocircpetition.org

Pritchard, Colin

The Child Abusers
New York: Open University Press, 2004

R

Radin, Dean

The Conscious Universe
The Scientific Truth of Psychic Phenomena
San Francisco: Harper & Row, 1997

Entangled Minds
Extrasensory Experiences in a Quantum Reality
New York: Paraview Pocket Books, 2006

Raknes, Ola

Wilhelm Reich and Orgonomy
Oslo: Universitetsforlaget, 1970

Wilhelm Reich und die Orgonomie
Eine Einführung in die Wissenschaft von der Lebensenergie
Frankfurt/M: Nexus, 1983

Randall, Neville

Life After Death
London: Robert Hale, 1999

Rank, Otto

Art and Artist
With Charles Francis Atkinson and Anaïs Nin
New York: W.W. Norton, 1989
Originally published in 1932

Published by Sirius-C Media Galaxy LLC, 2010

The Significance of Psychoanalysis for the Mental Sciences
New York: BiblioBazaar, 2009
First published in 1913

Rausky, Franklin

Mesmer ou la révolution thérapeutique
Paris, 1977

Redfield, James

The Tenth Insight
Holding the Vision
New York: Warner Books, 1996

The Celestine Prophecy
New York: Warner Books, 1995

Die Vision von Celestine
Berlin: Ullstein, 2004

Reich, Wilhelm

A Review of the Theories, dating from The 17th Century, on the Origin of Organic Life
by Arthur Hahn, Literature Assistant at the Institut für Sexualökonomische Lebensforschung, Biologisches Laboratorium, Oslo, 1938
©1979 Mary Boyd Higgins as Director of the Wilhelm Reich Infant Trust
XEROX Copy from the Wilhelm Reich Museum

Children of the Future
On the Prevention of Sexual Pathology
New York: Farrar, Straus & Giroux, 1984
First published in 1950

CORE (Cosmic Orgone Engineering)
Part I, Space Ships, DOR and DROUGHT
©1984, Orgone Institute Press
XEROX Copy from the Wilhelm Reich Museum

Der Einbruch der sexuellen Zwangsmoral
Frankfurt/M: Fischer, 1981

Die Entdeckung des Orgons II
Der Krebs
Frankfurt/M: Fischer, 1981
Köln: Kiepenheuer & Witsch, 1984

Die Funktion des Orgasmus
Sexualökonomische Grundprobleme der biologischen Energie
Köln: Kiepenheuer & Witsch, 1987

Die Massenpsychologie des Faschismus
Frankfurt/M: Fischer, 1974

Die sexuelle Revolution
Frankfurt/M: Fischer, 1966

Early Writings 1
New York: Farrar, Straus & Giroux, 1975

Ether, God & Devil & Cosmic Superimposition
New York: Farrar, Straus & Giroux, 1972
Originally published in 1949

Frühe Schriften 1
Aus den Jahren 1920-1925
Frankfurt/M: Fischer, 1983

Frühe Schriften 2
Genitalität in der Theorie und Therapie der Neurose
Frankfurt/M: Fischer, 1985

Genitality in the Theory and Therapy of Neurosis
©1980 by Mary Boyd Higgins as Director of the Wilhelm Reich
Infant Trust

Leidenschaften der Jugend
Köln: Kiepenheuer & Witsch, 1984

Published by Sirius-C Media Galaxy LLC, 2010

L'irruption de la morale sexuelle
Paris: Payot, 1972

Menschen im Staat
Frankfurt/M: Nexus, 1982

People in Trouble
©1974 by Mary Boyd Higgins as Director of the Wilhelm Reich
Infant Trust

Record of a Friendship
The Correspondence of Wilhelm Reich and A. S. Neill
New York, Farrar, Straus & Giroux, 1981

Selected Writings
An Introduction to Orgonomy
New York: Farrar, Straus & Giroux, 1973

The Bioelectrical Investigation of Sexuality and Anxiety
New York: Farrar, Straus & Giroux, 1983
Originally published in 1935

The Bion Experiments
reprinted in *Selected Writings*
New York: Farrar, Straus & Giroux, 1973

The Cancer Biopathy (The Orgone, Vol. 2)
New York: Farrar, Straus & Giroux, 1973

The Function of the Orgasm (The Orgone, Vol. 1)
Orgone Institute Press, New York, 1942

The Invasion of Compulsory Sex Morality
New York: Farrar, Straus & Giroux, 1971
Originally published in 1932

The Leukemia Problem: Approach
©1951, Orgone Institute Press
Copyright Renewed 1979
XEROX Copy from the Wilhelm Reich Museum

The Mass Psychology of Fascism
New York: Farrar, Straus & Giroux, 1970
Originally published in 1933

The Orgone Energy Accumulator
Its Scientific and Medical Use
©1951, 1979, Orgone Institute Press
XEROX Copy from the Wilhelm Reich Museum

The Schizophrenic Split
©1945, 1949, 1972 by Mary Boyd Higgins as Director of the
Wilhelm Reich Infant Trust
XEROX Copy from the Wilhelm Reich Museum

The Sexual Revolution
©1945, 1962 by Mary Boyd Higgins as Director of the
Wilhelm Reich Infant Trust

Zeugnisse einer Freundschaft
Der Briefwechsel zwischen Wilhelm Reich und A.S.
Neill (1936-1957)
Köln: Kiepenheuer & Witsch, 1986

Reid, Daniel P.
The Tao of Health, Sex & Longevity
A Modern Practical Guide to the Ancient Way
New York: Simon & Schuster, 1989

Guarding the Three Treasures
The Chinese Way of Health
New York: Simon & Schuster, 1993

Reps, Paul
Zen Flesh, Zen Bones
Rutland: Tuttle Publishing, 1989

Published by Sirius-C Media Galaxy LLC, 2010

Rhodes, Richard

The Making of the Atomic Bomb
New York, Simon & Schuster, 1995

Richardson, Justin

Everything You Never Wanted Your Kids to Know About Sex
With Mark. A. Schuster
New York: Three Rivers Press, 2003

Richet, Charles

Metapsychical Phenomena
Methods and Observations
Kessinger Publishing Reprint Edition, 2004
Originally published in 1905

Riso, Don Richard & Hudson, Russ

The Wisdom of the Enneagram
The Complete Guide to Psychological and Spiritual Growth
For The Nine Personality Types
New York: Bantam Books, 1999

Robbins, Anthony

Awaken The Giant Within
New York: Simon & Schuster, 1991

Unlimited Power
The New Science of Personal Achievement
New York: Free Press, 1997

Roberts, Jane

The Nature of Personal Reality
New York: Amber-Allen Publishing, 1994
First published in 1974

Die Natur der Persönlichen Realität
Ein neues Bewusstsein als Quelle der Kreativität
München: Kailash Verlag, 2007

The Nature of the Psyche
Its Human Expression
New York, Amber-Allen Publishing, 1996
First published in 1979

Die Natur der Psyche
Ihr menschlicher Ausdruck in Kreativität, Liebe, Sexualität
Genf: Ariston Verlag, 1985

Die Natur der Psyche
Ihr menschlicher Ausdruck in Kreativität, Liebe, Sexualität
München: Kailash Verlag, 2008

Roman, Sanaya

Opening to Channel
How To Connect With Your Guide
New York: H.J. Kramer, 1987

Zum Höheren Selbst Erwachen
Das Herz dem Bewusstsein des Lichts öffnen
Genf: Ansata Verlag, 2003

Rosen, Sydney (Ed.)

My Voice Will Go With You
The Teaching Tales of Milton H. Erickson
New York: Norton & Co., 1991

Rosenbaum, Julius

The Plague of Lust
New York: Frederick Publications, 1955

Rossman, Parker

Sexual Experiences between Men and Boys
New York, 1976

Published by Sirius-C Media Galaxy LLC, 2010

Rothschild & Wolf

Children of the Counterculture
New York: Garden City, 1976

Rousseau, Jean-Jacques

Émile ou de l'Éducation, 1762
Reprint, Paris: Garnier, 1964

The Social Contract
And Later Political Writings
Cambridge, MA.: Cambridge University Press, 1997

Rudhyar, Dane

Astrology of Personality
A Reformulation of Astrological Concepts and Ideals in
Terms of Contemporary Psychology and Philosophy
New York: Aurora Press, 1990

An Astrological Triptych
Gifts of the Spirit, The Way Through, and The Illumined Road
New York: Aurora Press, 1991

Astrological Mandala
New York: Vintage Books, 1994

L'astrologie de la transformation
Paris: Rocher, 1984

Ruiz, Don Miguel

The Four Agreements
A Practical Guide to Personal Freedom
San Rafael, CA: Amber Allen Publishing, 1997

The Mastery of Love
A Practical Guide to the Art of Relationship
San Rafael, CA: Amber Allen Publishing, 1999

The Voice of Knowledge
A Practical Guide to Inner Peace
With Janet Mills
San Rafael, CA: Amber Allen Publishing, 2004

Ruperti, Alexander

Cycles of Becoming
The Planetary Pattern of Growth
New York: CRCS Publications, 1978

La roue de l'expérience individuelle
Paris: Librairie de Médicis, 1991

Rush, Florence

The Best Kept Secret
Sexual Abuse of Children
New Jersey: Prentice-Hall, 1980

Das bestgehütete Geheimnis
Sexueller Kindesmissbrauch
Berlin: Sub-Rosa Frauenverlag, 1984

S

Saint-Simon, Claude-Henri de

De la réorganisation de la société européenne
Avec Auguste Thierry
Paris, 1814
Lausanne: Centre de Recherches Européennes, 1967

Salomé, Jacques

Si je m'écoutais, je m'entendrais
Avec Sylvie Galland
Paris: Éditions de l'Homme, 1990

Published by Sirius-C Media Galaxy LLC, 2010

Sandfort, Theo

The Sexual Aspect of Pedophile Relations
The Experience of Twenty-five Boys
Amsterdam: Pan/Spartacus, 1982

SantoPietro, Nancy

Feng Shui, Harmony by Design
How to Create a Beautiful and Harmonious Home,
New York: Putnam-Berkeley, 1996

Satinover, Jeffrey

Homosexuality and the Politics of Truth
New York: Baker Books, 1996

The Quantum Brain
New York: Wiley & Sons, 2001

Satprem

Sri Aurobindo ou l'aventure de la conscience
Paris: Buchet/Castel, 1970

Scarro A. M., Jr. (Ed.)

Male Rape
New York: Ams Press, 1982

Schérer, René

Co-ire
Album systématique de l'enfance
Avec Guy Hocquenghem
Recherches No. 22
Paris: E.S.F., 1976

Émile perverti, ou des rapports entre l'éducation et la sexualité
Paris: Robert Laffont, 1974
Paris, Désordres, 2006
Nouvelle Édition

Le corps interdit
Avec Georges Lapassade
Paris: E.S.F., 1976

Une érotique puérile
Paris: Éditions Galilée, 1978

Schlipp, Paul A. (Ed.)

Albert Einstein
Philosopher-Scientist
New York: Open Court Publishing, 1988

Schonberg, Harold

The Great Pianists
From Mozart to the Present
New York: Simon and Schuster (Fireside), 2006
Originally published in 1963

Schrenck-Notzing, Albert von

Phenomena of Materialization
A Contribution to the Investigation of Mediumistic Teleplastics
Perspectives in Psychical Research
New York: Kegan Paul, 1920

Schultes, Richard Evans, et al.

Plants of the Gods
Their Sacred, Healing, and Hallucinogenic Powers
New York: Healing Arts Press
2nd edition, 2002

Die Pflanzen der Götter
Die magischen Kräfte der Rausch- und Giftgewächse
München: AT Verlag, 1998

Schumacher, E.F.

Small is Beautiful
Economics as if People Mattered
San Francisco: Harper Perennial, 1989

Schwartz, Andrew E.

Guided Imagery for Groups
Fifty Visualizations That Promote Relaxation, Problem-Solving,
Creativity, and Well-Being
Whole Person Associates, 1995

Senf, Bernd

Die Wiederentdeckung des Lebendigen
Aachen: Omega, 2003
Erstmals veröffentlicht 1996 mit Zweitausendeins Verlag in Frankfurt/M

Nach Reich: Neue Forschungen zur Orgonenergie
Sexualökonomie / Die Entdeckung der Orgonenergie
Herausgegeben zusammen mit Professor James DeMeo,
Ashland, Oregon, USA
Frankfurt/M: Zweitausendeins Verlag, 1997

Sepper, Dennis L.

Goethe Contra Newton
Polemics and the Project of a New Science of Color
Cambridge: Cambridge University Press, 1988

Shalabi, Ahmad

Islam
Cairo, 1970

Sharaf, Myron

Fury on Earth
A Biography of Wilhelm Reich
London: André Deutsch, 1983

Wilhelm Reich
Der heilige Zorn des Lebendigen
Berlin: Simon & Leutner, 1994

Sheldrake, Rupert

A New Science of Life
The Hypothesis of Morphic Resonance
Rochester: Park Street Press, 1995

Das Schöpferische Universum
Die Theorie des morphogenetischen Feldes
Neue und erweiterte Auflage
Berlin: Ullstein, 2009

Sher, Barbara & Gottlieb, Annie

Wishcraft
How to Get What You Really Want
2nd edition
New York: Ballantine Books, 2003

Shone, Ronald

Creative Visualization
Using Imagery and Imagination for Self-Transformation
New York: Destiny Books, 1998

Simonton, O. Carl et al.

Getting Well Again
Los Angeles: Tarcher, 1978

Singer, June

Androgyny
New York: Doubleday Dell, 1976

Smith, C. Michael

Jung and Shamanism in Dialogue
London: Trafford Publishing, 2007

Published by Sirius-C Media Galaxy LLC, 2010

Spiller, Jan

Astrology for the Soul
New York: Bantam, 1997

Spock, Benjamin

Dr. Spock's Baby and Child Care
8th Edition
New York: Pocket Books, 2004

Säuglings- und Kinderpflege
Berlin: Ullstein, 1986

Spretnak, Charlene

Green Politics
Rochester, VT: Inner Traditions, 1986

Stein, Robert M.

Redeeming the Inner Child in Marriage and Therapy
in: Reclaiming the Inner Child
ed. by Jeremiah Abrams
New York: Tarcher/Putnam, 1990, 261 ff.

Steiner, Rudolf

Theosophy
An Introduction to the Spiritual Processes in Human Life
and in the Cosmos
New York: Anthroposophic Press, 1994

Die Erziehung des Kindes
Dornach: Rudolf Steiner Verlag, 2003
First published in 1907

Stekel, Wilhelm

Auto-Eroticism
A Psychiatric Study of Onanism and Neurosis
Republished, London: Paul Kegan, 2004

Patterns of Psychosexual Infantilism
New York, 1959 (reprint edition)

Psychosexueller Infantilismus
Die seelischen Kinderkrankheiten der Erwachsenen
Berlin: Urban & Schwarzenberg, 1922

Sadism and Masochism
New York: W.W. Norton & Co., 1953

Sex and Dreams
The Language of Dreams
Republished
New York: University Press of the Pacific, 2003

Störungen des Trieb- und Affektlebens
Bände I & II
Berlin: Urban & Schwarzenberg, 1921

Stiene, Bronwen & Frans

The Reiki Sourcebook
New York: O Books, 2003

The Japanese Art of Reiki
A Practical Guide to Self-Healing
New York: O Books, 2005

Stone, Hal & Stone, Sidra

Embracing Our Selves
The Voice Dialogue Manual
San Rafael, CA: New World Library, 1989

Du bist viele
Das 100fache Selbst und seine Entdeckung durch
die Voice-Dialogue Methode
München: Heyne Verlag, 1994

Published by Sirius-C Media Galaxy LLC, 2010

Strassman, Rick

DMT: The Spirit Molecule
A doctor's revolutionary research into the biology of near-death
and mystical experiences
Rochester: Park Street Press, 2001

Sun Tzu (Sun Tsu)

The Art of War
Special Edition
New York: El Paso Norte Press, 2007

Die Kunst des Krieges
Hamburg: Nikol Verlag, 2008

Suryani, Luh Ketut & Jensen, Gorden D.

The Balinese People
A Reinvestigation of Character
New York: Oxford University Press, 1993

Symonds, John Addington

A Problem in Greek Ethics
New York: M.S.G. House, 1971

Szasz, Thomas

The Myth of Mental Illness
New York: Harper & Row, 1984

T

Talbot, Michael

The Holographic Universe
New York: HarperCollins, 1992

Das holographische Universum
Die Welt in neuer Dimension
München: Droemer Knaur, 1994

Tansley, David V.

Chakras, Rays and Radionics
London: Daniel Company Ltd., 1984

Targ, Russell & Katra, Jane

Miracles of Mind
Exploring Nonlocal Consciousness and Spiritual Healing
Novato, CA: New World Library, 1999

Tarnas, Richard

Cosmos and Psyche
Intimations of a New World View
New York: Plume, 2007

The Passion of the Western Mind
Understanding the Ideas that have Shaped Our World View
New York: Ballantine Books, 1993

Tart, Charles T.

Altered States of Consciousness
A Book of Readings
Hoboken, N.J.: Wiley & Sons, 1969

Tatar, Maria M.

Spellbound: Studies on Mesmerism and Literature
Princeton, N.Y., 1978

Tchouang-tseu

Oeuvre complète
Paris: Gallimard/Unesco, 1969

Temple, Robert

The Sirius Mystery
New Scientific Evidence of Alien Contact 5000 Years Ago
Rochester: Destiny Books, 1998

Textor, R. B.

A Cross-Cultural Summary
New Haven, Human Relations Area Files (HRAF)
Press, 1967

The Advent of Great Awakening

A Course in Miracles
Text Workbook and Manual for Teachers
New York: New Christian Church of Full Endeavor, 2007

The Tibetan Book of the Dead

The Great Liberation through Hearing in the Bardo
Translated with commentary by Francesca
Fremantle & Chögyam Trungpa
Boston: Shambhala Dragon Editions, 1975

The Ultimate Picasso

New York: Harry N. Abrams, 2000

Thorsson, Edred

Futhark
A Handbook of Rune Magic
San Francisco: Weiser Books, 1984

Tiller, William A.

Conscious Acts of Creation
The Emergence of a New Physics
Associated Producers, 2004 (DVD)

Psychoenergetic Science
New York: Pavior, 2007

Conscious Acts of Creation
New York: Pavior, 2001

Tischner, Rudolf

F.A. Mesmer
München, 1928

Todaro-Franceschi, Vidette

The Enigma of Energy
Where Science and Religion Converge
New York: Crossroad Publishing, 1991

Toffler, Alvin

Powershift
Knowledge, Wealth, and Violence at the Edge of the 21st Century
New York: Bantam, 1991

Revolutionary Wealth
How it will be created and how it will change our lives
New York: Broadway Business, 2007

The Third Wave
New York: Bantam, 1984

Tolle, Eckhart

The Power of Now
A Guide to Spiritual Enlightenment
Novato, CA: New World Library, 2004

Jetzt! Die Kraft der Gegenwart
Ein Leitfaden zum spirituellen Erwachen
Bielefeld: Kamphausen Verlag, 2000

A New Earth
Awakening to Your Life's Purpose
New York: Michael Joseph (Penguin), 2005

Eine neue Erde
Bewusstseinssprung anstelle von Selbstzerstörung
München: Goldmann, 2005

Published by Sirius-C Media Galaxy LLC, 2010

Too, Lillian

Feng Shui
Kuala Lumpur: Konsep Books, 1994

U

Unlawful Sex

Offences, Victims and Offenders in the Criminal Justice System
of England and Wales
The Report of the Howard League Working Party
London: Waterloo Publishers Ltd., 1985

V

Van Gelder, Dora

The Real World of Fairies
A First-Person Account
Wheaton: Quest Books, 1999
First published in 1977

Vanguard, Thorkil

Phallós
A Symbol and its History in the Male World
New York: International Universities Press, 2001

Villoldo, Alberto

Healing States
A Journey Into the World of Spiritual Healing and Shamanism
With Stanley Krippner
New York: Simon & Schuster (Fireside), 1987

Dance of the Four Winds
Secrets of the Inca Medicine Wheel
With Eric Jendresen
Rochester: Destiny Books, 1995

Die Macht der vier Winde
Eine Reise ins Reich der Schamanen
München: Goldmann, 2009

Shaman, Healer, Sage
How to Heal Yourself and Others with the Energy Medicine
of the Americas
New York: Harmony, 2000

Hüter des alten Wissens
Schamanisches Heilen im Medizinrad
Darmstadt: Schirner Verlag, 2007

Healing the Luminous Body
The Way of the Shaman with Dr. Alberto Villoldo
DVD, Sacred Mysteries Productions, 2004

Mending The Past And Healing The Future with Soul Retrieval
New York: Hay House, 2005

Seelenrückholung: die Vergangenheit schamanistisch erkunden
Die Zukunft heilen
München, Goldmann, 2006

Vitebsky, Piers
The Shaman
Voyages of the Soul, Trance, Ecstasy and Healing from
Siberia to the Amazon
New York: Duncan Baird Publishers, 2001
Originally published in 1995

Von Riezler, Sigmund
Geschichte der Hexenprozesse in Bayern
Stuttgart: Magnus Verlag, 1983

Published by Sirius-C Media Galaxy LLC, 2010

W

Walker & Walker

The English Legal System
6th Edition, by R.J. Walker
London: Butterworths, 1985

Ward, Elizabeth

Father-Daughter Rape
New York: Grove Press, 1985

Watts, Alan W.

The Way of Zen
New York: Vintage Books, 1999

This Is It
And Other Essays on Zen and Spiritual Experience
New York: Vintage, 1973

Wee Chow Hou

The 36 Strategies of the Chinese
Adapting Ancient Chinese Wisdom to the Business World
New York: Addison-Wesley, 2007

Weiss, Jess E.

The Vestibule
New York: Ashley Books, 1979

West's Encyclopedia of American Law

Second Edition
New York: Gale Group, 2008

Wharton

Wharton's Criminal Law
14th ed. by Charles E. Torcia
Vol. II, §§99-282
Rochester, New York: The Lawyers Cooperative Publishing Co., 1979

What the Bleep Do We Know!?

See Arntz, William

Whiteman

Digest of International Law
Vol. 6
Washington, D.C.: Department of State Publication 8350, 1968

Whitfield, Charles L.

Healing the Child Within
Deerfield Beach, Fl: Health Communications, 1987

Whiting, Beatrice B.

Children of Six Cultures
A Psycho-Cultural Analysis
Cambridge: Harvard University Press, 1975

Wiener, Jon

Gimme Some Truth: The John Lennon FBI Files
Los Angeles: University of California Press, 1999

Wilber, Ken

Sex, Ecology, Spirituality
The Spirit of Evolution
Boston: Shambhala, 2000

Quantum Questions
Mystical Writings of The World's Greatest Physicists
Boston: Shambhala, 2001

Published by Sirius-C Media Galaxy LLC, 2010

Wild, Leon D.

The Runes Workbook
A Step-by-Step Guide to Learning the Wisdom of the Staves
San Diego: Thunder Bay Press, 2004

Wilhelm Helmut

The Wilhelm Lectures on the Book of Changes
Princeton: Princeton University Press, 1995

Wilhelm, Richard

The I Ching or Book of Changes
With C. Baynes
3rd Edition, Bollingen Series XIX
Princeton, NJ: Princeton University Press, 1967

Williams, Strephon Kaplan

Dreams and Spiritual Growth
With Patricia H. Berne and Louis M. Savary
New York: Paulist Press, 1984

Durch Traumarbeit zum eigenen Selbst
Die Jung-Senoi Methode
Interlaken: Ansata Verlag, 1987

Dream Cards
Understand Your Dreams and Enrich Your Life
New York: Simon & Schuster (Fireside), 1991

Wing, R. L.

The I Ching Workbook
Garden City, N.Y.: Doubleday, 1984

Das Arbeitsbuch zum I Ching
Mit Chinesischen Orakel Münzen
München: Goldmann, 2004

Het I Tjing Werkboek
Baarn: Bigot & Van Rossum, 1986

Woerly, Franz
Esprit Guide
Entretiens avec Karlfried Dürckheim
Paris: Albin Michel, 1985

Wolf, Fred Alan
Taking the Quantum Leap
The New Physics for Nonscientists
New York: Harper & Row, 1989

Der Quantensprung ist keine Hexerei
Frankfurt/M: Fischer Verlag, 1990

Parallel Universes
New York: Simon & Schuster, 1990

The Dreaming Universe
A Mind-Expanding Journey into the Realm Where
Psyche and Physics Meet
New York: Touchstone, 1995

The Eagle's Quest
A Physicist Finds the Scientific Truth At the Heart of the
Shamanic World
New York: Touchstone, 1997

Die Physik der Träume
Frankfurt/M: DTV Verlag, 1997

Mind into Matter
A New Alchemy of Science and Spirit
New York: Moment Point Press, 2000

Published by Sirius-C Media Galaxy LLC, 2010

Words and Phrases Legally Defined

Ed. By John Saunders
2nd Edition
London: Butterworths, 1969

Wydra, Nancilee

Feng Shui
The Book of Cures
Lincolnwood: Contemporary Books, 1996

Y

Yang, Jwing-Ming

Qigong, The Secret of Youth
Da Mo's Muscle/Tendon Changing and Marrow/Brain Washing Classics
Boston, Mass.: YMAA Publication Center, 2000

The Root of Chinese Qigong
Secrets for Health, Longevity, & Enlightenment
Roslindale, MA: YMAA Publication Center, 1997

Yates, Alayne

Sex Without Shame
Encouraging the Child's Healthy Sexual Development
New York, 1978
Republished Internet Edition

Yeats, William Butler

Irish Fairy and Folk Tales
New York: Modern Library, 2003

Mythologies
New York: Simon & Schuster, 1998
Author Copyright 1959, Renewed 1987 by Anne Yeats

Ywahoo, Dhyani

Voices of Our Ancestors
Cherokee Teachings from the Wisdom Fire
New York: Shambhala, 1987

Am Feuer der Weisheit
Lehren der Cherokee Indianer
Zürich: Theseus Verlag, 1988

Z

Znamenski, Andrei A.

Shamanism
Critical Concepts in Sociology
New York: Routledge, 2004

Zinker, Joseph

Se créer par la Gestalt
Montréal: Les Éditions de l'Homme, 1981

Zukav, Gary

The Dancing Wu Li Masters
An Overview of the New Physics
New York: HarperOne, 2001

Die tanzenden Wu Li Meister
Der östliche Pfad zum Verständnis der modernen Physik
Vom Quantensprung zum schwarzen Loch
Berlin: Rowohlt, 2000

Zweig, Stefan

Die Heilung durch den Geist
Mesmer, Mary Baker-Eddy, Freud
Frankfurt/M: Fischer Verlag, 1982
Originally published in 1931

Published by Sirius-C Media Galaxy LLC, 2010

Zyman, Sergio

The End of Marketing as We Know It
New York: HarperCollins, 2000

Das Ende der Marketing Mythen
Erfolgsrezepte des Aya-Cola für Umsatz und Profit
Berlin: Econ Verlag, 2000

FROM THE SAME AUTHOR

A Bibliography

You can search publications from here:
http://ipublica.com/books/

For audio books and music, you can start here:
http://ipublica.com/audio/

All paperbacks, audio downloads, audio book compact discs, music downloads and music compact discs, as well as Kindle books, are referenced on the site.

For free podcasts search iTunes under my author name.

For quoting my publications, please use the following form:
Pierre F. Walter, [Title]: [Subtitle], Newark: Sirius-C Media Galaxy LLC, 2010

Web Presence

Pierre F. Walter on the Web

Sites

http://authoryourlife.com

http://ipublica.com

http://ipublica.net

http://ipublica.org

http://ipublica.tv

Video Channels

http://youtube.com/user/ipublica

http://youtube.com/user/authoryourlife

http://vimeo.com/pierrefwalter/channels

http://ipublica.blip.tv/

http://authoryourlife.blip.tv/

http://emosexuality.blip.tv/

http://pierrefwalter.blip.tv/

Our real criminals are not populating our prisons and jails but our oval offices and national parliaments, for they have the real power for *power abuse* which is the only form of abuse our media never talk about.

And for good reason, for they need convenient scapegoats and poison containers for their repressed and perverse desires that, unfortunately so, are the lot of the majority in a society that has since long lost its genuine morality, replacing it by compulsive moralism which is violent, abrasive and destructive, and which could lead, between 2012 and 2020 to the annihilation of the entire human race.

I have written this book in the optimism that this will not happen because of the corrective powers of nature and the divine realm that is not inhuman, but more-than-human, thereby being fully human.

The problem of humanity is that it always considered nature as impossible and violent and moralism as possible and nonviolent – while of course, the exact opposite is true.
– Pierre F. Walter

NOTES

Annotations

[1] See Albert Hofmann, *LSD: My Problem Child (1980/2009)* as well as Stanislav Grof, LSD: Doorway to the Numinous (2009) and *Realms of the Human Unconscious: Observations from LSD Research (1976).*

[2] Terence McKenna, *The Archaic Revival (1992)*, p. 45.

[3] Mircea Eliade, Michael Harner, Richard Schultes, Ralph Metzner, Adam Gottlieb and Terence McKenna.

[4] Dean Radin, *The Conscious Universe (1997)*, p. 263.

[5] The character Tao 道 means path or way, but in Chinese religion and philosophy it has abstract meanings.

[6] Pierre F. Walter, *The Idiot Guide to Leadership (2010)*, Chapter Three ('Right Action') and *The I Ching's Perennial Pro-Life Code: An Analysis of Pattern, Audio Book (2010).*

[7] See, for example, Mircea Eliade, *Shamanism (1964)*, Piers Vitebsky, *Shamanism (2001)*, Ralph Metzner (Ed.), *Ayahuasca (1999)*, Michael Harner, *Ways of the Shaman (1990)*, Jeremy Narby, *The Cosmic Serpent (1999)*, Richard E. Schultes et al., *Plants of the Gods (2002)*, Terence McKenna, *The Invisible Landscape (1994)*, *True Hallucinations (1998)*, *The Archaic Revival (1992)*, *Food of the Gods (1993)*, Robert Forte (Ed.), *Entheogens and the Future of Religion (2000)*, Luis Eduardo Luna, Pablo Amaringo, *Ayahuasca Visions (1999)*, Adam Gottlieb, *Peyote and Other Psychoactive Cacti (1997)*, Aldous Huxley, *The Doors of Perception and Heaven and Hell (1954)*, Rick Strassman, *DMT: The Spirit Molecule (2001)*, Josep M. Fericla, *Al trasluz de la Ayahuasca (2002).*

[8] See, for example, Sheila Ostrander & Lynn Schroeder, *Superlearning 2000 (1994).*

[9] Please allow me to use the masculine preposition 'he' here and in the following, while I am aware of the fact that in the oldest and most powerful shamanic traditions, the shaman was not a male but a *female*. My intention here is motivated by aesthetic reasons, simply to avoid rendering the script unnecessarily disrupted by the endless repetition of 'he or she'.

[10] Holger Kalweit, *Shamans, Healers and Medicine Men (2000)*, p. 3.

[11] Id., p. 54.

Published by Sirius-C Media Galaxy LLC, 2010

[12] Id., pp. 228-229.

[13] See, for example, Richard Gerber, *Vibrational Medicine (2002)*.

[14] Mircea Eliade, *Shamanism: Archaic Techniques of Ecstasy (1964)*, p. 329.

[15] Id., p. 364.

[16] Alberto Villoldo, *Shaman, Healer, Sage (2000)*, p. 9.

[17] Id., pp. 13-14.

[18] Id. pp. 22-23.

[19] Alberto Villoldo explains in *Shaman, Healer, Sage (2000)*, pp. 42-43 what the luminous energy field is:

> We all possess a *Luminous Energy Field* that surrounds our physical body and informs our body in the same way that the energy fields of a magnet organize iron filings on a piece of glass. Our Luminous Energy Field has existed since before the beginning of time. It was one with the unmanifest light of Creation, and it will endure / throughout infinity. It dwells outside of time but manifests in time by creating new physical bodies lifetime after lifetime.

[20] See Carlos Castaneda, *The Teachings of Don Juan (1985)*, *Journey to Ixtlan (1991)*, *Tales of Power (1991)*, *The Second Ring of Power (1991)*.

[21] Alberto Villoldo, Shaman, Healer, Sage (2000), pp. 48-49.

[22] Id., p. 46.

[23] See Erika Nau, *Self-Awareness through Huna (1981)*.

[24] Max Long, *The Secret Science at Work (1958/1995)* and *Growing into Light (1955)*.

[25] See Wilhelm Reich, *Cosmic Superimposition (1949/1972)*, Ola Raknes, *Wilhelm Reich and Orgonomy (1970)*. See also Pierre F. Walter, *Energy Science and Vibrational Healing, Monograph (2010)* and *The Science of Orgonomy, Monograph (2010)*.

[26] See, for example, Ong Hean Tatt, *Amazing Scientific Basis of Feng Shui (1997)*.

[27] See also the remarkable study by Valerie Hunt, *Infinite Mind (2000)*.

[28] Alberto Villoldo, *Shaman, Healer, Sage (2000)*, p. 50.

[29] See, for example, the revealing study by Shafica Karagulla and Dora van Gelder, *The Chakras (1989)*.

[30] Alberto Villoldo, *Shaman, Healer, Sage (2000)*, p. 53.

[31] Id., p. 56.

[32] Id.

[33] Id., pp. 56-57.

[34] Id., p. 57.

[35] Id., p. 61.

[36] See O. Carl Simonton, *Getting Well Again (1978)*.

[37] Id., p. 113.

[38] Alberto Villoldo, Shaman, *Healer, Sage (2000)*, p. 159.

[39] See, for example, Rupert Sheldrake, *A New Science of Life (1995)*.

[40] Alberto Villoldo, Shaman, *Healer, Sage (2000)*, p. 157.

[41] In German, there is a very pointed expression for such kind of attitude, it is called *'die Zeit totschlagen'* (slaughtering time).

[42] Stanley Krippner, Antonio Villoldo, *Healing States (1984)*, p. 85.

[43] Terence McKenna, *The Archaic Revival (1992)*, p. 15.

[44] See further, Pierre F. Walter, *The Idiot Guide to Love, Awareness Guide (2010)* and *The Idiot Guide to Sanity, Awareness Guide (2010)*. See also Pierre F. Walter, *Love or Laws?, Audio Book (2010)* and *The Legal Split in Child Protection, Audio Book (2010)*.

[45] See, for example, *Fritjof Capra, The Tao of Physics (1975), The Turning Point (1982), The Web of Life (1996), The Hidden Connections (2002), Steering Business Toward Sustainability (1995)*. There are many further references in all of Capra's books.

[46] Fritjof Capra, *The Tao of Physics (1975/1984)*, p. 8.

Published by Sirius-C Media Galaxy LLC, 2010

[47] Fritjof Capra, *The Turning Point (1987)*, p. 66.

[48] See, for example Thomas Moore, *Care of the Soul (1994)*, Michael Talbot, *The Holographic Universe (1992)* and Michael Murphy, *The Future of the Body (1992)*, Part 2, (21) and (22), pp. 464-527.

[49] See Michael Talbot, *The Holographic Universe (1992)*, Part 2 (5), pp. 119 ff, Michael Murphy, *The Future of the Body (1992)*, Part 2 (21), pp. 464 ff.

[50] See Piers Vitebsky, *The Shaman (1995/2001)*.

[51] See, for example, Alberto Villoldo, *Mending The Past And Healing The Future with Soul Retrieval (2005)*.

[52] Andrei A. Znamenski, *Shamanism (2004)*, 314, 321.

[53] See, for example, Susanne Cho, *Kindheit und Sexualität im Wandel der Kulturgeschichte (1983)*; Larry L. & Joan M. Constantine, *Treasures of the Island (1976)* and *Where are the Kids? Children in Alternative Life-Styles (1977)*.

[54] Bronislaw Malinowski, *The Sexual Life of Savages in North West Melanesia (1929)* and *Sex and Repression in Savage Society (1927)*.

[55] V. Elwin, *The Muria and their Ghotul (1947)*, Richard Currier, *Juvenile Sexuality in Global Perspective (1981)*, 9 ff.

[56] Id.

[57] Margaret Mead, *Sex and Temperament in Three Primitive Societies (1935)*.

[58] See Pierre F. Walter, *Natural Order, Monograph (2010)*, Chapter One.

[59] C. Michael Smith, *Jung and Shamanism in Dialogue (2007)*, 82-83.

[60] Id., p. 96.

[61] Id., p. 97.

[62] Pierre F. Walter, *Eight Dynamic Patterns of Living, Audio Book (2010)*.

[63] Terence McKenna, *The Archaic Revival (1992)*, p. 13.

[64] Id., p. 144.

[65] Sallie Nichols, *Jung and Tarot (1986)*.

[66] Dr. Joseph Murphy, *The Power of Your Subconscious Mind (1963)*, *The Miracle of Mind Dynamics (1964)* and *Think Yourself Rich (2001)*.

[67] Dr. Joseph Murphy, *The Power of Your Subconscious Mind*, p. 165, Pierre F. Walter, *Creative Prayer, Audio Book (2010)*.

[68] Joseph Murphy, *The Power of Your Subconscious Mind*, pp. 29-30.

[69] Manly P. Hall, *The Secret Teachings of All Ages (2003)*, p. 347.

[70] Ervin Laszlo, *Science and the Akashic Field (2004)*, p. 55.

[71] See, for example, Edward de Bono, *The Use of Lateral Thinking (1967)*, *The Mechanism of Mind (1969)*, *Sur/Petition (1993)*, *Tactics (1993)*, *Serious Creativity (1996)*. See also Pierre F. Walter, *Edward de Bono and Serious Creativity, 110 Book Reviews (2010)*.

[72] See, for example, J. Krishnamurti, *Freedom from the Known (1969)*, *Education and the Significance of Life (1978)* or *Beyond Violence (1985)*. See also Pierre F. Walter, *Krishnamurti, Education and the Significance of Life, 110 Book Reviews (2010)*.

[73] Albert Hofmann, *LSD: My Problem Child (1979/2009)*, pp. 15, 57.

[74] Terence McKenna, *The Archaic Revival (1992)*, p. 219.

[75] See, for more details, Richard Evans Schultes, et al., *Plants of the Gods (1992)*, pp. 124 ff.

[76] See, for example, Mircea Eliade, *Shamanism (1972)*, Piers Vitebsky, *Shamanism (2001)*, Ralph Metzner (Ed.) *Ayahuasca (1999)*, Michael Harner, *Ways of the Shaman (1990)*, Jeremy Narby, *The Cosmic Serpent (1999)*, Richard Evans Schultes et al, *Plants of the Gods (2002)*, Terence McKenna, *The Invisible Landscape (1994)*, *True Hallucinations (1998)*, *The Archaic Revival (1992)*, *Food of the Gods (1993)*, Robert Forte (Ed.), *Entheogens and the Future of Religion (2000)*, Luis Eduardo Luna & Pablo Amaringo, *Ayahuasca Visions (1999)*, Adam Gottlieb, *Peyote and Other Psychoactive Cacti (1997)*, Rick Strassman, *DMT (2001)*.

[77] Rick Strassman, *DMT (2001)*.

[78] Jeremy Narby, *The Cosmic Serpent (2003)*.

[79] Id., pp. 127-128.

Published by Sirius-C Media Galaxy LLC, 2010

[80] Id., p. 129.

[81] Alberto Villoldo, *Shaman, Healer, Sage (2000)*, p. 41.

[82] Dora van Gelder, *The Real World of Fairies (1999)*, p. 4.

[83] Charles W. Leadbeater, *Astral Plane (1894)*, pp. 54-55.

[84] Id., pp. 55-56.

[85] Charles W. Leadbeater, *Astral Plane (1894)*, p. 61.

[86] Dr. William A. Tiller, *Conscious Acts of Creation (DVD)*.

[87] Charles W. Leadbeater, *Astral Plane (1894)*.

[88] See, for example, Rupert Sheldrake, *A New Science of Life (1995)*.

[89] I use the expression *koan* in my books in order to denote the paradoxes that are the outcome of our intellectual, conceptual look at life. However, please be aware that I am convinced that life or nature is not per se paradoxical, but bound together in infinite harmony. The paradoxical nature of life is an appearance of reality made up through our limited, concept-based and thought-created image of reality - and not reality itself.

[90] Jeremy Narby, *The Cosmic Serpent (2003)*, p. 42.

[91] See Aldous Huxley, *The Doors of Perception and Heaven and Hell (1954)*.

[92] In fact, the first title of this paper had been *The Eight Principles of Living*.

[93] Id., p. 80.

[94] Id., p. 81.

[95] Id.

[96] Fritjof Capra, *The Hidden Connections (2002)*, p. 9.

[97] See Fritjof Capra, *The Web of Life (1996)*, p. 37.

[98] Pierre F. Walter, *Eight Dynamic Patterns of Living, Audio Book (2010)*.

[99] Interestingly, neither Bronislaw Malinowski nor Margaret Mead found sexual paraphilias present in Melanesia's Trobriand culture where children enjoy utmost emotional and sexual freedom.

[100] See, for example, *Walter's Encyclopedia, Academic Edition (2010)* and *The Idiot Guide to Emotions (2010)* as well as *The Idiot Guide to Science (2010)*.

[101] Wilhelm Reich, *The Mass Psychology of Fascism (1933)*.

[102] More information and references on this research topic are to be found in Pierre F. Walter, *The Idiot Guide to Sanity (2010)* and *The Idiot Guide to World Peace (2010)*.

[103] See Lloyd DeMause, *Foundations of Psychohistory (1982)*.

[104] See Lloyd DeMause, *The Psychohistory Journal* (Periodical).

[105] Françoise Dolto, *Séminaire de Psychanalyse d'Enfants, Tome 2, (1985)*, p. 21 (translation mine).

[106] See Ralph Metzner in Ralph Metzner (Ed.), *Ayahuasca (1999)*, p. 24.

[107] Terence McKenna, *The Archaic Revival (1992)*, p. 13.

[108] Id., p. 144.

[109] See already notes 1 and 73 supra.

[110] Id., p. 219.

[111] Dennis McKenna & Terence McKenna, *The Invisible Landscape (1993)*.

[112] Mircea Eliade, *Shamanism (1989)*.

[113] See, for example, W.Y. Evans-Wentz, *The Fairy Faith in Celtic Countries (1911)*.

[114] Thomas Moore, *Care of the Soul (1994)*.

[115] See Pierre F. Walter, *The Idiot Guide to Science (2010)*.

[116] See Johann Wolfgang von Goethe, *The Theory of Colors (1970)*, and Frederick Burwick, *The Damnation of Newton (1986)*.

[117] Terence McKenna, *Food of the Gods (1992)*, p. 7.

Published by Sirius-C Media Galaxy LLC, 2010

[118] See *Walter's Encyclopedia, Academic Edition (2010)*, 'Oedipal Culture' and my audio book *Oedipal Hero (2010)*.

[119] See *Walter's Encyclopedia, Academic Edition (2010)*, 'Essenes'.

[120] Id, 'Direct Perception'.

[121] Schultes, Hofmann and Rätsch report that mushrooms are among the most archaic forms of living on earth and have therefore assimilated an infinite amount of knowledge about evolution on earth that is not per se accessible to a human being without access to entheogenic substances. See Richard Evans Schultes, Albert Hofmann, Christian Rätsch, *Plants of the Gods (1992)*.

[122] See Sheila Ostrander & Lynn Schroeder, *Superlearning 2000 (1994)*.

[123] Terence McKenna, *Food of the Gods (1992)*, p. 52.

[124] My study of the age-old Chinese science of Feng Shui from 1996 has revealed me clearly that already as a child I had the faculty of directly sensing the bioenergetic charge emanating from people and being present in certain locations. I still have this faculty today.

[125] See, for example, Robert Temple, *The Sirius Mystery (1998)*, Part One, pp. 53 ff.

[126] Terence McKenna, *The Archaic Revival*, San Francisco: Harper & Row, 1992, p. 45.

[127] Carlos Castaneda, *The Teachings of Don Juan (1985)*, *Journey to Ixtlan (1991)*, *Tales of Power (1991)*, *The Second Ring of Power* (1991).

[128] Ralph Metzner (Ed.), *Ayahuasca (1999)*.

[129] Id., p. 72 (emphasis mine).

[130] Id., p. 123 (emphasis mine).

[131] Id., p. 282 (emphasis mine).

[132] Id., p. 283 (emphasis mine).

[133] William A. McCarey M.D., *In Search for Healing (1996)*.

[134] Id., p. 4.

[135] See also Pierre F. Walter, *Consciousness and Shamanism, Audio Book (2010)*.

[136] Robert M. Stein, *Redeeming the Inner Child in Marriage and Therapy (1990)*, 261 ff.

[137] See *Walter's Encyclopedia, Academic Edition (2010)*, 'Repression'.

[138] Literally: *green house*.

[139] See, for example, Françoise Dolto, *La Cause des Enfants (1985)*, *La Cause des Adolescents (1988)*. See also Pierre F. Walter, *Françoise Dolto and Language, Great Minds Series (2010)*.

[140] See, for example, Wilhelm Reich, *The Invasion of Compulsory Sex-Morality (1971)*.

[141] See, for example, Stone & Stone, *Embracing Our Selves (1989)*.

[142] See, for example, Joseph Campbell, *The Hero With A Thousand Faces (1973)*, *Occidental Mythology (1973)*, and *The Power of Myth (1988)*.

[143] See Riane Eisler, *The Chalice and the Blade (1995)*, *Sacred Pleasure (1996)*.

[144] Larry L. & Joan M. Constantine, *Treasures of the Island (1976)*, *Where are the Kids? (1977)*.

[145] J. P. Alston & F. Tucker, *The Myth of Sexual Permissiveness (1973)*.

[146] See for a definition of what is understood under matriarchal culture the extensive studies of Johann Jakob Bachofen, in: *Gesammelte Werke, Band II, Das Mutterrecht (1948)*.

[147] Bronislaw Malinowski, *The Sexual Life of Savages in North West Melanesia (1929)*.

[148] Bronislaw Malinowski, *Sex and Repression in Savage Society (1927)*, p. 76.

[149] V. Elwin, *The Muria and their Ghotul (1947)*, Richard L. Currier, *Juvenile Sexuality in Global Perspective (1981)*.

Published by Sirius-C Media Galaxy LLC, 2010

[150] I speak here about the pre-Hellenic cultures such the *Minoan Civilization* on the island of Crete that was a highly developed civilization with a natural focus on the senses an on beauty, free sexuality and a matriarchal worldview, respecting the female and female children. It had a low crime rate, no slavery, no male god but venerating goddesses, and low degree of violence, a culture that however was virtually raped and eaten up by the patriarchal invader tribes.

[151] Bronislaw Malinowski, *The Sexual Life of Savages in North West Melanesia (1929), Sex and Repression in Savage Society (1927).*

[152] In most Western nations today, especially the USA, divorce rate is around 70 to 75%.

[153] See Erich Fromm, *To Have or To Be (1976).*

[154] Riane Eisler suggests to *completely abandon* the dichotomy 'matriarchal-patriarchal' replacing it by 'egalitarian-dominator', thus avoiding endless discussions if or not in matriarchal cultures males were oppressed by females. The question in fact is not who dominates whom, but if a culture in general runs on a *dominator paradigm* or on a sharing or *egalitarian paradigm*. It is now shared by the majority of scientists that 'matriarchal' cultures were clearly *more egalitarian* than the subsequent patriarchal or dominator cultures. Thus, a way back to love obviously will have to consider a sort of *archaic revival*, to speak with Terence McKenna.

[155] H.J. Campbell, *The Pleasure Areas (1973).*

[156] *Grant's Method of Anatomy (1980)*, p. 61.

[157] Ashley Montagu, *Touching: The Human Significance of the Skin (1971).*

[158] Id., pp. 15 ff.

[159] Id., p.18.

[160] Id., p. 234.

[161] See for example Frederick Leboyer, *Birth Without Violence (1975).*

[162] Michel Odent, *La Santé Primale (1986)*, p. 24 (Translation mine).

[163] R.B. Textor, *A Cross-Cultural Summary (1967).*

[164] Id., pp. 10, 11.

[165] Id., p. 10.

[166] Id.

[167] Id., p. 12.

[168] Id.

[169] Id., p. 13.

[170] See James W. Prescott, *Deprivation of Physical Affection as a Primary Process in the Development of Physical Violence (1979)*, pp. 77, 78.

[171] Alice Miller, *The Drama of the Gifted Child: In Search for the True Self (1996)*.

[172] Alice Miller, *Thou Shalt Not Be Aware (1998)*, and Alexander Lowen, *Narcissism: Denial of the True Self (1997)*.

[173] Françoise Dolto, *La Cause des Enfants (1985)*, p. 13. See also Ronald David Laing, *Divided Self (1991)*, Bruno Bettelheim, *A Good Enough Parent (1988)* and *The Uses Of Enchantment (1989)*.

[174] Françoise Dolto, *La Cause des Enfants (1985)*, pp. 28, 29, citing Ariès, *The Childhood of French King Louis XIII*.

[175] Id. (Translation mine).

[176] Jean-Jacques Rousseau, *Émile ou de l'Éducation (1762)*. See also critically René Schérer, *Émile Perverti ou des rapports entre l'éducation et la sexualité (1974)* and John Locke, *Some Thoughts Concerning Education (1690)*, Vol. IX., pp. 6-205.

[177] Alexander S. Neill, *Summerhill (1961)* and *Neill! Neill! Orange Peel! (1972)*.

[178] Alexander S. Neill, *Summerhill (1961)*, pp. 139 ff. und pp. 301 ff.

[179] Id., p. 207.

[180] See also Floyd M. Martinson, *The Sex Education of Young Children (1981)*, pp. 51 ff.

[181] See A. S. Neill, *Summerhill (1961)*, p. 29.

[182] Id.

Published by Sirius-C Media Galaxy LLC, 2010

[183] Ashley Montagu, *Touching (1971)*.

[184] Id., p. 15.

[185] Id.

[186] Id., p. 234.

[187] Id., pp. 20, 21 and Michel Odent, *Birth Reborn (1986)*.

[188] Id., p. 15.

[189] Frederick Leboyer, *Loving Hands (1977)*.

[190] James W. Prescott, *Body Pleasure and the Origins of Violence (1975)*.

[191] Herbert James Campbell, *The Pleasure Areas (1973)*.

[192] James W. Prescott, *Body Pleasure and the Origins of Violence (1975)*, p. 13.

[193] See for example Susanne Cho, *Kindheit und Sexualität im Wandel der Kultur-geschichte (1983)* or Floyd M. Martinson, *The Sex Education of Young Children (1981)*, pp. 51 ff.

[194] This fact is corroborated by descriptions of the habits of the Royal Family in France, as reported by the doctor of Louis XIII, Héroard, see: J. Héroard, *Journal de Jean Héroard sur l'Enfance et la Jeunesse de Louis XIII (1868)*. See also: Lloyd deMause (Ed.), *The History of Childhood (1974)*, p. 23 and Philippe Ariès, *Centuries of Childhood (1962)*.

[195] Jean Liedloff, *The Continuum Concept (1986)*.

[196] Joseph Campbell, *Occidental Mythology (1991)*, p. 70.

[197] See Pierre F. Walter, *Joseph Campbell and the Lunar Bull, Great Minds Series (2010)* and *The Lunar Bull, Audio Book (2010)*.

[198] See, for example, Riane Eisler, *The Chalice and the Blade (1995)* and *Sacred Pleasure (1996)*.

[199] Originally in German language, Johann Jakob Bachofen, *Das Mutterrecht (1948)*.

[200] See Lloyd deMause, *Foundations of Psychohistory (1982)*.

[201] Joseph Campbell, *Occidental Mythology (1973,)* p. 21.

[202] Joseph Campbell, *The Power of Myth (1988)*, p. 54.

[203] Id., p. 178.

[204] See, for example, Sallie Nichols, *Jung And Tarot (1986).*

[205] Joseph Campbell, *Power of Myth (1988)*, pp. 215, 216.

[206] Id., p. 248.

[207] Joseph Campbell, *Occidental Mythology (1973)*, p. 27.

[208] I refer here to the courageous research of Professor Noam Chomsky, from MIT, as I honestly did not know about this fact. See an overview of his most important video lectures here, and see for yourself what you think of America's claim to be the 'most moral culture of the world':

http://ipublica.com/noam-chomsky/

[209] See Pierre F. Walter, *The Idiot Guide to Consciousness (2010)* and *Eight Dynamic Patterns of Living, Audio Book (2010).*

[210] See Erich Neumann, *The Great Mother (1972)*, p. 141, with many reproductions of ancient sculptures featuring both types of bull. See also the recent study of Michael Balter, *The Goddess and the Bull (2006).*

[211] See Joseph Campbell, *The Hero with A Thousand Faces, (1973)*, pp. 258, 259, note 5:

> This recognition of the secondary nature of the personality of whatever deity is worshipped is characteristic of most of the traditions of the world. (...) In Christianity, Mohammedanism, and Judaism, however, the personality of the divinity is taught to be final – which makes it comparatively difficult for the members of these communions to understand how one may go beyond the limitations of their own anthropomorphic divinity. The result has been, on the one hand, a general obfuscation of the symbols, and on the other, a god-ridden bigotry such as is unmatched elsewhere in the history of religion.

[212] See, for example, Lynn Schroeder, Sheila Ostrander & Nancy Ostrander, *Superlearning 2000 (1997).*

[213] See, for example, Joseph Campbell, *The Hero With A Thousand Faces* (1973), *Occidental Mythology (1973/1991)*, and *The Power of Myth (1988).*

Published by Sirius-C Media Galaxy LLC, 2010

[214] See Wilhelm Reich, *The Irruption of Compulsive Sex-Morality, (1971).*

[215] See Joseph Campbell, *The Power of Myth (1988)*, p. 248.

[216] Joseph Campbell, *Occidental Mythology (1973/1991)*, p. 75.

[217] The great majority of authors before the new perspective brought up by Riane Eisler were founding their terminology on Bachofen's study about *Das Mutterrecht* (The Matrilineal Order) that was first published as early as in the 1920s and similar studies by Bronislaw Malinowski and Wilhelm Reich in the 1930s. All these authors spoke from a shift of matriarchy to patriarchy.

[218] In *Occidental Mythology (1973/1991)*, p. 70, Joseph Campbell remarks:

> I am taking pains in this work to place considerable stress upon the world age and symbolic order of the goddess; for the findings both of anthropology and of archeology now attest not only to a contrast between the mythic and social systems of the goddess and the later gods, but also to the fact that in our own European culture that of the gods overlies and occludes that of the goddess – which is nevertheless effective as a counterplayer, so to say, in the unconscious of the civilization as a whole.

[219] See R.E.L. Masters, *Forbidden Sexual Behavior and Morality (1951)*, p. 387 who observes:

> It is ironic that where the desire to defile, humiliate or otherwise sadistically abuse children is concerned, it is so often the very notion of the child's purity and innocence that leads to the violation.

[220] In Greek mythology, Persephone, daughter of the earth goddess *Demeter* became the queen of the underworld after her abduction by *Hades*.

[221] As I show in *The Idiot Guide to Love (2010)*, it was in the Victorian Era that the virgin cult became a real sexual obsession or mania, a cult for well-to-do men to rape and defile young girls, when those were available, for example, in the worker classes of the poor quarters in industrialized London and other major cities of that time of the early *Industrial Revolution.*

[222] See, Florence Rush, *The Best Kept Secret (1980)*, and Riane Eisler, *The Chalice and The Blade (1995).*

[223] Id.

[224] Joseph Campbell writes in *The Power of Myth (1988)*, p. 217:

> The virgin birth comes into Christianity by way of the Greek tradition. When you read the four gospels, for example, the only one in which the virgin birth appears is the Gospel According to Luke, and Luke was a Greek.

[225] See Pierre F. Walter, *The Idiot Guide to Love, Awareness Guide (2010)*.

[226] According to Joseph Campbell, the grail symbolizes the respect not of abstract rules and regulations (sexual or other ones) but respect of nature, of creation.

[227] See Joseph Campbell, *Occidental Mythology (1973)*, p. 86.

[228] Joseph Campbell, *The Masks of God (1962)*, p. 4.

[229] See Joseph Campbell, *The Power of Myth (1988)*, p. 245.

[230] Id., p. 46.

[231] Id., p. 245.

[232] Id., p. 254.

[233] Thomas Moore, *Care of the Soul (1994)*, p. 17.

[234] Id., pp. 239, 240.

[235] Wilhelm Reich, *Children of the Future (1950)*, p. 44.

[236] See, in more detail, Pierre F. Walter, *Erós and Agapé: A Case of Healing the Split, Monograph (2010)*.

[237] Id., p. 204.

[238] Id., p. 197.

[239] Id., p. 208.

[240] Id., pp. 214, 215.

[241] Id., p. 215.

[242] Joseph Campbell, *Occidental Mythology (1973)*, pp. 21, 22.

243 Id., p. 70.

244 Thomas Moore, *Care of the Soul (1994)*, p. 17.

245 Id., p. 18.

246 Regarding systems theory and the transformation of a traditionally mechanistic science into a *systemically intelligent* society, see for example, Fritjof Capra, *The Web of Life (1996)* and *The Hidden Connections (2002)*.

247 See, for example, Daniel Goleman, *Emotional Intelligence (1995)*.

248 Thomas Moore, *Care of the Soul (1994)*, p. 20.

249 Carl Gustav Jung, *Religious and Psychological Problems of Alchemy (1993)*, p. 541.

250 Thomas Moore, *Care of the Soul (1994)*, p. 67.

251 Id., p. 62.

252 Id., p. 13.

253 Ralph Metzner (Ed.), *Ayahuasca (1999)*, p. 34.

254 See Jeremy Narby, *The Cosmic Serpent (1999)*, p. 66, with further references.

255 Id., pp. 46-57.

256 Id., p. 48.

257 Id., pp. 76-85.

258 Id., p. 83.

259 Id., pp. 129-131.

260 Id., p. 130.

261 See, for example, Sigmund Freud, *The Interpretation of Dreams (1980)* and Carl Gustav Jung, *The Meaning and Significance of Dreams 1991*.

606 | The Science of Shamanism

[262] This is the prevailing view in shamanic cultures. Interestingly, this is paralleled with what has been received as information from extrasensorial dimensions through the practice of channeling.

[263] See Pierre F. Walter, *The Idiot Guide to Intuition, Awareness Guide (2010)*.

[264] See Pierre F. Walter, *The Idiot Guide to Soul Power, Awareness Guide (2010)*.

[265] See Pierre F. Walter, *The Idiot Guide to Science, Awareness Guide (2010)*.

[266] See Pierre F. Walter, *The Idiot Guide to World Peace, Awareness Guide (2010)*.

[267] See, for example, Dean M. Herman, *A Statutory Proposal to Prohibit the Infliction of Violence upon Children*, 19 FAMILY LAW QUARTERLY, 1986, 1-52.

[268] See Pierre F. Walter, *The Idiot Guide to Sanity, Awareness Guide (2010)*.

www.ingramcontent.com/pod-product-compliance
Lightning Source LLC
Chambersburg PA
CBHW031809170526
45157CB00001B/17

9 781456 585853